Modern Semiconductor Device Physics

Modern Semiconductor Device Physics

Editor:

Nelson Bolívar, Ph.D

ArclerPress

www.arclerpress.com

Modern Semiconductor Device Physics

Editor: Nelson Bolívar, Ph.D

708 3rd Avenue, 6th Floor
New York
NY 10017
United States of America
www.arclerpress.com

ISBN: 978-1-68094-399-3

Library of Congress Control Number: 2016951049

About the Editor

Nelson Bolívar, Ph.D

Nelson Bolivar has a PhD. in physics from the University of Lorraine in France finished in 2014. His expertise is in quantum systems and condensed matter. His interest includes spintronic devices and correspondences between general relativity and condensed matter. He is currently an associate professor at the Central University of Venezuela

Contents

List of Contributors

Jin-Seong Park
Division of Materials Science and Engineering, Hanyang University, 222 Wangsimni-ro Seongdon-gu, Seoul 133-791, Republic of Korea

H. Kim
Department of Materials Science and Engineering, Korea Advanced Institute of Technology, 291 Daehak-ro, Yuseong-gu, Daejeon 305-701, Republic of Korea

Il-Doo Kim
Department of Materials Science and Engineering, Korea Advanced Institute of Technology, 291 Daehak-ro, Yuseong-gu, Daejeon 305-701, Republic of Korea

H.X. Xu
Department of Electronic Science & Technology, Huazhong University of Science and Technology, Wuhan
Department of Physics & Electronic Technology, Huanggang Normal University, Huangzhong

J.P. Xu
Department of Electronic Science & Technology, Huazhong University of Science and Technology, Wuhan

C.X. Li
Department of Electrical & Electronic Engineering, The University of Hong Kong, Pokfulam Road, Hong Kong, Hong Kong

C.L. Chan
Department of Electrical & Electronic Engineering, The University of Hong Kong, Pokfulam Road, Hong Kong, Hong Kong

P.T. Lai
Department of Electrical & Electronic Engineering, The University of Hong Kong, Pokfulam Road, Hong Kong, Hong Kong

CHEN Bin
Key Laboratory of Ministry of Education for Wide Band-Gap Semiconductor Materials and Devices, School of Microelectronics, Xidian University, Xi'an 710071, China;
Xi'an Electronic Engineering Research Institute, Xi'an 710100, China;

YANG YinTang
Key Laboratory of Ministry of Education for Wide Band-Gap Semiconductor Materials and Devices, School of Microelectronics, Xidian University, Xi'an 710071, China;

XIE XuanRong
No. 771 Institute of Microelectronics Technology, Xi'an 710054, China

WANG Ning
Key Laboratory of Ministry of Education for Wide Band-Gap Semiconductor Materials and Devices, School of Microelectronics, Xidian University, Xi'an 710071, China;

MA ZhenYang
Key Laboratory of Ministry of Education for Wide Band-Gap Semiconductor Materials and Devices, School of Microelectronics, Xidian University, Xi'an 710071, China;

SONG Kun
Key Laboratory of Ministry of Education for Wide Band-Gap Semiconductor Materials and Devices, School of Microelectronics, Xidian University, Xi'an 710071, China;

ZHANG XianJun
Key Laboratory of Ministry of Education for Wide Band-Gap Semiconductor Materials and Devices, School of Microelectronics, Xidian University, Xi'an 710071, China;

Faten Adel Ismail Chaqmaqchee
School of Computer Science and Electronic Engineering, University of Essex, Colchester CO4 3SQ, UK

Simone Mazzucato
School of Computer Science and Electronic Engineering, University of Essex, Colchester CO4 3SQ, UK

Murat Oduncuoglu
School of Computer Science and Electronic Engineering, University of Essex, Colchester CO4 3SQ, UK
Department of Physics, Faculty of Science and Art, University of Kilis 7 Aralik, Kilis, Turkey

Naci Balkan
School of Computer Science and Electronic Engineering, University of Essex, Colchester CO4 3SQ, UK

Yun Sun
School of Computer Science and Electronic Engineering, University of Essex, Colchester CO4 3SQ, UK

Mustafa Gunes
School of Computer Science and Electronic Engineering, University of Essex, Colchester CO4 3SQ, UK

Maxime Hugues
Department of Electronic and Electrical Engineering, University of Sheffield, Sheffield S1 3JD, UK

Mark Hopkinson
Department of Electronic and Electrical Engineering, University of Sheffield, Sheffield S1 3JD, UK

Jagan Singh Meena
Department of Electronics Engineering and Institute of Electronics, National Chiao Tung University, Hsinchu 30010, Taiwan

Simon Min Sze
Department of Electronics Engineering and Institute of Electronics, National Chiao Tung University, Hsinchu 30010, Taiwan

Umesh Chand
Department of Electronics Engineering and Institute of Electronics, National Chiao Tung University, Hsinchu 30010, Taiwan

Tseung-Yuen Tseng
Department of Electronics Engineering and Institute of Electronics, National Chiao Tung University, Hsinchu 30010, Taiwan

Chao Chen
Key Laboratory of Artificial Micro- and Nano-structures of Ministry of Education, School of Physics and Technology, Wuhan University, Wuhan 430072, People's Republic of China.

Ti Wang
Key Laboratory of Artificial Micro- and Nano-structures of Ministry of Education, School of Physics and Technology, Wuhan University, Wuhan 430072, People's Republic of China.

Hao Wu
Key Laboratory of Artificial Micro- and Nano-structures of Ministry of Education, School of Physics and Technology, Wuhan University, Wuhan 430072, People's Republic of China.

He Zheng
Center for Electron Microscopy, School of Physics and Technology, Wuhan University, Wuhan 430072, People's Republic of China.

Jianbo Wang
Center for Electron Microscopy, School of Physics and Technology, Wuhan University, Wuhan 430072, People's Republic of China.

Yang Xu
Key Laboratory of Artificial Micro- and Nano-structures of Ministry of Education, School of Physics and Technology, Wuhan University, Wuhan 430072, People's Republic of China.

Chang Liu
Key Laboratory of Artificial Micro- and Nano-structures of Ministry of Education, School of Physics and Technology, Wuhan University, Wuhan 430072, People's Republic of China.

Dmitri B. Strukov
University of California Santa Barbara, Santa Barbara, CA 93106, USA

R. Stanley Williams
Hewlett-Packard Laboratories, 1501 Page Mill Road, MS1123, Palo Alto, CA 94304, USA

Gang WANG
School of Physical Electronics, University of Electronic Science and Technology of China, Chengdu, 610054, China

Shuqiang CHEN
School of Physical Electronics, University of Electronic Science and Technology of China, Chengdu, 610054, China

Huajun YANG
School of Physical Electronics, University of Electronic Science and Technology of China, Chengdu, 610054, China

Jose Antonio Ayllon
Chemistry Department, Universitat Autonòma de Barcelona, Campus UAB, 08193 Bellaterra, Spain

Monica Lira-Cantu
Centro de Investigaciò en Nanociencia i Nanotecnologia, CIN2 (CSIC-ICN), Campus UAB, Edifici ETSE 2nd Floor, 08193 Bellaterra, Barcelona, Spain

Sudeshna Samanta
Department of Condensed Matter Physics and Material Sciences, S N Bose National Centre for Basic Sciences

Kaustuv Das
Department of Condensed Matter Physics and Material Sciences, S N Bose National Centre for Basic Sciences

Arup Kumar Raychaudhuri
Department of Condensed Matter Physics and Material Sciences, S N Bose National Centre for Basic Sciences

Jaakko Huikari
Department of Electrical Engineering, Electronics Laboratory, University of Oulu, Linnanmaa, 90014 Oulu, Finland

Eugene Avrutin
Department of Electronics, University of York, York YO10 5DD, UK

Boris Ryvkin
Department of Electrical Engineering, Electronics Laboratory, University of Oulu, Linnanmaa, 90014 Oulu, Finland
A.F. Ioffe Physico-Technical Institute, 194021 St. Petersburg, Russia

Juha Kostamovaara
Department of Electrical Engineering, Electronics Laboratory, University of Oulu, Linnanmaa, 90014 Oulu, Finland

Lara F dos Santos
Physics Department, Federal University of São Carlos, São Carlos, Brazil

Yara Galvão Gobato
Physics Department, Federal University of São Carlos, São Carlos, Brazil

Márcio D Teodoro
Physics Department, Federal University of São Carlos, São Carlos, Brazil

Victor Lopez-Richard
Physics Department, Federal University of São Carlos, São Carlos, Brazil

Gilmar E Marques
Physics Department, Federal University of São Carlos, São Carlos, Brazil

Maria JSP Brasil
Physics Institute, UNICAMP, Campinas, Brazil

Milan Orlita
Grenoble High Magnet Field Laboratory, Grenoble, France
Institute of Physics, Charles University, Ke Karlovu 5, 121 16 Praha 2, Czech Republic

Jan Kunc
Grenoble High Magnet Field Laboratory, Grenoble, France
Institute of Physics, Charles University, Ke Karlovu 5, 121 16 Praha 2, Czech Republic

Duncan K Maude
Grenoble High Magnet Field Laboratory, Grenoble, France

Mohamed Henini
School of Physics and Astronomy, Nottingham Nanotechnology and Nanoscience Centre, University of Nottingham, Nottingham, NG7 2RD, UK

Robert J Airey
EPSRC National Centre for III-V Technologies, The University of Sheffield, Sheffield, UK

Olle Inganäs
Biomolecular and Organic Electronics, IFM, Linköpings Universitet, 58183 Linköping, Sweden

Villy Sundström
Biomolecular and Organic Electronics, IFM, Linköpings Universitet, 58183 Linköping, Sweden

Abdulrahman Al-Muhanna
King Abdulaziz City for Science and Technology, KACST, National Nanotechnology Research Center, Riyadh, Saudi Arabia

Abdullah Alharbi
King Abdulaziz City for Science and Technology, KACST, National Nanotechnology Research Center, Riyadh, Saudi Arabia

Abdelmajid Salhi
King Abdulaziz City for Science and Technology, KACST, National Nanotechnology Research Center, Riyadh, Saudi Arabia

Samir Labiod
Electronic Department, Faculty of Engineering Sciences, Mentouri University Constantine, Constantine, Algeria;

Saida Latreche
Electronic Department, Faculty of Engineering Sciences, Mentouri University Constantine, Constantine, Algeria;

Mourad Bella
Electronic Department, Faculty of Engineering Sciences, Mentouri University Constantine, Constantine, Algeria;

Christian Gontrand
Institute of Nanotechnology of Lyon (INL), University of Lyon, Lyon, France.

Kazunori Shinohara
Advanced Power Electronics Project, Kanagawa Academy of Science and Technology, Kawasaki City, Japan

Qiang Yu
Department of Mechanical Engineering, Yokohama National University, Yokohama, Japan

Supriya Karmakar
Intel Corporation, Hillsboro, USA;

Faquir C. Jain
Department of Electrical & Computer Engineering, University of Connecticut, Storrs, USA.

Yoshitaka Takahashi
Department of Electronic Engineering, Graduate School of Engineering, Gunma University, Kiryu, Japan

Shinji Sekiya
Package Material Production Division, Hitachi Cable, Ltd., Hitachi, Japan

Tatsuro Suemune
Personal Solutions Business Unit, NEC Corp., Kawasaki, Japan

K. H. Al-Mossawi
Scholarships & Cultural Affairs, Ministry of Higher Education & Scientific Research, Iraq

Petr Bilik
Faculty of Electrical Engineering and Computer Science, VSB – Technical University of Ostrava, Czech Republic

Jan Zidek
Faculty of Electrical Engineering and Computer Science, VSB – Technical University of Ostrava, Czech Republic

Vaclav Kus
Department of Electromechanics and Power Electronics, University of West Bohemia, Pilsen, Czech Republic

Tereza Josefova
Department of Electromechanics and Power Electronics, University of West Bohemia, Pilsen, Czech Republic

Preface

Semiconductor devices such as diodes, transistors and integrated circuits are found everywhere in our daily lives, in televisions, automobiles, washing machines and computers. We have come to rely on them in every aspect and we are expecting higher performance at lower cost.

Personal computers are the corner stone on this subject. The complexity and performance of the today's personal computers vastly exceeds that of their old computer and that for about an equivalent purchase price.The essential fact which has driven the successful growth of this technological industry is that through industrial skill and technological advances. These devices deliver year after year better performance while consuming less power and because of their smaller size they can also be manufactured at a lower cost per device.Technology of the twenty-first century is relying heavily on the development of new materials and their structures in which the technology of semiconductors is the foundation of modern electronics, including transistors, solar cells, light-emitting diodes, quantum dots, and digital and analog integrated circuits A new era has just begun, where people's life changes dramatically with every step the technology takes and the science is being developed faster with each advance. A mutual relation that only goes towards benefits in an endless cycle of growing. This book takes some interesting aspects and advances of devices that relies on semiconductor physics, guiding the reader deeper into both the physical basis and their applications.

Editor
Nelson Bolívar, Ph.D

INTRODUCTION

Semiconductor devices are electronic components that exploit the electronic properties of semiconductor materials, principally silicon, germanium, and gallium arsenide, as well as organic semiconductors. Semiconductor devices have replaced thermionic devices (vacuum tubes) in most applications. They use electronic conduction in the solid state as opposed to the gaseous state or thermionic emission in a high vacuum.

Semiconductor devices are manufactured both as single discrete devices and as integrated circuits (ICs), which consist of a number—from a few (as low as two) to billions—of devices manufactured and interconnected on a single semiconductor substrate, or wafer.

Semiconductor materials are useful because their behavior can be easily manipulated by the addition of impurities, known as doping. Semiconductor conductivity can be controlled by the introduction of an electric or magnetic field, by exposure to light or heat, or by the mechanical deformation of a doped monocrystalline grid; thus, semiconductors can make excellent sensors. Current conduction in a semiconductor occurs via mobile or "free" electrons and holes, collectively known as charge carriers. Doping a semiconductor such as silicon with a small amount of impurity atoms, such as phosphorus or boron, greatly increases the number of free electrons or holes within the semiconductor. When a doped semiconductor contains excess holes it is called "p-type", and when it contains excess free electrons it is known as "n-type", where p (positive for holes) or n (negative for electrons) is the sign of the charge of the majority mobile charge carriers. The semiconductor material used in devices is doped under highly controlled conditions in a fabrication facility, or fab, to control precisely the location and concentration of p- and n-type dopants. The junctions which form where n-type and p-type semiconductors join together are called p–n junctions.

A semiconductor diode is a device typically made from a single p–n junction. At the junction of a p-type and an n-type semiconductor there forms a depletion region where current conduction is inhibited by the lack of mobile charge carriers. When the device is forward biased (connected with the p-side at higher electric potential than the n-side), this depletion region is diminished, allowing for significant conduction, while only very small current can be achieved when the diode is reverse biased and thus the depletion region expanded.

Exposing a semiconductor to light can generate electron–hole pairs, which increases the number of free carriers and thereby the conductivity. Diodes optimized to take advantage of this phenomenon are known as photodiodes. Compound semiconductor diodes can also be used to generate light, as in light-emitting diodes and laser diodes.

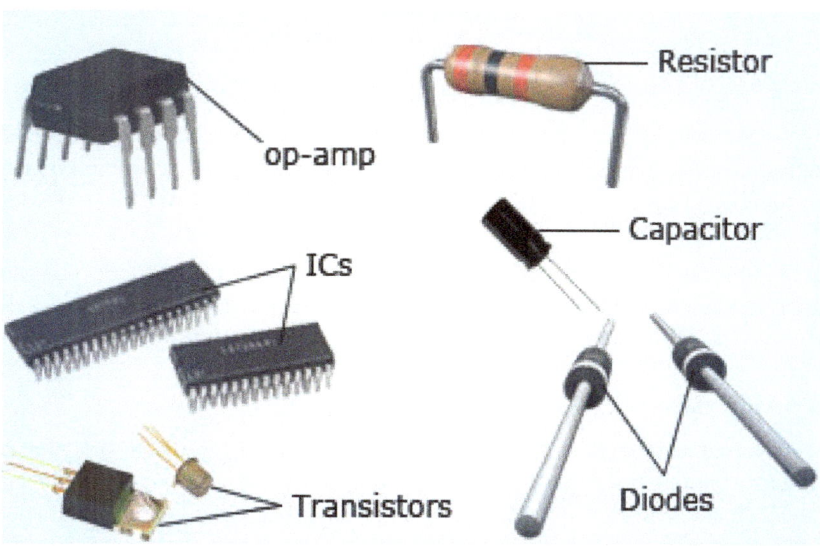

Diode

A semiconductor diode is a device typically made up of a single p-n junction. The junction of a p-type and n-type semiconductor forms a depletion region where current conduction is reserved by the lack of mobile charge carriers. When the device is forward biased, this depletion region is reduced, allowing for significant conduction, when the diode is reverse biased, the only less current can be achieved and the depletion region can be extended. Exposing a semiconductor to light can produce electron hole pairs, which increases the number of free carriers and thereby the conductivity. Diodes optimized to take advantage of this phenomenon is known as photodiodes. Compound semiconductor diodes are also being used to generate light, light-emitting diodes and laser diodes.

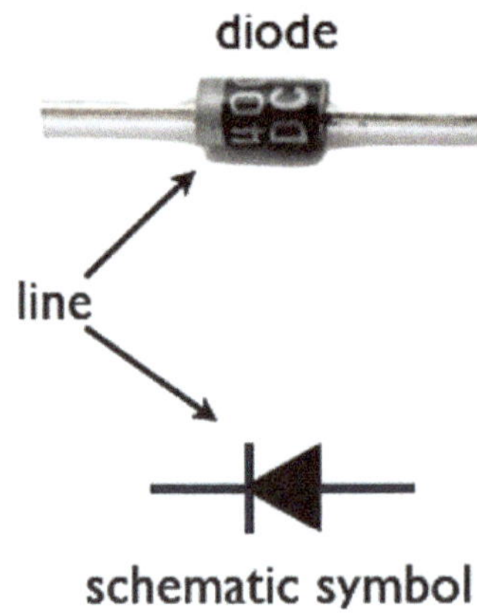

Transistor

Bipolar junction transistors are formed by two p-n junctions, in either p-n-p or n-p-n configuration. The middle or base, the region between the junctions is typically very narrow. The other regions, and their related terminals, are known as the emitter and collector. A small current injected through the junction between the base and emitter change the properties of the base collector junction so it can be conduct current even though it is reverse biased. This creates a larger current between the collector and emitter, and controlled by the base-emitter current.

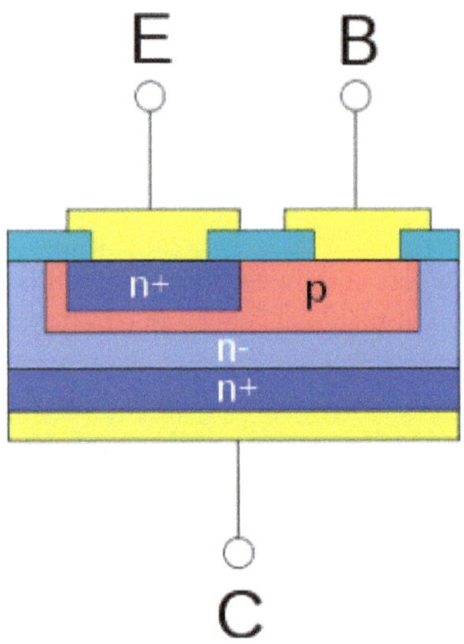

Another type of transistor named as field-effect transistor, it operates on the principle that semiconductor conductivity can increased or decreased by the presence of an electric field. An electric field can increase the number of electrons and holes in a semiconductor, thus changing its conductivity. The electric field may be applied by a reverse-biased p-n junction, and it's forms a junction field-effect transistor (JFET) or by an electrode insulated from the bulk material by an oxide layer, and it forms a metal-oxide semiconductor field-effect transistor (MOSFET).

Now a day's most used in the MOSFET, a solid-state device, and semiconductor devices. The gate electrode is charged to produce an electric field that can control the conductivity of a "channel" between two terminals, is called the source and drain. Depending upon the type of carrier in the channel, the device may be n-channel (for electrons) or p-channel (for holes) MOSFET.

Semiconductor Device Materials

The silicon (Si) is most widely used material in semiconductor devices. It's having lower raw material cost and relatively simple process. Its useful temperature range makes it currently the best compromise among the various competing materials. Silicon used in semiconductor device manufacturing is presently fabricated into bowls that are large enough in diameter to allow the manufacture of 300 mm (12 in.) wafers. Germanium (Ge) was a widely used in early semiconductor material, but its thermal sensitivity makes less useful than silicon. Nowadays, germanium is often alloyed with (Si) silicon for use in very-high-speed SiGe devices; IBM is a main producer of such devices. Gallium arsenide (GaAs) is also widely used with high-speed devices, but so far, it has been difficult to form large-diameter bowls of this material, limiting the wafer diameter sizes significantly smaller than silicon wafers thus making mass production of Gallium arsenide (GaAs) devices significantly more expensive than silicon.

List of Common Semiconductor Devices

The list of common semiconductor devices mainly includes two terminals, three terminals and four terminal devices.

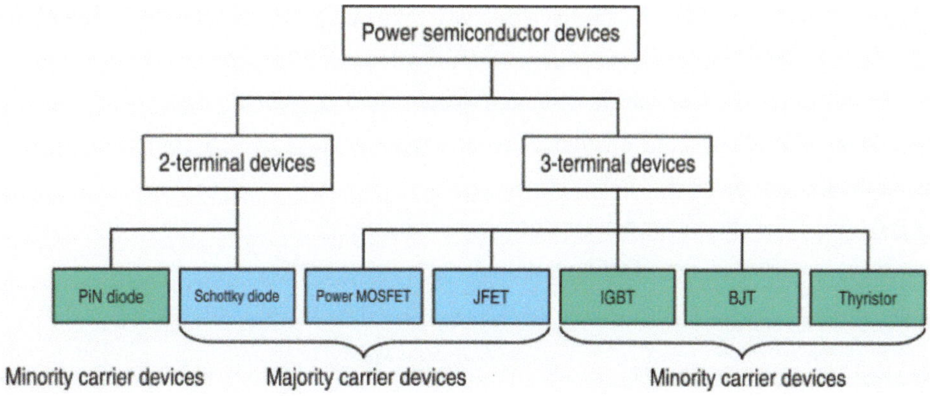

The two-terminal devices are

- Diode (rectifier diode)
- Gunn diode
- IMPATT diode
- Laser diode
- Zener diode
- Schottky diode
- PIN diode
- Tunnel diode
- Light-emitting diode (LED)
- Photo transistor
- Photocell
- Solar cell
- Transient-voltage-suppression diode
- VCSEL

Three-terminal devices are

- Bipolar transistor
- Field-effect transistor
- Darlington transistor
- Insulated-gate bipolar transistor (IGBT)
- Unijunction transistor
- Silicon-controlled rectifier
- Thyristor
- TRIAC

Four-terminal devices are

- Photo coupler (Optocoupler)
- Hall effect sensor (magnetic field sensor)

Semiconductor Device Applications

All types of transistor can be used as the building blocks of logic gates, which is useful to design of digital circuits. In digital circuits like as microprocessors, transistors so which is acting as a switch (on-off); in the MOSFET, for example, the voltage applied to the gate determines whether the switch is on or off.

The transistors are used for analog circuits do not act as switches (on-off); relatively, they respond to a continuous range of input with a continuous range of output. Common analog circuits include oscillators and amplifiers. The circuits that interface or translate between analog circuits and digital circuits are known as the mixed-signal circuits.

Advantages of Semiconductor Devices

- As semiconductor devices have no filaments, hence no power is needed to heat them to cause the emission of electrons.
- Since no heating is required, semiconductor devices are set into operation as soon as the circuit is switched on.
- During operation, semiconductor devices do not produce any humming noise.
- Semiconductor devices require low voltage operation as compared to vacuum tubes.
- Owing to their small sizes, the circuits involving semiconductor devices are very compact.
- Semiconductor devices are shock proof.
- Semiconductor devices are cheaper as compared to vacuum tubes.
- Semiconductor devices have an almost unlimited life.
- As no vacuum has to be created in semiconductor devices, they have no vacuum deterioration trouble.

Disadvantages of Semiconductor Devices

- The noise level is higher in semiconductor devices as compared to that in the vacuum tubes.
- Ordinary semiconductor devices cannot handle as more power as ordinary vacuum tubes can do.
- In high frequency range, they have poor responder.

Thus, this is all about different types of semiconductor devices include two terminals, three terminals and four terminal devices. We hope that you have got a better understanding of this concept.

Semiconductor Detectors

Semiconductor detectors are sensing devices that measure radiation levels or particle counts. They may use one element, a line of elements, or a two-dimensional array of detecting elements. The semiconductor detecting material used is only 0.10 mm thick. Semiconductor detectors have very important measurement and imaging applications in different fields, such as special imaging and particle physics. A silicon detector is commonly used for measuring intensity of radiation or number of ionizing particles in a specific location. The basic principle involved uses a diode, which is a positive (P)-type and negative (N)-type (P-N) junction in reverse bias. The very low idle current of the non-conducting diode becomes the idle reference. When radiation hits the detector, there will be an increase in reverse current that can trigger counters or particle frequency meters that may indicate particle ionization rates. Microstrip detectors are silicon detectors arranged linearly, and are an expansion of the point detection in diode-type detectors. They are usually made on a base N-type material. By aligning the particle flow with the longitudinal span of the microstrip detector, the path of particles of interest can be confirmed.

Picture elements or pixel detectors are widely used for two-dimensional imaging. Before the invention of pixel detectors, the only way to make use of the ability of X-ray to penetrate and pass through the human body was a barium compound plate, which is commonly known as an X-ray plate. With modern technological capability, pixel detectors allow real-time display of two-dimensional imaging of a part of interest in a medical surgical procedure. Another use could be a real-time display of otherwise hidden material parts, such as in quality assurance.

Particle detectors are available as solid-state semiconductor particle detectors. A particle detector is able to sense the presence of radiation or an ionizing particle. Early particle detectors made use of a sealed tube filled with an ionizing gas. When radiation is present, the gas ionizes and produces a brief moment of increased conductivity. Security enforcement increasingly relies upon semiconductor detectors. Simple metal detectors that use an electromagnetic field to detect dangerous weapons may be also incorporated with a radiation detector that detects dangerous radioactive materials at security checkpoints. Conveyor belt X-ray machines have increased in importance in rapid inspection of baggage for all types of passengers. The rapid clearing of people entering high-security facilities has also been made possible by the use of semiconductor detectors.

CHAPTER 1

ELECTROCERAMIC MATERIALS FOR OXIDE SEMICONDUCTOR THIN FILM TRANSISTORS

Jin-Seong Park, H. Kim & Il-Doo Kim

Keywords: Electroceramics, Oxide semiconductors, Transistors, Displays, Gate insulators

INTRODUCTION

Oxide semiconductors have, in recent years, become the focus of extreme interest for emerging electronic applications, such as flexible and transparent displays [1–5]. In particular, several display industries have intensively researched and demonstrated advanced flat-panel display (FPD) technology utilizing oxide semiconductors including electronic paper (e-paper), organic light-emitting diode displays (OLEDs) and liquid crystal displays (LCDs) [6, 7]. Figure 1depicts the history of R&D trends relative to oxide-semiconductors and prototype display panels. Although efforts dedicated towards the research and development of oxide semiconductors have been relatively recent (see Fig. 1), by 2012, many FPD manufacturers (e.g., Samsung, LG, Sharp, and AUO) had already announced major investments in large AM (Active Matrix) OLED-production based on oxide-semiconductor devices. In 2013, a 55-in. AMOLED TV (LG Electronics), with options for flat or curved screens, reached commercial markets [8].

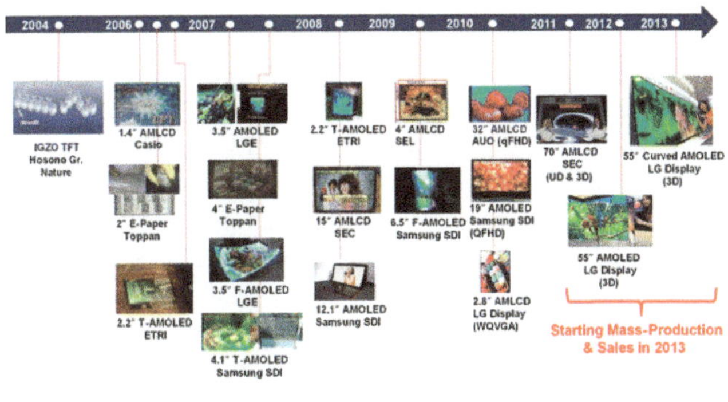

Figure. 1 Historical overview of developments in oxide-semiconductor based TFT-based devices.

Citation: Jin-Seong Park & H. Kim & Il-Doo Kim, electroceramic materials for oxide semiconductor thin film transistors, DOI 10.1007/s10832-013-9858-0.

The first version of an oxide semiconductor TFT, which utilized evaporated SnO_2 as the channel layer in a bottom-gate-staggered configuration on a glass substrate, was reported in 1964 [9]. The earliest SnO_2 TFT exhibited poor electrical performance, including lack of saturation, positive curvature of the output characteristics, an almost negligible amount of transfer curve, and the inability to turn the device off. In 1968, the first ZnO-based TFT was reported which also did not saturate and could not be turned off by application of a gate voltage. Since then, metal oxide TFTs have been continually improved, so that they now exhibit desired characteristics [10].

LCD and OLED flat-panel-display products have traditionally been driven by amorphous silicon (a-Si) and low temperature poly-silicon (LTPS) thin-film-transistors, respectively. While commercial markets have been demanding high-end products with high resolution (4000 × 2000 pixels), high frame rates (over 240 Hz), and large size (more than 70 in.), conventional silicon-based TFTs have been unable to deliver them. These TFTs have inherent limitations, with a-Si TFTs exhibiting relatively low mobility ($<1\ cm^2/V \cdot s$) and inferior bias stability, while LTPS TFTs suffer from non-uniform performance and high processing costs.

After the Hosono group reported the 'amorphous indium gallium zinc oxide' (a-IGZO) semiconductor in 2004 [11], amorphous oxide semiconductors (AOSs) were rapidly adopted as the channel layer in flexible and transparent TFT-based electronic devices. They have since been taking the place of a-Si and LTPS, given several key qualities, including high electrical mobility, amorphicity, and high optical transparency. Moreover, AOSs have a lower density-of-states (DOS), including tail and deep level states, compared to that of a-Si, resulting in superior electrical stability [12].

In this review, we report on recent developments in AOS TFT devices that address several major issues including the nature of the semiconducting oxides, gate insulators, and device instability. We begin by describing the historical development of oxide semiconductors from binary materials to multi-component materials, optimized for practical applications. Then we move on to introduce various gate insulators for producing high performance AOS TFTs, from their physical structures to their properties. Improvements in AOS TFTs by use of high-k gate insulators are also discussed. Finally, we introduce recent progress in minimizing device instability of AOS TFTs under gate bias, temperature, and illumination.

OXIDE SEMICONDUCTOR MATERIALS

Operation of the TFTs and important device parameters

Before focusing on the issues related to the choice and optimization of the oxide semiconductor and dielectric materials, we briefly review the TFT and its operating characteristics to aid readers less familiar with this technology. TFTs are three terminal field-effect devices, as illustrated in Fig. 2(a). The TFT structure is assembled by the stacking of the gate, gate insulator, channel (oxide semiconductor), and source/drain electrodes (both top and bottom gate designs are illustrated) onto the substrate. During operation, the current flowing within the semiconductor channel between the "Source" and "Drain" electrodes are modulated by the electric field imposed between the gate and Drain. To insure that no current flows between Gate and Drain, when the Gate voltage is applied Gate electrode, a highly insulating layer is placed between the semiconductor and the Gate. Furthermore, a high relative dielectric constant of the Gate insulator insures efficient field modulation of the source to drain current even for small applied voltages.

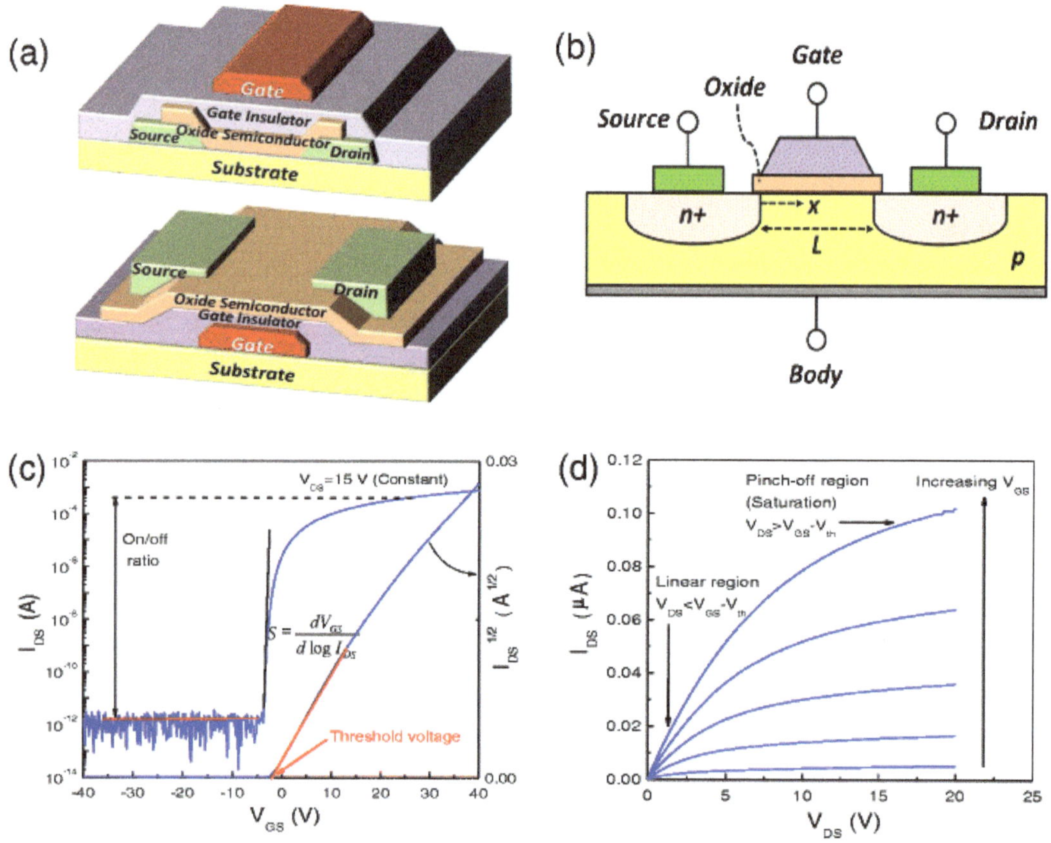

Figure. 2 (**a**) Common TFT device geometries used to fabricate oxide TFTs, which can be classified as top gate and bottom gate structure. (**b**) A schematic image of n-channel MOSFET device structure. Typical transfer (**c**) and output (**d**) electrical characteristics of oxide TFTs. Analysis of TFT characteristics are also shown including threshold voltage (V_{th}), subthreshold swing (S), and on/off ratio

Since carrier flow within the channel can be adjusted by application of an electric field at the insulator/semiconductor interface, these devices are known as "Field-Effect" devices. Confusion often exists about the distinction between TFTs and MOSFETs [Metal Oxide Semiconductor Field-Effect Transistor, the well-known device used in microelectronics, see Fig. 2(b)]. There are two important differences. The first relates to how the semiconductor layer is formed and/or deposited. Semiconductors in TFTs are "thin films", deposited by various thin film methods. But, MOSFETs are constructed on single-crystal (bulk) silicon wafers, requiring high temperature processes (exceeding 1000 °C). TFTs, on the other hand, are readily fabricated by low temperature processes on various inexpensive substrates such as glasses and plastics. The second difference relates to the carrier flow mode. The conductance in TFTs relies on an accumulation layer, while that in MOSFET is controlled by inversion layer formed in the interface between semiconductor and insulator. Thus, the n-type TFTs have n-type semiconductor layers but the n-type MOSFETs have p-type silicon substrates.

To understand the operation of TFTs, there are several important parameters [on/off ratio, field-effect mobility (μ_{fe}), threshold voltage (V_{th}), and subthreshold swing (S)], extracted from the transfer and output characteristics shown in Fig. 2(c) and (d).

On/off ratio

This is defined as the ratio of the maximum to the minimum I_{DS} (drain current). As shown in Fig. 2(c), the minimum I_{DS} is given by the noise level of the measurement equipment or by the gate leakage current during the off position of the TFT device. The lower this value, the lower the leakage current and therefore power dissipation during the off state. Thus the on/off ratio of TFTs represents the ideality of the electronic switch characteristic.

Field-effect mobility (μ_{fe})

The mobility reflects the efficiency of charge carrier transport across the semiconductor layer, depending on device configurations (width, length, and thickness of semiconductor) and the nature and significance of the various scattering mechanisms (lattice vibration, ionized impurity, surface roughness, and other defects). In TFTs, the following two mobilities can be extracted depending on V_{DS}: linear field-effect mobility and saturation mobility.

At low V_{DS} ($V_{DS} \sim 0 \ll V_{GS}$), defined as linear field-effect mobility (μ_{lin}):

$$I_{DS} = \frac{W}{L}\mu_{lin} C_i \left[(V_{GS} - V_{th}) V_{DS} - \frac{V_{DS}^2}{2} \right] \tag{1}$$

At high V_{DS} ($V_{DS} \gg V_{GS} - V_{th}$), defined as saturation mobility (μ_{sat}):

$$I_{DS} = \frac{W}{2L}\mu_{sat} C_i (V_{GS} - V_{th})^2 \tag{2}$$

where W denotes the channel width, L the channel length, C_i the capacitance per unit area of the dielectric layer, V_{DS} the drain to source voltage, and V_{GS} the gate to source voltage.

Threshold voltage (V_{th}): The V_{th} is the gate voltage, at which significant current flow initiates (conducting channel or accumulation layer) between source and drain via the channel formed at the semiconductor/insulator interface,. The value is usually extracted from the $I_{DS}^{1/2}$ vs. V_{GS}characteristics, as shown in Fig. 2(c). The plot of $I_{DS}^{1/2}$ vs V_{GS} intersects the x-axis (V_{GS}) at the value of V_{th}.

Subthreshold swing (S): The inverse of the maximum slope of the transfer characteristics: This represents how rapidly I_{DS} increases by one decade under an increasing applied V_{GS}, i.e.,

$$S = \left(\left. \frac{d\log(I_{DS})}{dV_{GS}} \right|_{Max} \right)^{-1} \tag{3}$$

Small values of S (generally, 0.1 ~ 0.5 V/decade) reflect low power consumption for the microelectronic devices.

The output characteristics representative TFT operation,, where I_{DS} is plotted against V_{DS} for various V_{GS}, are shown in Fig. 2(d). As V_{DS} increases under constant V_{GS}, I_{DS} increases linearly and then saturates to a value (called "hard saturation"). Both characteristics are representative of TFT behavior and reflect the relative performance of the TFT.

BINARY OXIDE SEMICONDUCTORS: ZNO, IN_2O_3, SNO_2, AND TIO_X

The key binary oxide semiconductors (ZnO, In_2O_3, SnO_2, and TiO_x etc.) of interest have wide band gaps (>3.0 eV) that allow the transmission of visible light [13]. Often even when undoped, they exhibit much higher electrical conductivities ($10^{-2}\,\Omega^{-1}\,cm^{-1}$ to $10^2\,\Omega^{-1}\,cm^{-1}$) than other wide gap metal-oxide materials due to inherent high levels of n-type conductivity. This indicates that native shallow defects must exist including either oxygen vacancies, cation interstitials, or substitutional or interstitial hydrogen [14–16]. Because of such donors, these binary oxides exhibit relatively high carrier concentrations (>10^{18} cm^{-3}) and electron mobilities (>10 $cm^2/V \cdot s$) even when deposited at room temperature. As a consequence, they have undergone intensive research as transparent conducting materials for various electronic devices (e.g., LCDs, OLEDs, and solar cells).

In 2003, following initial development of oxide semiconductors and their applications for display purposes, several active research groups revisited ZnO-based TFTs. Masuda et al. [17] reported a ZnO channel layer with bottom-gate structure, exhibiting low carrier concentration (<5×10^{16} cm^{-3}). The ZnO layer was deposited onto a silicon substrate by pulsed-laser deposition (PLD) at a substrate temperature of 450 °C, under oxygen atmosphere. The ZnO TFTs showed hard saturation and ideal saturation output curves (drain current—drain voltage characteristics). The field-effect mobility (μ_{fe}) was about 1 $cm^2/V \cdot s$ for a depletion mode device. Hoffman et al. [18] also reported the electrical performance of ZnO TFTs prepared by the ion-beam sputter method. The device had a μ_{fe} of 2.5 $cm^2/V \cdot s$ and high drain current, on/off ratio of ~10^7 after annealing at 800 °C in oxygen. This enhancement in performance was attributed to the improvement of the ZnO crystallinity. However, the high deposition and annealing temperature was still a significant problem for other electronic applications. Carcia et al. [19] suggested the potential suitability of room temperature grown ZnO films as the active channel layer of oxide TFTs prepared by radio-frequency (rf) magnetron sputtering. They fabricated ZnO TFTs with precise control of oxygen partial-pressure during sputter deposition, with the products exhibiting ideal output curves (μ_{fe} of >2 $cm^2/V \cdot s$), and a drain current on/off ratio >10^6.

After publication of these results, many ZnO TFTs were reported with improved performance and with lowered deposition temperatures suitable for flexible and transparent electronics applications [20–23]. As shown in Fig. 3(a) and (b), Fortunato et al. [20] reported room temperature fabricated ZnO TFTs with invert-staggered structure (high mobility value: approximately 50 $cm^2/V \cdot s$) by carefully optimizing oxygen partial pressure during ZnO growth. In fact, the control of oxygen partial pressure suggested a path to the fabrication of highly efficient and reliable TFTs, opening new doors towards the realization of transparent electronic circuits.

SnO_2 and In_2O_3 TFTs have also been investigated as channel layers by several groups. Presley et al. [24] demonstrated a staggered, bottom-gate SnO_2 TFT using $AlTiO_x$ as the gate insulator, and ITO (InSnOx) as the gate, source, and drain contact. The SnO_2 channel was deposited by rf magnetron sputtering and then subjected to rapid thermal annealing in oxygen at 600 °C. Based on optimizing the SnO_2 channel thickness (10–20 nm) on TFTs, the device performance exhibited a μ_{fe} of 2.0 $cm^2/V \cdot s$ and a drain current on/off ratio of 10^5. Unfortunately, SnO_2TFTs did not appear extensively in binary oxide TFTs, because satisfactory n-type electrical properties (of SnO_2) were difficult to achieve at low deposition temperatures, unlike other binary oxide semiconductors.

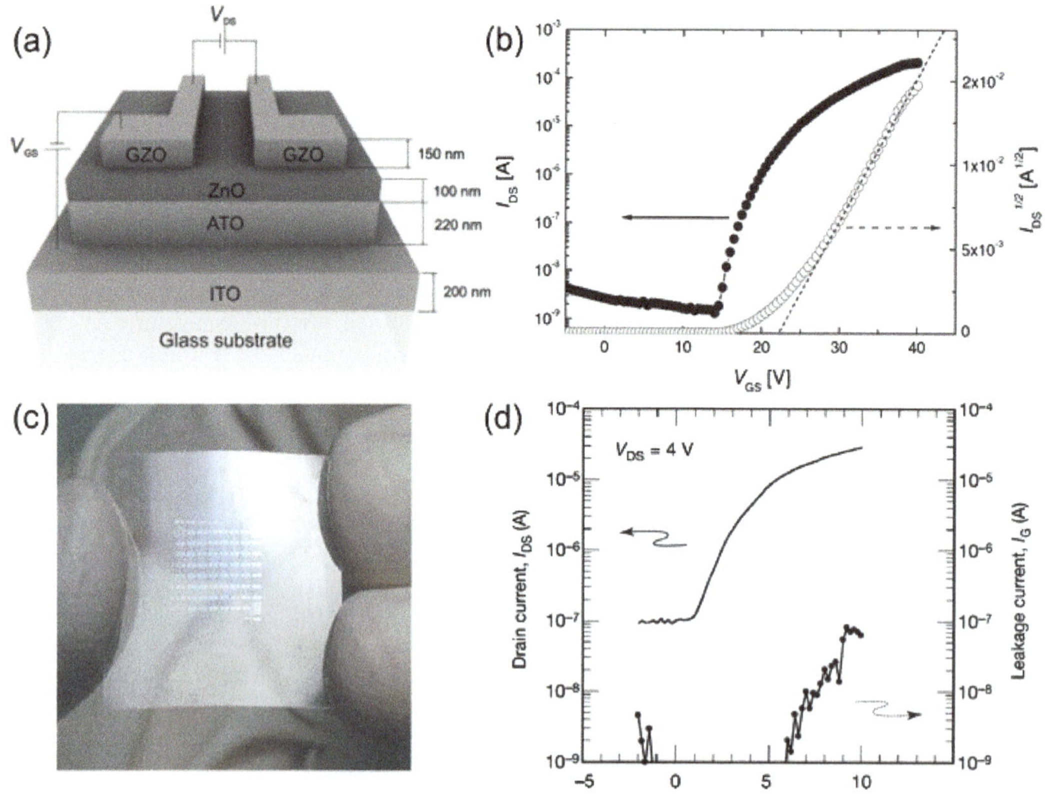

Figure. 3 (**a**) Schematic illustration of the ZnO-based TFT structure. The ZnO channel layer and the gate insulator, consisting of a superlattice of Al_2O_3 and TiO_2 (ATO,) are 100 and 220 nm in thickness, respectively. (**b**) Typical ZnO TFT transfer characteristics for V_{DS}=20 V. The ZnO TFT operates in the enhancement mode with a threshold voltage of 21 V and a saturation mobility of 20 $cm^2/V \cdot s$. Reprinted with permission from [20]. Copyright 2005, Wiley-VCH Verlag GmbH. (**c**) A photograph of the flexible *a*-IGZO TFT sheet. (**d**) Typical transfer characteristics of *a*-IGZO TFT before bending (threshold voltage of ~1.6 V and saturation mobility of ~8.3 $cm^2/V \cdot s$). Reprinted with permission from [11]. Copyright 2004, Macmillan Publishers Ltd

The development of In_2O_3 TFTs was launched late because indium oxide (In_2O_3), while a well-known transparent conductive oxide (TCO) with high carrier concentration (>10^{19} cm^{-3}), shows some difficulty in having its carrier concentrations reduced to semiconducting levels (<10^{18} cm^{-3}). Lavareda et al. [25] reported InO_x TFTs with bottom gate structure. They optimized the rf power and oxygen pressure during InO_x deposition by 'radio-frequency plasma enhanced reactive thermal evaporation', resulting in broad ranges of electrical resistivity from 13.7 to 1.7×10^7 Ωcm. The InO_x TFTs showed reasonable electrical performance, including μ_{fe} of 0.02 $cm^2/V \cdot s$ and drain current on/off ratio of 10^4. Following that, Vygranenko et al. [26] improved the electrical performance of In_2O_3 TFTs by using In_2O_3 channel layers deposited by ion-beam-assisted evaporation. They fabricated a reliable In_2O_3 TFT at room temperature, showing μ_{fe} of 3.3 $cm^2/V \cdot s$, subthreshold swing (S) of 0.5 V/decade, and drain current on/off ratio of 10^6. They also suggested that achieving a high performance In_2O_3 device is dependent on precise control of the oxygen ion beam flux and interface defects (In_2O_3/SiO_x). Moreover, Dhanajay et al. [27] investigated In_2O_3 TFTs fabricated at a substrate temperature of 100 °C as a function of active channel thickness (5–20 nm). They found that there is a reduction in grain boundary density in thicker films leading to a higher mobility (~34 $cm^2/V \cdot s$). This suggests that grain size control of the active layer is a key factor in obtaining better electrical performance with regard to mobility.

Recently, much more abundant and inexpensive titanium oxide (TiO_2) emerged as an alternative to that of indium oxide, given the known scarcity and expense of In.. In 2006, Katayama et al. [28] first reported a TiO_2 TFT with the rutile structure. A top-gate transistor structure with an MgO insulating buffer between TiO_2 semiconductor and amorphous $LaAlO_3$gate insulator, was fabricated on ultra-smooth, rutile, single crystal substrate. The TiO_2 active layer was deposited by PLD and annealed at 700 °C in air. The device exhibited an on-to-off ratio exceeding 10^4 and μ_{fe} of 0.08 $cm^2/V \cdot s$. To fabricate a TiO_x TFT at lower temperature, Park et al. [29] used 'metal-organic chemical-vapor-deposition (MOCVD)-grown TiO_2 active layer' or 'plasma-enhanced atomic-layer-deposition (PEALD)-grown TiO_x active layer', obtained at a deposition temperature of 250 °C, leading to reasonable TFT characteristics. TFTs optimized with N_2O plasma treatment showed μ_{fe} of 1.64 $cm^2/V \cdot s$, S of 1.86 V/decade, and drain current on/off ratio of 4.7×10^5. Also, Choi et al. [30] fabricated more practical TiO_x TFTs by direct-current (dc) magnetron sputtering. The device exhibited μ_{fe} of 0.69 $cm^2/V \cdot s$, subthreshold swing (S) of 2.45 V/decade, and drain current on-to-off ratio of 2.04×10^7 after rapid thermal annealing (RTA) in nitrogen gas. Very recently, Park et al. [31] reported that the conduction behavior of TiO_x materials is very different from that of ZnO materials due to electronic d-orbital ordering in TiO_x.

Table 1 summarizes the electrical properties and deposition conditions of binary oxide semiconductor based TFTs. Unfortunately, although there have been many efforts, over an extended period, to improve the electrical performance of these devices, and to improve the fabrication processes for binary oxide semiconductors, they have received little attention. This is largely due to the fact that these oxides are polycrystalline, and are thereby inappropriate for use as the active layer in TFTs due to poor uniformity. Therefore, the search for amorphous oxide semiconductors with high electrical performance for large-area production has remained a challenging issue, in terms of practical fabrication approaches.

Table 1 Summary of key process and property characteristics of binary oxide semiconductor thin film transistors [RF/DC (radio frequency/direct current) sputter, PE/AP ALD (plasma enhanced/atmospheric pressure atomic layer deposition), PLD (pulsed laser deposition), R.T. (room temperature), ND (no data)]

Channel Materials	**Deposition Method**	**Mobility ($cm^2/V \cdot s$)**	**Subthreshold Swing (V/decade)**	**Process Temperature (°C)**	**Ref.**
ZnO	rf sputter	>2	3	R.T.	[19]
ZnO	PLD	0031–0.97	ND	450	[17]
ZnO	Ion-beam sputter	0.3–2.5	ND	600–800	[18]
ZnO	ALD	4	ND	200	[21]
ZnO	AP ALD	10	ND	200	[22]
ZnO	rf sputter	27	1.39	R.T.	[23]
InO_x	Reactive evaporation	0.02	ND	R.T.	[25]
In_2O_3	Ion-beam evaporation	3.3	0.5	R.T.	[26]
In_2O_3	Reactive evaporation	34	ND	100	[27]
SnO_2	rf sputter	2	ND	600	[24]
TiO_x	PLD	0.08	ND	700	[28]
TiO_x	PEALD	1.64	1.86	250	[29]
TiO_x	dc sputter	0.69	2.45	400	[30]

MULTICOMPONENT OXIDE SEMICONDUCTORS

An historic event in the early development of oxide semiconductors was the demonstration by Nomura et al. of amorphous "a-IGZO" TFTs on flexible substrates [11], as illustrated in Fig. 3(c)and (d). The channel layer, a 30 nm thick a-IGZO, was deposited by PLD at room temperature onto a polyethylene terephthalate (PET) substrate. These TFTs exhibited high field effect mobility (μ_{fe}~8.3 $cm^2/V \cdot s$). This report made a significant impact not only among academic researchers but also industrial engineers, given thermal stability of the amorphous phase of IGZO in air up to ~500 °C, and the ability to lower the carrier concentration below <10^{17} cm^{-3}without decreasing the electron mobility. Interestingly, the electronic configurations of In, Ga, and Zn are $(n-1)$ $d^{10}ns^0$, where n is the principle quantum number of the cations ($n \geq 5$). Those vacant s-orbitals have isotropic properties and spherical symmetry, enabling electronic conduction via direct overlap of the s-orbitals in neighboring cations even within the structural disorder characteristic of amorphous oxides. Moreover, the decrease in carrier concentration may be attributed to the Ga^{3+} ions, which tightly bind to oxygen ions and thereby suppress the formation of oxygen vacancies. After this discovery, multi-component oxide semiconductors have been rapidly developed with the dual strategy of forming amorphous structures together with controlled carrier concentrations. Table 2 summarizes historically important multi-component oxide semiconductors and their TFT properties.

Table 2 Summary of key process and property characteristics of multi-component oxide semiconductor thin film transistors [*a*-/*p*-(amorphous/poly-crystal structure), MOCVD (metal organic chemical vapor deposition)]

Channel Materials	Deposition Method	Mobility ($cm^2/V \cdot s$)	Subthreshold Swing (V/decade)	Process Temperature (°C)	Ref.
a-ZnON	Reactive sputter	10	0.8	350	[32]
p-ZnON	ALD	6.7	0.67	150	[33]
MgZnO	MOCVD	40	0.25	450	[34]
a-InZnO	Sputter	20	1.2	R.T.	[35]
a-InZnO	Sputter	4.5	0.87	40	[36]
a-InZnO	Sputter	27 (19)	ND	600 (200)	[37]

a-InGaZnO	PLD	9	ND	R.T.	[11]
a-InGaZnO	Sputter	12	0.2	R.T.	[38]
a-InGaZnO	Sputter	35.9	0.59	350	[39]
ZrInZnO	Sputter	3.9	0.98	350	[40]
a-HfInZnO	Sputter	10	0.23	200	[41]
a-SnInZnO	Sputter	24.6	0.12	300	[42]
a-AlSnInZnO	Sputter	31.4	0.14	250	[43]
SiInZnO	Sputter	21.6	1.52	150	[44]
a-ZnSnO	Sputter	5–15 (20–50)	ND	300 (600)	[45]
a-ZnSnO	PLD	10	1.4	450	[46]
ZnSnO	Sputter	14	1.6	250	[47]
a-AlZnSnO	Sputter	10.1	0.6	180	[46]
GaZnSnO	Sputter	24.6	0.38	300	[47]
ZrZnSnO	Sputter	8.9	0.7	350	[48]

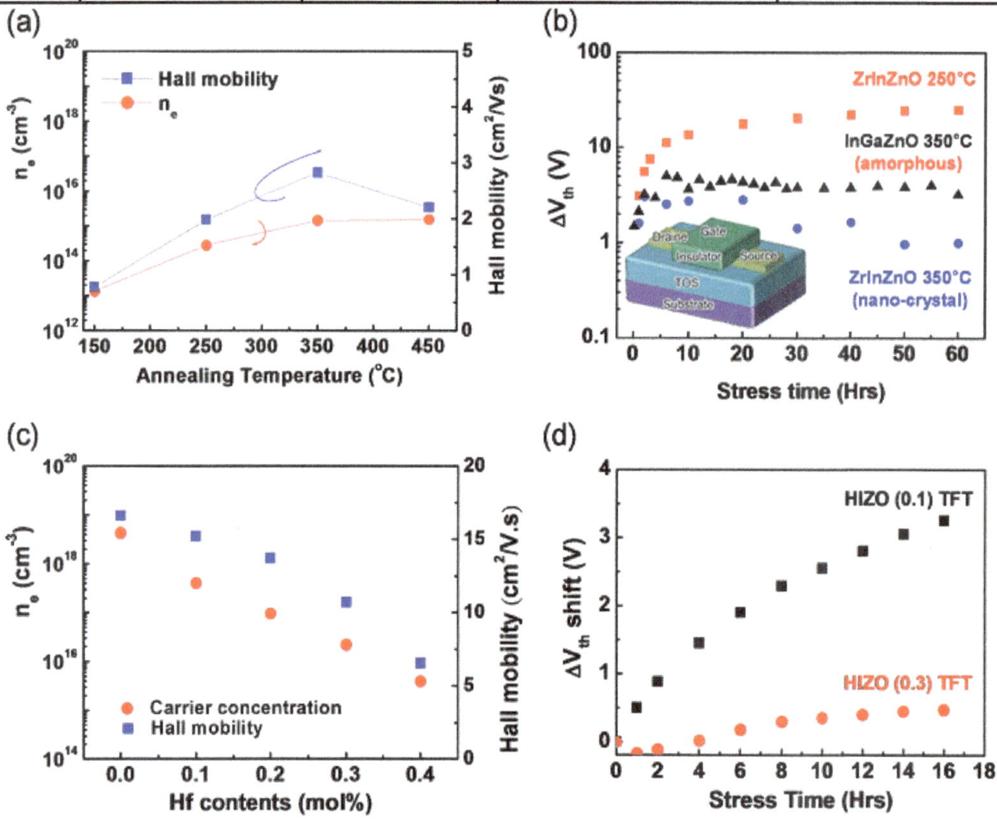

Fig. 4 (**a**) Carrier concentration and Hall mobility of ZrInZnO films for different annealing temperatures. (**b**) The variation of V_{th} for ZrInZnO TFTs annealed at 250 and 350 °C as a function of applied bias stress time. The stability data for *a*-InGaZnO TFT annealed at 350 °C is also presented. Reprinted with permission from [40]. Copyright 2009, Wiley-VCH Verlag GmbH. (**c**) Carrier concentration and Hall mobility as a function of Hf content (HfInZnO, HIZO) (**d**) The variation of V_{th} for the HIZO (Hf=0.1 mol%) and (Hf=0.3 mol%) TFTs as a function of bias temperature stability (BTS) time. The following condition of BTS was applied to both TFTs: V_{GS} of 20 V, V_{DS} of 10 V, and substrate temperature of 60 °C. The TFTs were kept under the BTS condition for 16 h. Reprinted with permission from [41]. Copyright 2009, American Institute of Physics

Another approach has been focused on exchanging Ga with other elements such as Zr [40], Hf [41], La [49], Mg [50], and Si [44] to form amorphous In(Zr, Hf, La, Mg, Si)ZnO-matrices since these elements can act as carrier-suppressors within InZnO as well as network stabilizers. These carrier-suppressors normally have relatively high electro-negativities for binding oxygen strongly within a crystalline structure. Indeed, various multi-component, oxide semiconductors are being examined with the aim of improving the stability of devices by proper selection of carrier-suppressors. Park et al. [40], for example, reported the performance of a ZrInZnO-semiconductor device with its characteristics shown in Fig. 4(a) and (b) including a comparison with that of a-IGZO TFTs. The ZrInZnO films, produced by co-deposition by rf sputtering from InZnO and ZrO_2 targets had nanoscale ZrInZnO crystallites dispersed in an amorphous-phase matrix. The ZrInZnO TFTs exhibited μ_{fe} of 3.9 cm^2/V · s, S of 0.98 V/decade, and drain current on-to-off ratio of 10^7 after a 350 °C anneal in a nitrogen atmosphere. In terms of device stability, the threshold voltage (V_{th}) shift after the application of biasing stress (drain current (I_{DS})=3 μA for 60 h) was only 0.99 V, better than those obtained for a-IGZO and ZnO TFT devices. In spite of the ZrInZnO investigation, it is still difficult to define the precise role of zirconium (Zr), in terms of influencing the electron mobility and device stability.

Kim et al. [41] reported the application of an amorphous HfInZnO thin-film as an oxide semiconductor layer [Fig. 4(c) and (d)]. They suggested that adding Hf (hafnium) could suppress growth of the columnar structure, and drastically decrease the carrier concentration and Hall mobility. Also, the TFTs exhibited a decrease of μ_{fe} and an improvement in device stability, as the amount of Hf increased in the HfInZnO matrix. Therefore, the role of the carrier-suppressor appears to influence carrier concentration, Hall mobility, and device stability due to its high oxygen-bonding ability.

Cost-efficient ZnSnO-matrix-based oxide semiconductors have been under investigation with the addition of a variety of carrier-suppressors due to the high-cost and scarcity of indium and gallium [47, 51]. Cho et al. [51] reported an amorphous AlZnSnO TFT, that exhibited μ_{fe} of 10.1 $cm^2/V \cdot s$, S of 0.6 V/decade, and drain current on/off ratio of 10^9 after annealing at 180 °C. Later, Fortunato et al. [52] developed an amorphous GaSnZnO channel layer, produced by rf magnetron co-sputtering using gallium zinc oxide (GZO) and tin (Sn) targets. The GaSnZnO TFTs exhibited high electrical performance (μ_{fe} of 24.6 $cm^2/V \cdot s$, S of 0.38 V/decade, and drain current on/off ratio of 8×10^7). The researchers suggested that high post-annealing (300 °C) treatment of a-GaSnZnO TFTs produced better electrical performance and stability than those of a-IGZO TFTs due to the Ga^{3+} and Sn^{4+} ions. There have been many similar demonstrations to date involving Si, Al, Zr, Hf, Ga, and Sn with InZnO and ZnSnO matrices, aimed at improving electrical performance (e.g., μ_{fe} and stability) [38–52].

Some researchers have raised concerns about such multi-component oxide semiconductors, considering the potential for achieving non-uniform composition distributions over large areas, and the narrow process-window zones during deposition. Thus, many still look to ZnO to solve both the crystal structure and electrical performance issues. Recently, several dopants (N, In, Sn and Hf) in a ZnO matrix were investigated leading to reasonable electrical performance (μ_{fe} over 10 $cm^2/V \cdot s$) and an amorphous structure; comparable with amorphous multi-component oxide semiconductors (e.g. a-InGaZnO) [32–37].

Ye et al. [32] developed zinc oxynitride semiconducting materials through reactive sputtering in which competition, between reactions responsible for the growth of hexagonal zinc oxide (ZnO) and cubic zinc nitride (Zn_3N_2), is promoted. Interestingly, the zinc oxynitride films formed an amorphous, or highly disordered, nanocrystalline structures depending on the process conditions. The Hall-mobilities of the ZnON films were 47 and 110 $cm^2/V \cdot s$; at 50 and 400 °C annealing temperatures, respectively. These TFTs also exhibited high electrical performance (saturation mobility of 10 $cm^2/V \cdot s$, S of 0.38 V/decade, and drain current on-to-off ratio of 8×10^7). Very recently, Kim et al. reported that the substitution of nitrogen in ZnO could improve the stability of the device on exposure to illumination, since the nitrogen in ZnON may deactivate oxygen vacancies by raising the valence band above the defect levels [37].

SOLUTION PROCESSED OXIDE SEMICONDUCTORS

As mentioned above, all of the oxide semiconductors discussed above were prepared as thin films under vacuum, given the focus on optimizing the electrical properties of semiconductor materials for incorporation into thin film transistor devices. However, alternative electronic modules, such as printable display applications, are rapidly emerging. In this regard, solution-process methods are attractive means of fabricating oxide semiconductor materials suitable for low-cost, large area production. Oxide semiconductors, as shown in Fig. 5, are generally fabricated either by spin coating or printing an aqueous solution, and subsequent annealing in air, or in an inert atmosphere. The aprotic solvent is highly volatile and does not dissociate the metal ligand precursor. The liquid thin-film formed by the precursor, readily loses the solvent and forms a uniform film, which then absorbs moisture from the ambient air. Finally, a solid metal-oxide-thin-film is formed by the substitution reaction between water and metal ligands. Solution-processes can therefore offer an innovative cost-reducing means for increasing substrate size, reducing the number of mask steps, and eventually improving production yield [53].

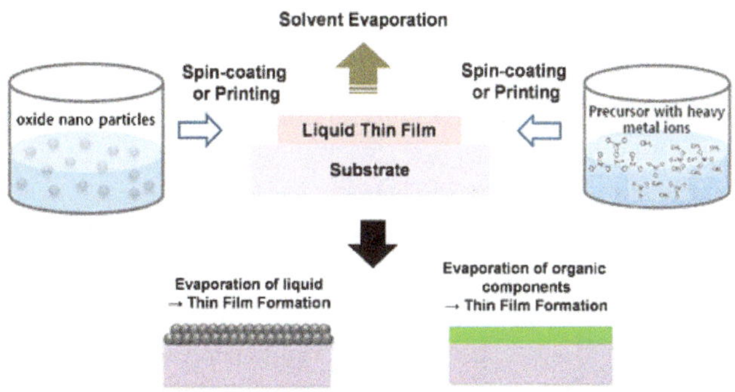

Figure. 5 A schematic illustration of solution-processed thin films based on nano-particle and precursor types

Solution-processed oxide TFTs were initially fabricated with spin-coated ZnO films using Zn-nitrate [$Zn(NO_3)_2$-$6H_2O$] in DI water. The electrical performance of the corresponding TFTs exhibited a mobility of 0.2 $cm^2/V \cdot s$ and an on/off ratio of 10^7 at an annealing temperature of 600 °C [54]. In terms of mobility, Ong et al. [55] were able to further increase the mobility of spin-coated ZnO-TFTs by controlled annealing at 500 °C to enhance the (002) preferred orientation of ZnO. Following this, the mobility of the TFT increased to a value as high as 5.25 $cm^2/V \cdot s$. However, the process temperature is also an important factor that is critical in determining whether solution-processing is advantageous compared to other semiconductor or conventional vacuum processes. Cheng et al. [56] grew a ZnO film using chemical bath deposition (CBD) and subsequently annealing at 100 °C. They obtained a mobility of 0.25 $cm^2/V \cdot s$ out of a bottom-gate TFT.

Multi-component oxide TFTs have also been intensively manipulated to achieve an amorphous structure, high mobility (over 10 $cm^2/V \cdot s$) and low process temperature (below 250 °C) since amorphous InGaZnO (a-IGZO), grown by conventional vacuum processes, exhibited excellent physical and chemical properties, and electrical performance as well. Lee et al. [57] explored a general route to realize printable high-mobility-oxide semiconductors by annealing at 600 °C, using $ZnCl_2$, $InCl_2$, and $SnCl_2$ precursors dissolved in acetonitrile. The mobility values of bottom-gated TFTs with InZnO (spin-coating) and InZnSnO (ink-jet printing) were 16 and 7.4 $cm^2/V \cdot s$, respectively. Although many researchers have studied various precursors, oxide materials, and device structures, the performance of the resulting TFTs was not adequate, and mobility values were far too low, to compete with conventional vacuum processes.

Recently, in terms of high mobility and low processing temperature, four remarkably effective solution-processed approaches have been reported: 'Sol–gel on Chip', 'Combustion', 'Aqueous route', and 'Photochemical activation of Sol–gel Oxide by DUV (Deep-Ultraviolet) irradiation'.

The first process by Banger et al. [58] reported the formation of amorphous metal oxide semiconducting thin films (InZnO and InGaZnO) using a 'sol–gel on chip' hydrolysis approach from soluble metal alkoxide precursors, which resulted in an unprecedented high mobility of 10 $cm^2/V \cdot s$, reproducible and stable threshold voltages (V_{th}~0 V) and high operational stability at maximum process temperatures as low as 230 °C. At lower temperatures, the non-hydrolysed IZO samples exhibited very poor TFT performance with low mobility (>0.1 $cm^2/V \cdot s$) and large hysteresis (35 V), indicative of significant charge trapping at the semiconductor/dielectric interface. In contrast, devices fabricated using the hydrolysed process exhibited high mobility, small hysteresis (<1.2 V) and small sub-threshold slopes (S <0.5 V/decade) at annealing temperatures as low as 230 °C.

The second process reported by Kim et al. [59] used combustion-processing as a new general route to solution-growth of diverse electronic metal oxide films at temperatures as low as 200 °C, using self-energy-generating combustion chemistry. The combustion-synthesis-based oxide TFTs exhibited high mobilities (3.37 $cm^2/V \cdot s$ in In_2O_3, 1.76 $cm^2/V \cdot s$ in ZnSnO, and 0.91 $cm^2/V \cdot s$ in InZnO) at low process temperatures (below 250 °C). This method therefore holds promise for achieving high-performance at low temperatures, with the opportunity to integrate amorphous oxide materials into novel electronic devices.

The third process is a simple and novel 'aqueous route' for fabricating oxide TFTs at annealing temperatures below 200 °C, suggested by several researchers [60–62]. The suggested aqueous route provides low-temperature oxide-formation and good stability by restricting the hydrolysis and condensation reactions within a solution state. An amorphous In_2O_3 thin film was obtained upon annealing at 175 °C, and the optimized TFT (200 °C) exhibited good uniformity and electrical performance (μ_{fe}=2.62±0.25 $cm^2/V \cdot s$, S=0.29±0.06 V/decade, and a turn-on voltage of 0 V) [60]. In addition, the starting Zn solutions were prepared by directly dissolving $Zn(OH)_2$ in aqueous ammonia solutions, to produce a so-called 'impurity-free precursor'. The use of zinc hydroxide allowed for low-temperature ZnO formation, indicating rapid, low-energy kinetics of metal-amine dissociation and simple dehydration and condensation reactions to form ZnO films [61, 62]. The undoped ZnO (140 °C under microwave-assisted annealing) and Li doped ZnO TFTs (300 °C annealing) exhibited remarkable performance: μ_{fe}~1.7 $cm^2/V \cdot s$ and 11.45 $cm^2/V \cdot s$ for undoped and Li-doped devices, respectively.

The fourth process is the DUV photochemical activation of sol–gel films. It plays an important role in forming high-performance, stable metal-oxide semiconductors because DUV irradiation induces efficient condensation and densification of oxide semiconducting films by photochemical activation at low temperature. As shown in Fig. 6, Kim et al. [63] reported that flexible a-IGZO TFTs, fabricated by DUV irradiation at room temperature, exhibited reasonably high mobility values of 14 and 7 $cm^2/V \cdot s$ with an Al_2O_3 gate insulator on glass and polymer substrates, respectively.

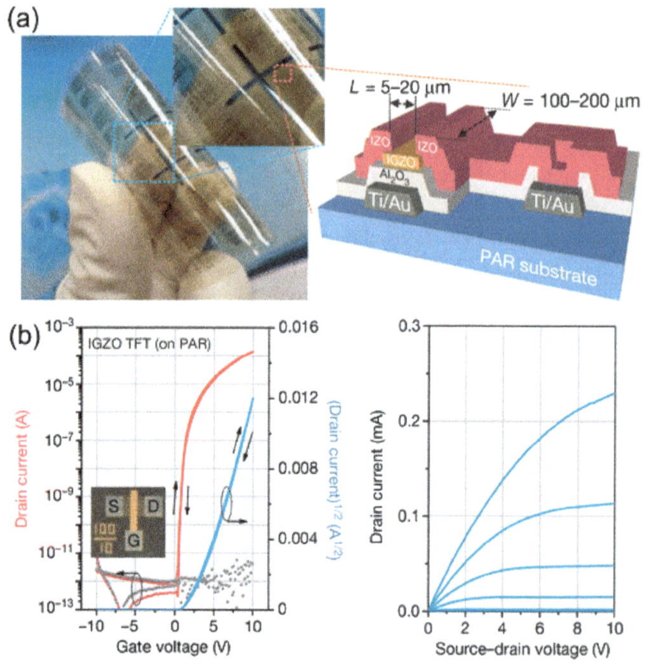

Figure. 6 (**a**) Optical micrographs and a schematic cross-section of photo-annealed IGZO TFTs on a polyarylate (PAR) substrate (**b**) Electrical characteristics (transfer and output curves) of photo-annealed IGZO TFTs (channel width and length are 10 and 100 μm, respectively) on a PAR substrate. Reprinted with permission from [63]. Copyright 2012, Macmillan Publishers Ltd

These four novel solution processes have mounted strong challenges to vacuum methods for achieving high-performance, flexible, printed metal-oxide thin-film electronic devices. Although it is still necessary to improve the mobility and stability of these TFTs, it seems reasonable that solution-processed oxide TFTs will ultimately exhibit desired properties following continued development. Table 3 summarizes approaches used in solution-processed oxide semiconductors [53–73].

Table 3 Listing of solution-processed oxide semiconductors including key process and property characteristics

Material	Coating	Solut. Type	μ_{fe} ($cm^2/V \cdot s$)	Temp. (°C)	Ref.		
ZnO	SC	Precursor	0.2	700	[54]		

ZnO	CBD	Precursor	0.25	100	[56]		
ZnO	SC	Precursor	5.25	500	[55]		
ZnSnO	SC/IJ	Precursor	16	600	[57]		
ZnInO	IJ	Precursor	7.4	600	[57]		
ZnO	IJ	Precursor	1.8	150	[64]		
ZnZrO	SC	Precursor	0.0042	300	[65]		
InZnO	SC	Precursor	7.3	500	[66]		
nc-IGZO	SC	Precursor	7.84	400	[67]		
IZTO	IJ	Precursor	30	600	[68]		
AIO	SC	Precursor	19.6	350	[69]		
a-ZnSnO	SC	Precursor	5	500	[70]		
a-InZnO	SC	Precursor	5	350	[71]		
ZnInSnO	SC	Precursor	10~100	400	[72]		
InGaZnO	SC	Precursor	2.3	95	[73]		
a-IZO	SC	Precursor	10	<250	[58]		
a-IGZO							
In_2O_3	SC	Precursor	3.37	<250	[59]		
a-InZnO						1.76	<250
In_2O_3	SC	Precursor	2.62	200	[60]		
ZnO	SC	Precursor	1.7	140	[61]		
Li-ZnO	SC	Precursor	11.45	300	[62]		
a-IGZO	SC	Precursor	14	<100	[63]		

SC spin coating, *IJ* ink jet, *CBD* chemical bath deposition

At present, in spite of promising research and frequent developments, the mobility and instability in oxide semiconductor TFTs must still be improved to realize the dream of achieving low cost non-silicon-based electronics. Fortunately, many researchers continue to search for more stable and efficient multi-oxide semiconductors, pursue an improved understanding of the mechanisms of instability, while considering how to achieve desired electrical properties, homogenous structures, lower-cost processes, process-adaptability for large areas, and bias-temperature-reliable materials.

GATE DIELECTRIC MATERIALS

High-k dielectric materials play an increasingly important role in the miniaturization of microelectronic components. The need to maintain a high capacity in spite of shrinking dimensions can no longer be accommodated by reductions in the thickness of SiO_2 and other low-k dielectrics, as structures are moving towards the nm thickness range. As described previously, thin film transistors (TFTs) are composed of a semiconductor channel, gate insulator, source, drain, and gate electrodes. Among them, the semiconductor channel is the most important element since it governs the major characteristics of the device (e.g., field-effect mobility, on/off current ratio, threshold voltage, and subthreshold swing). In addition, the interface properties between the semiconductor channel layer and gate insulator are considered key parameters for improving reliability as well as performance (e.g., charge trap density, threshold voltage, and sub-threshold swing) of the TFTs. In particular, the operation voltage of TFTs is determined by the capacitance level, which is directly related to the dielectric constant and the thickness of dielectric layers [74–76]. According to Eqs. 1 and 2 describing the TFT drain current I_{DS} [77], a viable approach towards increasing the drain current, while operating at low bias, is to increase the capacitance of the dielectrics; for a planar structure, $C_i = \varepsilon_0(kA/d)$, where k is the dielectric constant, A the area of the electrodes, and d the dielectric thickness.

In this regard, significant attention has been paid to gate insulators, especially optimized for oxide TFTs because their switching voltage can be markedly reduced using high-k dielectric layers or very thin gate dielectric films, leading to high capacitance values. Low-voltage operation in oxide-based TFTs result in cost savings by enabling use of lower-cost driver electronics and by low battery power consumption, making them suitable for portable applications. There are several types of dielectric layers which have been used in oxide-based TFTs. They include polymers [78, 79], high-k metal oxides [80, 81], SiO_2 [82, 83], Si_3N_4 [84,85], and hybrid dielectrics composed of multi-layered films [86, 87].

SiO_2 layers have to date been widely utilized in most types of TFTs as the gate insulator due to their easy fabrication by chemical vapor deposition (CVD) and their optimized film properties, including precise thickness control and uniform large-area coverage. However, the rather low dielectric constant (~3.9) and relatively high growth temperatures

(>300 °C) of SiO_2 layers are major limitations in the fabrication of SiO_2 dielectric layers on plastic substrates (i.e., polyethylene terephthalate or polycarbonate) with low glass-transition temperature. Although a high capacitance value can be obtained by the introduction of very thin SiO_2 dielectric layers (less than 20 nm), we note that pinhole formation and poor step-coverage on relatively rough plastic substrates are detrimental to the homogeneous coating of thin dielectric layers on plastic substrates [88, 89].

Similar problems are associated with polymer-based dielectric layers with relatively low dielectric constants. Although very low voltage operation has been successfully demonstrated through the use of several polymer dielectrics (i.e., SAM (self-assembled monolayer) [90, 91], SAS (self-assembled super-lattice) [92], CPB (cross-linked polymer blend) [93], PVA [poly(vinyl alcohol)] [94], PMMA [poly(methyl methacrylate)] [79], and PVP [poly(4-vinylphenol)] [95, 96]), good TFT performance with such layers has been achieved only on very smooth Si wafers or well-surface-treated, smooth-plastic substrates. Thus, inherent (bare) polymer substrates, characterized by rough surfaces (rms roughness ~3 nm), benefit from the use of high-k dielectrics given that high electric fields can be achieved with the use of thicker films (~200 nm) without need to increase the operating voltage.

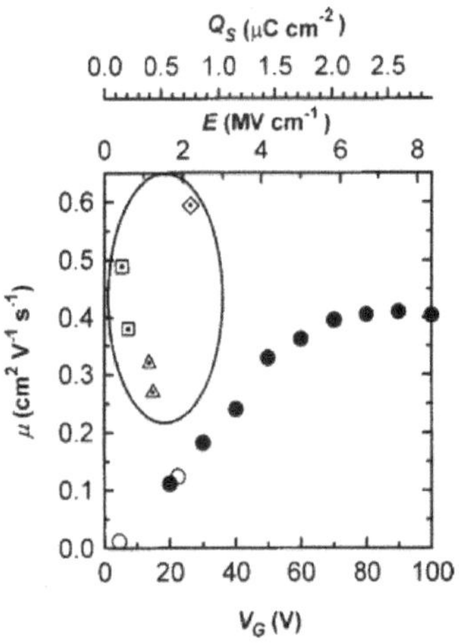

Figure. 7 Dependence of mobility on V_G, E, and Q_S. *Solid circles* refer to all three x axes and correspond to devices with 0.12-μm-thick SiO_2. *Open circles* refer only to E and Q_S axes (0.5-μm-thick SiO_2). *Dotted* symbols in ellipse refer to the E axis only. *Triangle*, BZT;*square*, BST; *diamond*, Si_3N_4 insulator. Reprinted with permission from [74]. Copyright 1999, AAAS

The use of such high-k oxide films with sufficient thickness (>100 nm), leads to the suppression of pinhole formation and minimization of problems associated with step coverage. Normally, high temperatures are required to obtain the crystalline form of high-k, dielectric-oxide thin films, given their typical refractory nature. In 1999, researchers at IBM first demonstrated room temperature grown amorphous Ba(Zr,Ti)O_3 (BZT) thin films as a gate insulator [74]. The rf-sputtering method was used to grow 122 nm thick BZT films (k=17.3). The pentacene-based organic TFTs with BZT gate insulators showed much lower voltage operation (<14 V) as compared to that (~100 V) of the pentacene TFTs using SiO_2 dielectric layers [Fig. 7]. Although the IBM group successfully demonstrated the potential suitability of room-temperature grown high-k gate insulators combined with organic semiconductors for the first time, the relatively low breakdown characteristics and high leakage current characteristics of the gate insulators prepared at room temperature were major drawbacks. These need to be solved to provide stable and reliable operation of oxide TFTs. Here, we provide an overview of various gate oxide insulators in terms of structural, compositional and dielectric properties. A number of promising gate insulators have been suggested and developed, particularly for low voltage operation of oxide TFTs.

BINARY OXIDE DIELECTRICS

A number of binary oxide-based gate insulators with relatively high dielectric constant values (k=10–40) have been investigated as alternative gate dielectrics to replace SiO_2. These medium or high-k binary oxides include Al_2O_3, HfO_2, TiO_2, ZrO_2, Y_2O_3, La_2O_3, and Ta_2O_5, and their material properties (i.e., optical band gap, crystal structure, dielectric constant, and breakdown strength) were summarized in Table 4 [97–117]. In particular, the potential feasibility of the usage of high-k binary oxides (Al_2O_3 (k=8.5) [33], HfO_2 (k=20.1) [118], TiO_2 (k=40) [119], ZrO_2 (k=25) [120], Y_2O_3 (k=14) [39], La_2O_3 (k=10.8) [121], Ta_2O_5 (k=20.3) [121]) has been demonstrated for application in ZnO-based oxide TFTs, as summarized in Table 5. However, a common limitation of near room temperature grown high-k dielectric films is their tendency to suffer from poor leakage current characteristics, which is detrimental to TFTs operation. In addition, the large polarizability of the high-k insulators has an adverse effect on their mobility [119]. These issues should thus be carefully addressed to insure widespread application of high-k gas insulators. In general, a binary oxide with a high dielectric constant typically possesses a smaller band gap, which is can be correlated with lower breakdown strength. In contrast, a binary oxide with a wide band gap typically has a smaller dielectric constant. Therefore, it is very important to understand that adoption of binary oxides involves a trade-off between breakdown strength and dielectric constant. As an example, Pei et al. [123] fabricated HfO_2-Al_2O_3 (HfAlO) composite insulators by rf co-sputtering HfO_2 and Al_2O_3 ceramic targets. As shown in Fig. 8, their optimized HfAlO films $[(HfO_2)_{0.86}(Al_2O_3)_{0.14}]$ showed larger band gap (~5.58 eV) and smaller dielectric constant (~16), compared to those (~5.31 eV and ~17.7, respectively) of pure HfO_2. Eventually, they were able to significantly reduce the leakage current density from $\sim 1\times10^{-2}$ A/cm^2 to below 3×10^{-10} A/cm^2, at an electric field strength of 0.25 MV/cm. Kim et al. [124] also proposed new types of $Mg_2Hf_5O_{12}$ gate insulators to improve leakage-current characteristics by combining relatively low-k MgO (k~10) and high-k HfO_2 (k=20.1), as shown in Fig. 9. Polycrystalline, single-phase $Mg_2Hf_5O_{12}$ thin films exhibited greatly enhanced leakage current characteristics ($<2\times10^{-7}$ A/cm^2) compared to that (2×10^{-5} A/cm^2) of HfO_2 thin films at 0.4 MV/cm and a high dielectric constant (k=22). The optical band gap of $Mg_2Hf_5O_{12}$ is ~0.16±0.02 eV larger than that (5.67±0.02 eV) of HfO_2. $InGaZnO_4$ TFTs using $Mg_2Hf_5O_{12}$ gate insulators, which were fabricated on plastic substrates, showed a high field-effect mobility of 27.32 cm^2/V · s and a current on/off ratio of 4.01×10^6. These reports highlight the importance of the manipulation of composition to achieve high dielectric constants as well as improved breakdown strength.

Table 4 Summary of material properties of binary oxide insulators

Binary Oxide	Al_2O_3	HfO_2	TiO_2	ZrO_2	Y_2O_3	La_2O_3	Ta_2O_5	
Optical Band gap, Eg (eV)	8.8 eV	5.5~6.0 eV	3.0~3.5 eV	5.0~7.0	6	4.3~6	4.4	
	[97]	[98–100]	[101]	[102]	[97]	[97, 103]	[97]	
Crystal Structure	Rhombohedral	Monoclinic: <1022 °C	Tetragonal:	Monoclinic: <1170 °C	Monoclinic: <500 °C	Hexagonal: <300 °C	Orthorhombic: 400~1450 °C	
	(Trigonal): >1000~1100 °C	Tetragonal: 1022~2422 °C	Anatase: <600 °C	Tetragonal: 1170~2370 °C	Cubic: >500 °C	Cubic: >300 °C	Tetragonal: >1450 °C	
	[104]	Cubic: >2422 °C	Rutile: >700 °C	Cubic: 2370~2680 °C	[109]	[110]	[111]	
			[105]	[106]				[107,108]
Relative Dielec-tric	8.8	22~25	Anatase: 30	22~24	13~17	27~30	22	
Constant	[112]	[99]	Rutile: 80~170 [106,113]	[102]	[114]	[103, 112]	[112]	
Breakdown	11.2~13.8	3.9~6.7	1.4~2.5	3.3~5.7	1.9~3.0	3.3~5.6	2.1~3.7	
Strength (MV/cm)	[115]	[115]	[106]	[115]	[116, 117]	[115]	[115]	

Table 5 Comparison of key processing and property characteristics of ZnO-based oxide TFTs with binary oxide insulators

Gate Insulator Materials	Deposition Method	Mobility ($cm^2/V \cdot s$)	Subthreshold Swing (V/decade)	Process Temperature (°C)	Ref.
Al_2O_3 (k=8.5)	ALD	6.7	0.67	150	[33]
HfO_2 (k=20.1)	rf sputter	3.53	0.095	300	[118]
TiO_2 (k=40)	ALD	9.9	0.22	250	[119]
ZrO_2 (k=25)	rf sputter	28	0.56	R.T.	[120]
Y_2O_3 (k=14)	rf sputter	12	0.2	140	[39]
La_2O_3 (k=10.8)	ALD	0.77	1.2	150	[121]
Ta_2O_5 (k=20.3)	rf sputter	40.5	0.23	R.T.	[122]

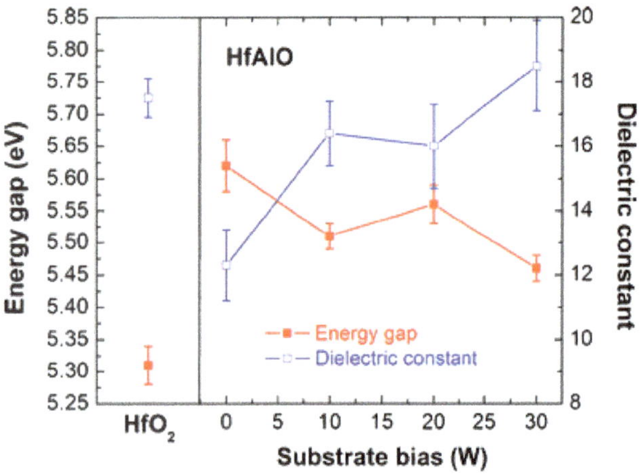

Figure. 8 Energy gap calculated from spectroscopic ellipsometry (SE) results and the dielectric constant determined from capacitance-voltage measurements in accumulation regime for HfAlO films deposited with different substrate biases. The values for pure HfO_2 are also presented. Reprinted with permission from [123]. Copyright 2009, The Electrochemical Society

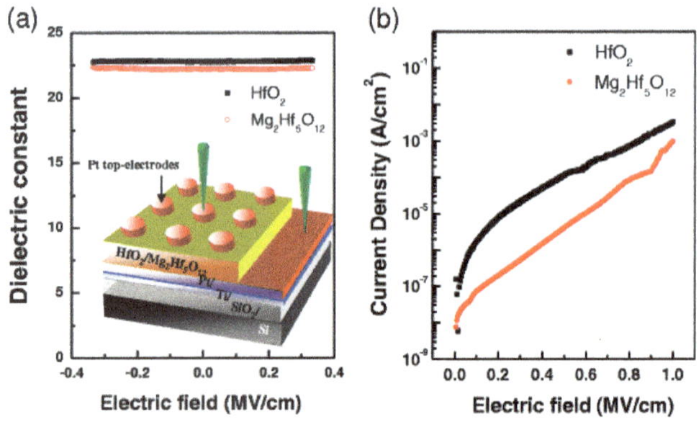

Figure. 9 (**a**) The relationship between the dielectric constant and the electric field in HfO_2 and $Mg_2Hf_5O_{12}$ thin films deposited at room temperature. (**b**) The relationship between the leakage current characteristics and the electric field applied to the HfO_2 and $Mg_2Hf_5O_{12}$thin films. Reprinted with permission from [124]. Copyright 2010, The Electrochemical Society

For routine application of high-k binary oxides as gate dielectrics, further investigation of their chemical stability, interfacial trap density distribution, and related device instability have to be performed. Kwon et al. [125] studied the effect of gate dielectric material on the device stability of a-HfInZnO TFTs. It was observed that the Al_2O_3 gated device exhibited a stability comparable to a conventional device with the SiO_2 gate dielectric (promising for future applications), while the HfO_2 device suffered from a huge negative threshold voltage shift (>11 V) during application of negative-bias-thermal illumination stress for 3 h [Fig. 10]. Their work suggests that HfO_2 normally contains a large number of trap sites or fixed charges, which have to be significantly reduced for its practical application. For example, Zou et al. [126] suggested that nitrogen incorporation into HfO_2 can effectively improve the interface quality and thus enhance the reliability of resulting a-IGZO TFTs with HfO_xN_y gate dielectrics. They demonstrated that the density of interface state (~5.2×10^{11} eV^{-1} cm^{-2}) of the HfO_xN_y/a-IGZO device is smaller than that (~1.1×10^{12} eV^{-1} cm^{-2}) of the reference HfO_2/a-IGZO, because nitrogen incorporation in the dielectric film can effectively passivate unsaturated dangling bonds, and hinder the diffusions of Hf and Ga.

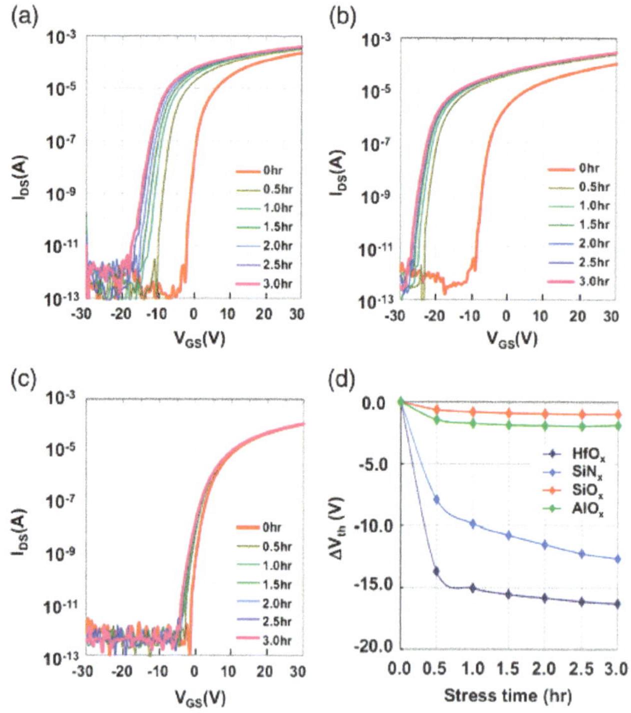

Fig. 10 (**a**) Evolution of the transfer characteristics for (**a**) SiNx/HIZO, (**b**) SiNx/HfOx/HIZO, and **c** SiNx/SiOx/HIZO devices as a function of the applied NBITS time for the HIZO TFT with various gate dielectric materials and structures. Reprinted with permission from [125]. Copyright 2010, American Institute of Physics

PEROVSKITE-BASED OXIDE DIELECTRICS

Among perovskite oxides, $(Ba,Sr)TiO_3$ (BST) has attracted much attention as a high-k dielectric material for application in storage capacitors of gigabit dynamic random-access memory (DRAM), ferroelectric random-access memory (FRAM), and electrically tunable microwave devices due to its superior properties (e.g., high dielectric constant, high dielectric breakdown strength, and low dissipation factor) [127–131]. In particular, the capacitive properties of BST can be tuned by more than 50 % at low bias levels (<5 V) resulting in similar percentage changes in frequency characteristics of tuned circuits [130, 131]. For the above-mentioned applications, BST films have been grown at high processing temperatures, typically above 600 °C, due to its refractory nature [128, 132]. However, layers of BST grown at high temperature are not suitable as gate insulator for oxide TFTs, particularly those fabricated on plastic substrates.

Dimitrakopolous et al. first proposed the potential use of room-temperature deposited barium strontium titanate BST film (k = 16) as a high-k gate dielectric for making low-voltage operating organic TFTs [74]. Following this technological breakthrough, a number of research groups have searched for BST-based high-k gate insulators for ZnO-based oxide TFTs, as summarized in Table 6. The voltage-dependent, dielectric-constant characteristic of high-temperature-grown BST dielectrics with polycrystalline nature is an attractive feature for application in tunable microwave devices. However, this behavior is less desirable when attempting to achieve low-voltage and high-frequency operating TFTs. Interestingly, room-temperature-grown BST films prepared by rf-sputtering exhibit an amorphous structure; thus the tunable dielectric behavior of BST films is not an issue [Fig. 11]. While the relative dielectric constant of the amorphous phase (k=28) is smaller than that of the polycrystalline phase (k >200), amorphous BST films with sufficiently high capacitance density possess a large voltage and frequency independent dielectric constant, which is desirable for the fabrication of room-temperature-processed, ZnO-based TFTs [inset of Fig. 11]. Kim et al. [133] fabricated high-performance top-contact bottom-gate a-IGZO TFTs utilizing a 170 nm thick amorphous BST as gate insulator. They demonstrated low-voltage operation (<3 V) with a high saturation mobility value of 10 $cm^2/V \cdot s$, an excellent sub-threshold swing of 60 mV/decade, and a low threshold voltage of 0.5 V, as shown in Fig. 12. Li-Ping et al. [139] also reported the fabrication of low-voltage, depletion-mode-operating, indium-tin-oxide (ITO) TFTs gated by an amorphous BST gate dielectric. Their BST films, with a thickness of 400 nm, showed a leakage current density of 6×10^{-8} A/cm^2 (at an applied electric field of 0.125 MV/cm) and a high dielectric constant of ~37. The ITO TFT devices exhibited a threshold voltage of −3.7 V, a sub-threshold swing of 0.5 V/decade, and a field-effect mobility of 3.2 $cm^2/V \cdot s$. While there have been some earlier promising results for high-performance TFTs with amorphous BST gate dielectrics, room-temperature grown BST films generally suffered from poor leakage current characteristics at voltages above 5 V, which is detrimental to stable operation of TFTs (see Fig. 13).

Table 6 Comparison of key processing and property characteristics of ZnO-based oxide TFTs with perovskite oxide insulators

Gate Insulator Materials	Deposition Method	Mobility ($cm^2/V \cdot s$)	Subthreshold Swing (V/decade)	Process Temperature (°C)	Ref.
BST (k=28)	rf sputter	10	0.06	325	[133]
Mg-BST (k=28)	rf sputter	16.3	0.4	R.T.	[134]
Mn-BST (k=24)	rf sputter	1.0	ND	R.T.	[135]
Ni-BST (k=26.5)	rf sputter	2.2	0.21	300	[136]
MgO-BST (k=18)	rf sputter	10.9	0.46	R.T.	[137]
PMMA/BST (k=19.5)	rf sputter	10.2	0.44	R.T.	[138]

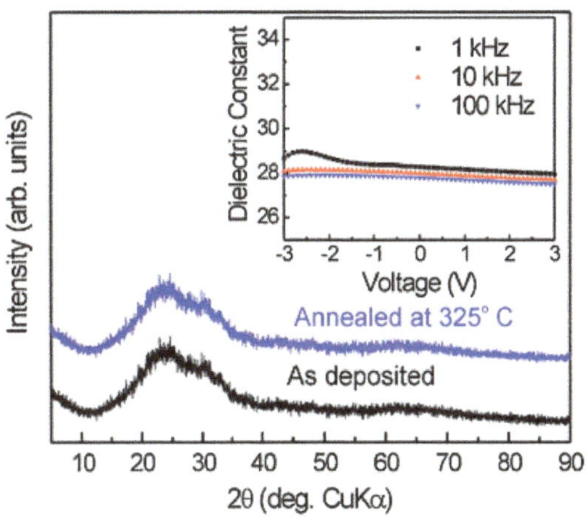

Figure. 11 XRD patterns of a-BST film on glass substrate as deposited at room temperature and after annealing at 325 °C for 1 h. *inset*: Dielectric constant measured on a Au/Ti/a-BST/ITO/glass sample as a function of bias voltage and at 1, 10, and 100 kHz. Reprinted with permission from [133]. Copyright 2008, American Institute of Physics

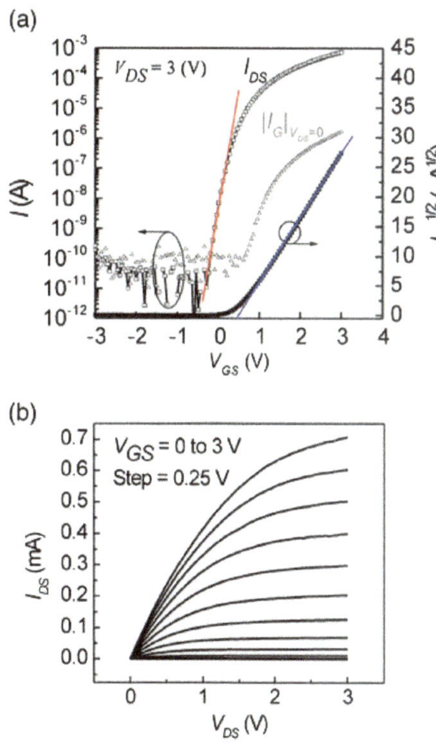

Figure. 12 Operation characteristics of sputtered a-BST/a-IGZO TFT with L=5 μm and W=1000 μm. (**a**) Transfer characteristics (*squares*) V_{GS} was swept from −3 to +3 V and leakage current (*triangles*) at V_{DS}=0 V. (**b**) Output characteristics. V_{DS} was swept from 0 to +3 V at each V_{GS} varied from 0 to 3 V at 0.25 V steps. Reprinted with permission from [133]. Copyright 2008, American Institute of Physics

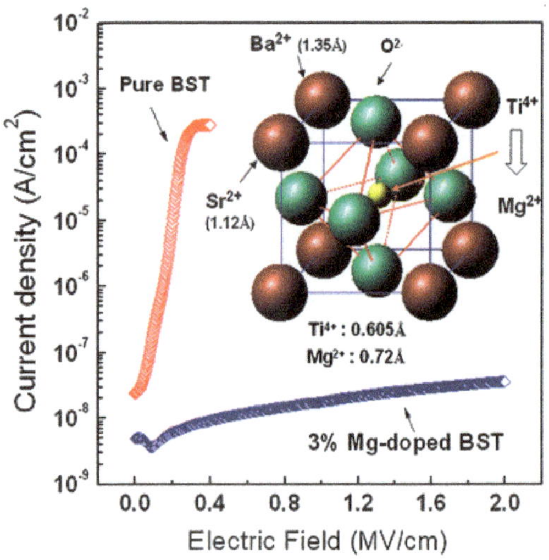

Figure. 13 Current-density-electric-field characteristics of pure BST and 3 % Mg-doped BST film. The *inset* shows a schematic structure for Mg substitution into Ti site in BST lattice. Reprinted with permission from [134]. Copyright 2007, American Institute of Physics

Table 7 Comparison of key processing and property characteristics of ZnO-based oxide TFTs with pyrochlore oxide insulators

Gate Insulator Materials	Deposition Method	Mobility ($cm^2/V \cdot s$)	Subthreshold Swing (V/decade)	Process Temperature (°C)	Ref.
BZN (k=51)	rf sputter	1.13	ND	R.T.	[81]
BZN (k=70)	PLD	42	0.24	100	[140]
MgO/BZN (k=31.5)	rf sputter	5.4	ND	R.T.	[86]
MgO-BZN (k=32)	rf sputter	37.2	0.46	R.T.	[141]

In order to improve leakage current characteristics (i.e., enhancement of breakdown strength, of room-temperature-grown BST thin films), doping the amorphous BST host with acceptor dopants (i.e., Mn, Ni, and Mg) which partially substitute for Ti on the B site of the $A^{2+}B^{4+}O^{2-}$perovskite structure, has been suggested [134–136]. On the basis of the similar ionic radii of Ti^{4+} (r_{eff}=0.605 Å) and Mg^{2+} (r_{eff}=0.72 Å) in six-fold oxygen coordination, we can assume that Ti^{4+} is replaced by Mg^{2+} in the BST lattice. For example, based on simple defect chemistry, negatively charged defects (Mg_{Ti}'') and a corresponding number of doubly ionized oxygen vacancies ($V_O^{\cdot\cdot}$) are simultaneously formed to satisfy site balance and charge neutrality. This can be described as:

$$MgO \xrightarrow{TiO_2} Mg_{Ti}'' O_O + V_O^{\cdot\cdot} \quad (3)$$

In this case, Mg behaves as an acceptor-type dopant, which can prevent reduction of Ti^{4+} to $Ti^{3+}$by neutralizing the donor action of the oxygen vacancies. For example, at the low oxygen partial pressures used to deposit BST film, reduction occurs with the formation of oxygen vacancies, resulting in n-type conductivity according to:

$$O_O \rightarrow V_O^{\cdot\cdot} + 2e' + 1/2\, O_2(g) \quad (4)$$

where O_O, $V_O^{\cdot\cdot}$, and e′ represent the oxygen ion on its normal site, the oxygen vacancy, and electron, respectively. By increasing in the oxygen vacancy concentration by Mg doping, reaction 4 is driven back to the left, causing a decrease in the concentration of electrons, and thereby an enhancement of the breakdown strength of Mg-doped BST thin films. In agreement with these arguments, the leakage current density ($<5\times10^{-8}$ A/cm^2 at 2 MV/cm) of 3 % Mg-doped BST film is markedly reduced compared to undoped BST films ($<3\times10^{-4}$ A/cm^2 at 0.4 MV/cm), as shown in Fig. 13. The valence state of an Mg ion is 2+, thus we can expect further reduction in the leakage current density in Mg-doped BST films compared to BST films doped by acceptors with the higher multi-valence states.. Kang et al. [134] suggested the potential suitability of a 3 % Mg-doped BST gate insulator for high field-effect mobility (16.3 cm^2/V · s) TFTs fabricated on PET substrates. Kim et al. [136] reported the successful integration of 1 % Ni-doped BST as a gate dielectric for ZnO TFTs exhibiting a very low operation voltage (4 V), field-effect mobility (2.2 cm^2/V · s), on/off current ratio (1.2×10^6), and subthreshold swing (0.21 V/decade). ZnO TFTs with 3 % Mn-doped BST showed a field-effect mobility of 1.0 cm^2/V · s and a low operating voltage of less than 7 V [135]. These results were based on a-BST small amounts of acceptor dopants (<5 %).

To further improve the leakage current properties of room-temperature-grown BST films, Kim et al. [137] fabricated MgO-BST composite thin films (i.e., excess MgO-loaded BST). With increasing MgO content, the leakage current was indeed significantly suppressed, as shown in Fig. 14. A remarkable reduction in the leakage current density (below ~1×10^{-7} A/cm^2 even up to an applied electric field of 0.5 MV/cm) could be achieved with a composition ratio of $MgO_{0.3}BST_{0.7}$. It is interesting to see the high resolution TEM image of $MgO_{0.3}BST_{0.7}$ film, suggesting a locally dispersed nanocrystalline MgO phase in an amorphous BST matrix (Fig. 15). The uniform distribution of MgO nanoparticles serves to suppress leakage current flow within the BST matrix due to their excellent insulating characteristics. Furthermore, low-voltage operation (4 V), high-performance a-IGZO TFT (μ_{fe} = 10.9 cm²/V · s) could be demonstrated by use of the $MgO_{0.3}BST_{0.7}$ gate dielectric on plastic substrates.

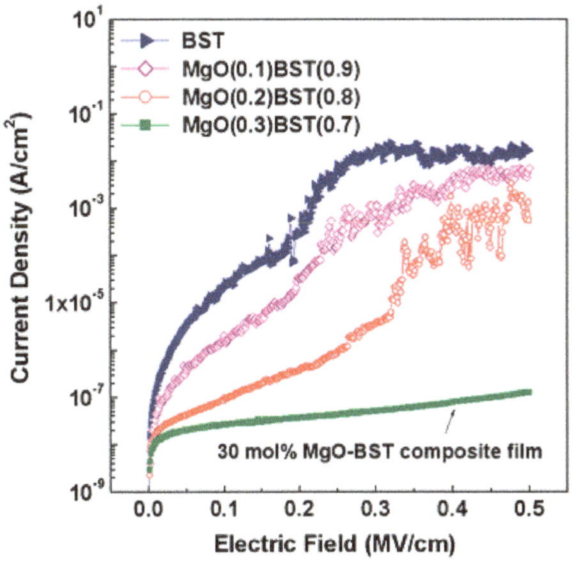

Figure. 14 Leakage current characteristics of pure BST and MgO-BST composite thin films deposited by PLD. Reprinted with permission from [137]. Copyright 2008, American Institute of Physics

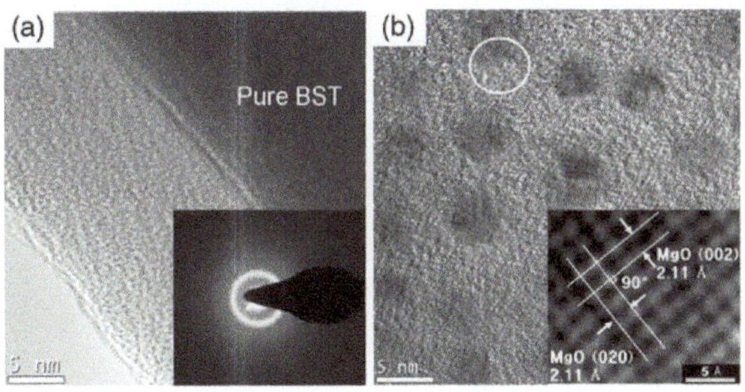

Figure. 15 (**a**) HR-TEM image of pure BST thin films. The inset exhibits a selected area diffraction pattern of the BST thin film. (**b**) HR-TEM image of MgO-BST composite thin film. The*inset* shows the lattice fringe of the MgO nanoparticles. Reprinted with permission from [137]. Copyright 2008, American Institute of Physics

PYROCHLORE BASED OXIDE DIELECTRICS

Bi-based pyrochlore materials with $A_2B_2O_7$ composition (e.g., $Bi_{1.5}Zn_{1.0}Nb_{1.5}O_7$, hereafter, BZN) have been widely investigated for various applications of voltage-tunable microwave devices and integrated decoupling capacitors [141–144]. Interestingly, the low dielectric loss tangent (tan δ) of 5×10^{-4} and the high resistivity (~3×10^{13} Ωcm) would normally make BZN films highly suitable for microwave regime applications, except for the fact that they exhibit low tunability [144]. However, for TFT operation, the low-voltage dependence of the dielectric constant ensures more predictable operation (Table 7). Kim et al. [81, 145] developed room-temperature-deposited, BZN gate insulators with enhanced leakage current characteristics (<10^{-7} A/cm² for voltages below 4 V) and high relative dielectric constant (~51); thereby leading to low-voltage operating (<4 V) ZnO TFTs. Also, it is noted that the BZN films exhibit the highest relative dielectric constant (above 50), among room-temperature-grown, high-k gate insulators prepared by physical vapor deposition such as sputtering. Fu et al. [140] also successfully fabricated high-performance ZnO nanowire (NW) TFTs combined with low-temperature (~100 °C) PLD-grown BZN (k=70) as gate insulator. The NW transistors exhibited a low operation voltage (<3 V), a high field-effect mobility (~42 cm²/V · s), and a steep subthreshold swing up to 0.24 V/decade. BZN films have a cubic, pyrochlore-like structure even for deposition at room temperature [see Fig. 16]. The BZN films exhibit a partially nanocrystalline morphology with a crystalline size distribution of 5–8 nm in diameter [inset of Fig. 17]. Therefore, the high dielectric constant achievable with BZN, even when processed at room temperature, can be understood on the basis of its crystalline-like structure (i.e., better short range order). However, room-temperature-deposited BZN films were shown still to be largely unreliable (>5×10^{-5} A/cm²) with respect to their leakage current characteristics at high applied electric fields (above 0.3 MV/cm) [86, 146]. Therefore, further optimization was necessary to improve breakdown strength and reduce leakage current for room-temperature-grown BZN films.

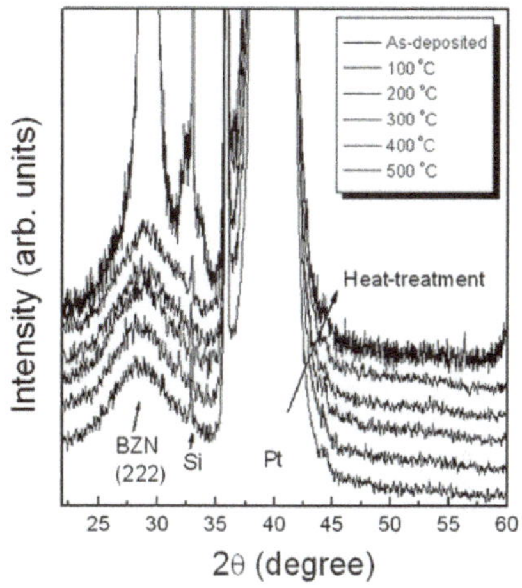

Figure. 16 X-ray diffraction pattern of BZN films as a function of annealing temperature. Reprinted with permission from [81]. Copyright 2006, American Institute of Physics

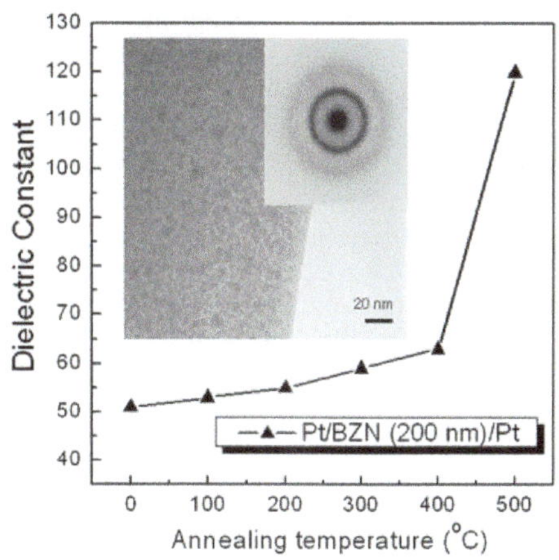

Figure. 17 Dielectric constant of BZN films as a function of annealing temperature. The *inset* shows diffraction pattern and TEM images of BZN films. Reprinted with permission from [81]. Copyright 2006, American Institute of Physics

MgO, with a relative dielectric constant of 9.96 and a high band gap of 7.3 eV, has been used as a protective or an intermediate buffer layer in various electronic devices due to its excellent insulation properties [147–149]. Lim et al. [86] demonstrated that the introduction of an insulating, MgO-capping layer onto the BZN films could improve the leakage current characteristics of BZN films (i.e., below 3×10^{-8} A/cm^2 at an applied electric field of 0.5 MV/cm). Recently, Cho et al. [146] reported the preparation of excess MgO-added BZN-composite thin films from a single composite target, rather than two-step thin-film deposition of BZN and MgO films. The two-step process is not desirable in terms of high processing cost and low manufacturing yield. The $MgO_{0.3}BZN_{0.7}$ composite gate insulators exhibited greatly enhanced leakage-current properties ($<2 \times 10^{-8}$ A/cm^2 at 0.3 MV/cm), while retaining an appropriate high relative dielectric constant of 32. Finally, the ZnO TFTs incorporating the MgO-BZN composite gate insulator showed superior TFT performance including a high field-effect mobility of 37.2 cm^2/V · s, a reasonable on/off ratio of 1.5×10^5, a sub-threshold swing of 0.46 V/decade, and a low threshold voltage of 1.7 V [see Fig. 18].

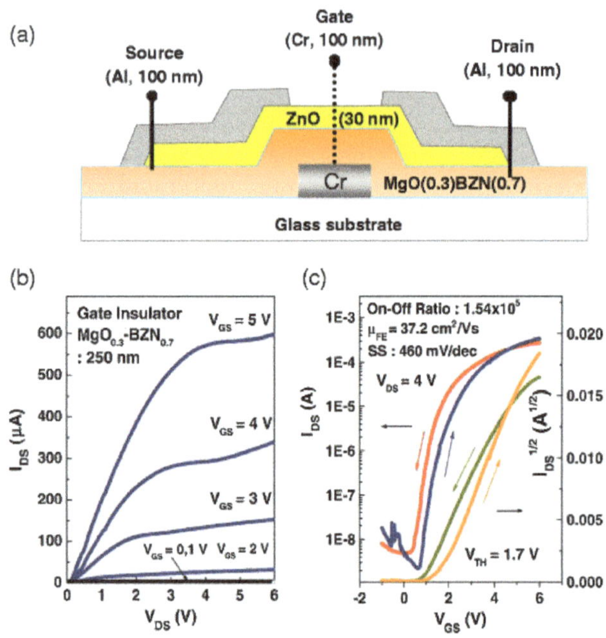

Figure. 18 (**a**) A schematic diagram of ZnO-TFTs with $MgO_{0.3}BZN_{0.7}$ gate insulator on glass substrate, (**b**) output characteristics of ZnO-TFTs using a $MgO_{0.3}BZN_{0.7}$ insulator. $V_{DS}$was swept from 0 V to +6 V in step of 0.02 V, (**c**) transfer characteristics of ZnO-TFTs using a $MgO_{0.3}BZN_{0.7}$ insulator. V_{GS} was swept from −1 V to +6 V and from −1 V at V_{DS} of 4 V [channel length (L) of 150 μm and channel width (W) of 2000 μm]. Reprinted with permission from [146]. Copyright 2010, Elsevier

As a concluding remark, for the application of BZN- and BST-based oxides as gate insulators in oxide TFTs, it is highly necessary in future research to investigate the interfacial trap density and long-term device stability (i.e., temperature, illumination, mechanical stress, voltage, or current stability) under real operating conditions.

CURRENT TFT OXIDE ISSUES

Oxide semiconductor-based TFTs have been successfully developed and integrated into flat panel displays. Oxide TFTs including ZnO and $InGaZnO_4$ semiconducting channel layers have also driven the creation of novel transparent electronics. However, many researchers and engineers still have concerns regarding several challenging issues that need to be overcome in order to realize mass production of oxide semiconductor-based TFTs. In particular, further improvements in terms of mobility and stability are required. For this purpose, all constituent layers in TFTs, including oxide semiconductor, gate insulator, gate/source-drain electrode, and passivation layer need to be carefully optimized. This work is summarized in Fig. 19.

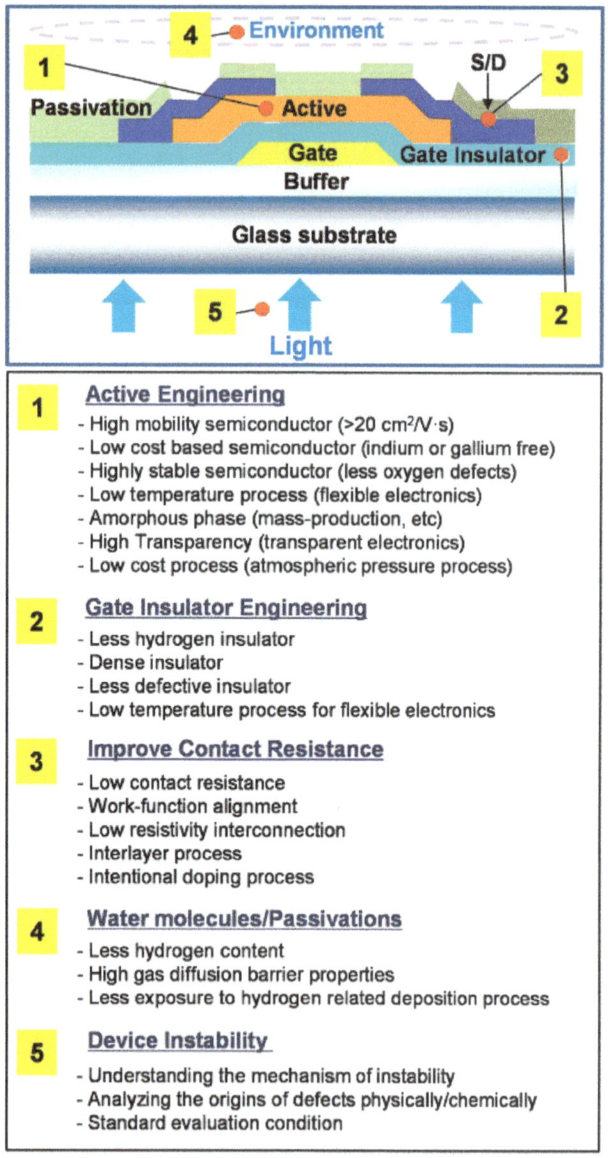

Figure. 19 Schematic design strategy including various consideration factors to achieve high-mobility and high-stability oxide TFTs

DEMAND FOR HIGHER MOBILITY

The current flat panel display (FPD) industry is being driven towards the development of high definition ("Ultra Definition"), large area (>70 in.), high frame rate (>240 Hz), and three-dimensional (3D) displays. In order to realize such high-end products such as 3D televisions that do not require specialized glasses, using either active matrix liquid crystal displays (AMLCDs) or active-matrix, organic light-emitting-diode (AMOLED) panels, high-performance thin film transistor (TFT) devices acting as switching or driving elements are necessary, preferably with field-effect mobility values exceeding 20 $cm^2/V \cdot s$ [12, 150]. For example, for large AMOLED TVs, the mobility requirement is generally predicted to be over 30 $cm^2/V \cdot s$ (depending on display resolution and pixel-circuit designs) because OLED pixels need high current in order to emit light through current injection. Many researchers have reported various oxide semiconductor materials and structures that can achieve high-mobility TFTs [37,151–153]. Interestingly, as we learn about the electronic nature of oxide semiconductors, we note that the mobility values are controllable in the range of 1–30 $cm^2/V \cdot s$, as long as device-instability is not a concern [152, 153]. Also, some promising results have recently been demonstrated with ultra-high mobility (above 50 $cm^2/V \cdot s$) in oxide-based TFTs, by adopting bilayer, active structures such as indium tin oxide (ITO)/IGZO or indium zinc oxide (IZO)/IGZO [87, 154]. Research on the selection and synthesis of tailored oxide-semiconducting layers, being proposed to obtain higher field effect mobility values, will be continued.

STABLE AND RELIABLE OXIDE TFTS

The instability of oxide semiconductor-based TFTs is perhaps the most critical issue that may block the practical application of oxide TFTs for AMLCDs or AMOLED displays. Therefore, recent efforts have been focused on understanding device instability and improving long-term stability. To more systematically understand the instability phenomena of oxide TFTs, many research groups have considered four different types of practical stress conditions: negative/positive gate-bias, temperature, illumination, and environment (e.g., humidity) [155,156]. Upon application of each stress, in general, the oxide TFT only exhibits a V_{th} shift without a significant change in mobility, as shown in Fig. 5. This may occur by either charge trapping at the channel/gate insulator interface, or by charge injection into the gate dielectric bulk. Occasionally, it has been found that the stressed devices spontaneously recover to their initial state after a relaxation period without any thermal annealing. Also, it is important to note that the V_{th} shift is significantly accelerated by applying two stress conditions simultaneously (e.g., negative gate-bias illumination stress) [157]. Although several groups have proposed a variety of possibilities such as oxygen vacancy, hole trapping, and electron injection, the origin of the related degradation mechanism is still unclear and under investigation.

In order to improve the stability of oxide TFTs, researchers have suggested various methods including optimized structures [82], suitable gate insulator materials [158], impermeable passivation layers [159], robust semiconductors [160], and post-annealing treatments [161]. For instance, by comparing bottom-gate TFTs with a back-channel-etch (BCE) or an etch-stop (ES) structure, researchers found that the stability of an ES-type device was superior to that of a BCE type device, which may be attributable to the formation of defective interfacial layers [82]. For gate insulators, the superior stability of SiO_2 or Al_2O_3 gated devices can be attributed to the suppression of hole injection, or trapping in the gate dielectric, owing to relatively large valence-band offset and lower hydrogen content [158]. In addition, low-permeable passivation layers (such as Al_2O_3 and $SiON_x$) resulted in better device stability, even under stress conditions [159]. Furthermore, it was revealed that robust oxide semiconductors normally contain higher oxygen content during the deposition process and post treatment [160, 161]. In summary, in terms of materials and structures, stable and reliable oxide TFTs have less hydrogen and higher oxygen composition to suppress V_{th} shifts when exposed to the four practical stress conditions.

Because the electronic conduction mechanisms in oxide semiconductors depend on the ability to control the types and concentration of defects such as oxygen vacancies and dissolved hydrogen, and the cation/oxygen stoichiometries, one is highly challenged to fabricate optimized and stable oxide TFTs using conventional processes, structures, and equipment. This follows from the unpredictable nature of defects that are generated or incorporated during the fabrication process. Thus the production of stable oxide TFTs that perform well in the final application remains challenging and will require very tight control of the processes, materials, and equipment as well as improvements in our ability to predict the impact of processing conditions on the defect and transport processes exhibited by the oxide semiconductor and dielectric layers.

CONCLUSION

Transistor circuits using low-temperature-processed oxide semiconductors such as ZnO and $InGaZnO_4$, have received intense attention for practical applications of high resolution AMLCD and AMOLED panels. In particular, low-voltage oxide transistors utilizing high-K dielectrics fabricated at or near room temperature have been successfully demonstrated. Room temperature processing of oxide semiconducting layers and gate insulators would render 3D integration of large stacks of active electronic device layers feasible. Low priced, flexible, oxide-TFT-based logic circuits such as inverters, ring oscillators and comparators, developed at low temperature, will serve as the key building blocks for future technology. To realize these technologies, more reproducible and robust oxide TFTs need to be developed and several key issues must be properly addressed. These issues include: (1) material tailoring of oxide semiconductors to achieve higher mobility and higher driving current, (2) interfacial control between high-k materials and oxide semiconductors, and (3) improved stability under bias, temperature, environmental and long-term use. In addition to recently commercialized IGZO-based TFTs, highly reliable and superior performance oxide TFTs would contribute to further progress in achieving products such as e-paper and e-books, and electronics integrated into textiles.

ACKNOWLEDGMENTS

I.D.K thanks Prof. Harry L. Tuller at Massachusetts Institute of Technology for very helpful discussion and thoughtful comments on the paper. Thanks to Prof. Tuller's careful suggestions, we could greatly improve our paper. This research was supported by the Development Program of Manufacturing Technology for Flexible Electronics with High Performance (SC0970) funded by Korea Institute of Machinery and Materials (KIMM). This work was supported by the National Research Foundation of Korea (NRF) grant funded by the Korea government (MEST) (No. 2012011730) and also supported by the IT R&D program of MKE/KEIT (Grant No. 10041416, the core technology development of light and space adaptable new mode display for energy saving on 7 in. and 2 W). This work was supported by the Center for Integrated Smart Sensors funded by the Ministry of Science, ICT & Future Planning as Global Frontier Project (CISS-2012M3A6A6054188).

REFERENCES

1. T. Kamiya, H. Hiramatsu, K. Nomura, H. Hosono, J. Electroceram. 17, 267 (2006)
2. J.F. Wager, D.A. Keszler, R.E. Presley, Transparent Electronics (Springer, New York, 2008). ISBN 978-0-387-72341-9
3. J.K. Jeong, Semicond. Sci. Technol. 26, 034008 (2011)
4. J.Y. Kwon, D.J. Lee, K.B. Kim, Electron. Mater. Lett. 7, 1 (2011)
5. J.S. Park, W.-J. Maeng, H.-S. Kim, J.-S. Park, Thin Solid Films 520, 1679 (2012)
6. J.-S. Park, T.-W. Kim, D. Stryakhilev, J.-S. Lee, S.-G. An, Y.-S. Pyo, D.-B. Lee, Y.G. Mo, D.-U. Jin, H.K. Chung, Appl. Phys. Lett. 95, 013503 (2009)
7. T. Rield, P. Gorrn, W. Kowalsky, J. Disp. Technol. 5, 501 (2009)
8. J.-S. Park, Inf. Disp. 29, 16 (2013)
9. H.A. Klasens, H. Koelmans, Solid State Electron. 7, 701 (1964)
10. G.F. Boesen, J.E. Jacobs, Proc. IEEE 56, 2094 (1968)
11. K. Nomura, H. Ohta, A. Takagi, T. Kamiya, M. Hirano, H. Hosono, Nature 432, 488 (2004)
12. T. Kamiya, K. Nomura, H. Hosono, Sci. Technol. Adv. Mater. 11, 044305 (2010)
13. T. Minami, Semicond. Sci. Technol. 20, S35 (2005)
14. R.B.H. Tahar, T. Ban, Y. Ohya, T. Takahashi, J. Appl. Phys. 83, 2631 (1998)
15. K. Ellmer, J. Phys. D. Appl. Phys. 34, 3097 (2001)
16. C.G. Van de Walle, Phys. Rev. Lett. 85, 1012 (2000)
17. S. Masuda, K. Kitamura, Y. Okumura, S. Miyatake, H. Tabata, T. Kawai, J. Appl. Phys. 93, 1624 (2003)
18. R.L. Hoffman, B.J. Norris, J.F. Wager, Appl. Phys. Lett. 82, 733 (2003)
19. P.F. Carcia, R.S. McLean, M.H. Reilly, G. Nunes, Appl. Phys. Lett. 82, 1117 (2003)
20. E.M.C. Fortunato, P.M.C. Barquinha, A.C.M.B.G. Pimentel, A.M.F. Gonçalves, A.J.S. Marques, L.M.N. Pereira, R.F.P. Martins, Adv. Mater. 17, 590 (2005)
21. S.-H.K. Park, C.-S. Hwang, M. Ryu, S. Yang, C. Byun, J. Shin, J.-I. Lee, K. Lee, M.S. Oh, S. Im, Adv. Mater. 21, 678 (2009)
22. D.H. Levy, D. Freeman, S.F. Nelson, P.J. Cowdery-Corvan, L.M. Irving, Appl. Phys. Lett. 92, 192101 (2008)
23. E.M.C. Fortunato, P.M.C. Barquinha, A.C.M.B.G. Pimentel, A.M.F. Gonçalves, A.J.S. Marques, R.F.P. Martins, L.M.N. Pereira, Appl. Phys. Lett. 85, 2541 (2004)
24. R.E. Presley, C.L. Munsee, C.-H. Park, D. Hong, J.F. Wager, D.A. Keszler, J. Phys. D. Appl. Phys. 37, 2810 (2004)
25. G. Lavareda, C.N. de Carvalho, E. Fortunato, A.R. Ramos, E. Alves, O. Conde, A. Amaral, J. Non-Cryst. Solids 352, 2311 (2006)
26. Y. Vygranenko, K. Wang, A. Nathan, Appl. Phys. Lett. 91, 263508 (2007)
27. S.-S. Dhanajay, C.-Y. Cheng, C.-W. Yang, Y.-C. Ou, M.C. Chuang, C.-W. Wu, J. Chu, Phys. D. Appl. Phys. 41, 092006 (2008)
28. M. Katayama, S. Ikesaka, J. Kuwano, Y. Yamamoto, H. Koinuma, Y. Matsumoto, Appl. Phys. Lett. 89, 242103 (2006)
29. J.-W. Park, D. Lee, H. Kwon, S. Yoo, J. Huh, IEEE Electron. Devices Lett. 30, 739 (2009)
30. K.-H. Choi, H.-K. Kim, Electrochem. Solid State Lett. 14, H314 (2011)
31. J. Park, K.-C. Ok, B.D. Ahn, J.H. Lee, J.-W. Park, K.-B. Chung, J.- S. Park, Appl. Phys. Lett. 99, 142104 (2011)
32. Y. Ye, R. Lim, J.M. Whiter, J. Appl. Phys. 106, 074512 (2009)
33. S.J. Lim, S.-J. Kwon, H. Kim, J.-S. Park, Appl. Phys. Lett. 91, 183517 (2007)
34. B. Yaglioglu, H.Y. Yeom, R. Beresford, D.C. Paine, Appl. Phys. Lett. 89, 062103 (2006)
35. Y.-L. Wang, F. Ren, W. Lim, D.P. Norton, S.J. Pearton, I.I. Kravchenko, J.M. Zavada, Appl. Phys. Lett. 90, 232103 (2007)
36. C.-J. Ku, Z. Duan, P.I. Reyes, Y. Lu, Y. Xu, C.-L. Hsueh, E. Garfunkel, Appl. Phys. Lett. 98, 123511 (2011)
37. H.-S. Kim, S.H. Jeon, J.S. Park, T.S. Kim, K.S. Son, J.-B. Seon, S.- J. Seo, S.-J. Kim, E. Lee, J.G. Chung, H. Lee, S. Han, M. Ryu, S.Y. Lee, K. Kim, Sci. Rep. 3, 1459 (2013)
38. M. Kim, J.H. Jeong, H.J. Lee, T.K. Ahn, H.S. Shin, J.-S. Park, J.K. Jeong, Y.-G. Mo, H.D. Kim, Appl. Phys. Lett. 90, 212114 (2007)
39. H. Yabuta, M. Sano, K. Abe, T. Aiba, T. Den, H. Kumomi, Appl. Phys. Lett. 89, 112123 (2006)
40. J.-S. Park, K.-S. Kim, Y.-G. Park, Y.-G. Mo, H.D. Kim, J.K. Jeong, Adv. Mater. 21, 329 (2009)
41. C.-J. Kim, S. Kim, J.-H. Lee, J.-S. Park, S. Kim, J. Park, E. Lee, J. Lee, Y. Park, J.H. Kim, S.T. Shin, U.-I. Chung, Appl. Phys. Lett. 95, 252103 (2009)
42. M.K. Ryu, S. Yang, S.-H.K. Park, C.-S. Hwang, J.K. Jeong, Appl. Phys. Lett. 95, 072104 (2009)
43. S. Yang, D.-H. Lee, M.K. Ryu, S.-H.K. Park, C.-S. Hwang, J. Jang, J.K. Jeong, Appl. Phys. Lett. 96, 213511 (2010)
44. E. Chong, Y.S. Chun, S.Y. Lee, Appl. Phys. Lett. 97, 102102 (2010)
45. H.Q. Chiang, J.F. Wager, R.L. Hoffman, J. Jeong, D.A. Keszler, Appl. Phys. Lett. 86, 013503 (2005)
46. W.B. Jackson, R.L. Hoffman, G.S. Herman, Appl. Phys. Lett. 87, 193503 (2005)
47. P. Görn, P. Hölzer, T. Riedl, W. Kowalsky, J. Wang, T. Weimann, P. Hinze, S. Kipp, Appl. Phys. Lett. 90, 063502 (2007)
48. B.S. Yang, M.S. Huh, S. Oh, U.S. Lee, Y.J. Kim, M.S. Oh, J.K. Jeong, C.S. Hwang, H.J. Kim, Appl. Phys. Lett.

98, 122110 (2011)
49. D.N. Kim, D.L. Kim, G.H. Kim, S.J. Kim, Y.S. Rim, W.H. Jeong, H.J. Kim, Appl. Phys. Lett. 97, 192105 (2010)
50. G.H. Kim, W.H. Jeong, B.D. Ahn, H.S. Shin, H.J. Kim, H.J. Kim, M.K. Ryu, K.B. Park, J.B. Seon, S.Y. Lee, Appl. Phys. Lett. 96, 163506 (2010)
51. D.-H. Cho, S. Yang, C. Byun, J. Shin, M.K. Ryu, S.-H.K. Park, C.- S. Hwang, S.M. Chung, W.-S. Cheong, S.M. Yoon, H.-Y. Chu, Appl. Phys. Lett. 93, 142111 (2008)
52. E.M.C. Fortunato, L.M.N. Pereira, P.M.C. Barquinha, A.M.B. do Rego, G. Gonçalves, A. Vilà, J.R. Morante, R.F.P. Martins, Appl. Phys. Lett. 92, 222103 (2008)
53. S.-Y. Han, D.-H. Lee, G.S. Herman, C.-H. Chang, J. Disp. Technol. 5, 520 (2009)
54. B.J. Norris, J. Anderson, J.F. Wager, D.A. Keszler, J. Phys. D. Appl. Phys. 36, L105 (2003)
55. B.S. Ong, C. Li, Y. Li, Y. Wu, R. Loutfy, J. Am. Chem. Soc. 129, 2750 (2007)
56. Y.-J. Chang, D.-H. Lee, G.S. Herman, C.-H. Chang, Electrochem. Solid State Lett. 10, H135 (2007)
57. D.-H. Lee, Y.-J. Chang, G.S. Herman, C.H. Chang, Adv. Mater. 19, 843 (2007)
58. K.K. Banger, Y. Yamashita, K. Mori, R.L. Peterson, T. Leedham, J. Rickard, H. Sirringhaus, Nat. Mater. 10, 45 (2011)
59. M.-G. Kim, M.G. Kanatzidis, A. Facchetti, T.J. Marks, Nat. Mater. 10, 382 (2011)
60. Y.H. Hwang, J.-S. Seo, J.M. Yun, H. Park, S. Yang, S.-H.K. Park, B.-S. Bae, NPG Asia Mater. 5, e45 (2013)
61. T. Jun, K. Song, Y. Jeong, K. Woo, D. Kim, C. Bae, J. Moon, J. Mater. Chem. 21, 1102 (2011)
62. S.Y. Park, B.J. Kim, K. Kim, M.S. Kang, K.-H. Lim, T.I. Lee, J.M. Myoung, H.K. Baek, J.H. Cho, Y.S. Kim, Adv. Mater. 24, 834 (2012)
63. Y.-H. Kim, J.-S. Heo, T.-H. Kim, S. Park, M.-H. Yoon, J. Kim, M.S. Oh, G.-R. Yi, Y.-Y. Noh, S.K. Park, Nature 489, 128 (2012)
64. S.T. Meyers, J.T. Anderson, C.M. Hung, J. Thompson, J.F. Wager, D.A. Keszler, J. Am. Chem. Soc. 130, 17603 (2008)
65. S.-C. Chang, C.-C. Yu, F.-W. Chang, S.-W. Liang, C.-H. Tsai, B.-C. Chuang, C.-Y. Tsay, SID Int. Symp. Dig. Tech. 39, 1188 (2008)
66. C.G. Choi, S.-J. Seo, B.-S. Bae, Electrochem. Soild State Lett. 11, H7 (2008)
67. G.H. Kim, B.D. Ahn, H.S. Shin, W.H. Jeong, H.J. Kim, H.J. Kim, Appl. Phys. Lett. 94, 233501 (2009)
68. D.H. Lee, S.-Y. Han, G.S. Herman, C.-H. Chang, J. Mater. Chem. 19, 3135 (2009)
69. Y.H. Hwang, J.H. Jeon, S.-J. Seo, B.-S. Bae, Electrochem. Soild State Lett. 12(H336) (2009)
70. S.K. Park, Y.-H. Kim, H.-S. Kim, J.-I. Han, Electrochem. Solid State Lett. 12, H256 (2009)
71. M.-K. Ryu, K.-B. Park, J.-B. Seon, S.Y. Lee, J. Soc. Inf. Display 18, 734 (2010)
72. M.-G. Kim, H.S. Kim, Y.-G. Ha, J. He, M.G. Kanatzidis, A. Facchetti, T.J. Marks, J. Am. Chem. Soc. 132, 10352 (2010)
73. Y.-H. Yang, S.S. Yang, C.-Y. Kao, K.-S. Chou, IEEE Electron. Device Lett. 31, 329 (2010)
74. C.D. Dimitrakopoulos, S. Purushothaman, J. Kymissis, A. Callegari, J.M. Shaw, Science 283, 822 (1999)
75. L.A. Majewski, R. Shreoder, M. Grell, Adv. Mater. 17, 192 (2005)
76. M. Halik, H. Klauk, U. Zschieschang, G. Schmid, C. Dehm, M. Schutz, S. Maisch, F. Effemberger, M. Brunnbauer, F. Stellacci, Nature 431, 963 (2004)
77. S.M. Sze, K.K. Ng, Physics of Semiconductor Devices (John Wiley and Sons, New York, 2007), ISBN 9780–471143239
78. F. Fleischhaker, V. Wloka, I. Hennig, J. Mater. Chem. 20, 6622 (2010)
79. D.H. Kim, S.-H. Choi, N.G. Cho, Y. Chang, H.-G. Kim, J.-M. Hong, I.-D. Kim, Electrochem. Solid State Lett. 12, H296 (2009)
80. K. Kang, M.-H. Lim, H.-G. Kim, Y. Choi, H.L. Tuller, I.-D. Kim, J.-M. Hong, Appl. Phys. Lett. 87, 242908 (2005)
81. I.-D. Kim, M.-H. Lim, K. Kang, H.-G. Kim, S.-Y. Choi, Appl. Phys. Lett. 89, 022905 (2006)
82. J.-Y. Kwon, K.S. Son, J.S. Jung, K.-H. Lee, J.S. Park, T.S. Kim, K.H. Ji, R. Choi, J.K. Jeong, B. Koo, S. Lee, Electrochem. Solid State Lett. 13, H213 (2010)
83. J.-S. Park, J. Electroceram. 25, 145 (2010)
84. J.-S. Park, C.W. Lee, J.J. Yoon, K.-B. Chung, J. Electroceram. 26, 63 (2011)
85. J.-S. Park, J. Electroceram. 28, 74 (2012)
86. M.-H. Lim, K. Kang, H.-G. Kim, I.-D. Kim, Y. Choi, H.L. Tuller, Appl. Phys. Lett. 89, 202908 (2006)
87. H.-S. Kim, J.S. Park, H.-K. Jeong, K.S. Son, T.S. Kim, J.-B. Seon, E. Lee, J.G. Chung, D.H. Kim, M. Ryu, S.Y. Lee, ACS Appl. Mater. Interfaces 4, 5416 (2012)
88. Y.W. Choi, I.D. Kim, H.L. Tuller, A.I. Akinwande, IEEE Trans. Electron. Dev. 52, 2819 (2005)
89. H. Kauk, H. Halik, F. Eder, G. Schmid, C. Dehm, U. Zschieschang, D. Rohde, R. Brederlow, S. Briole, S. Maisch, F. Effenberger, Tech. Digit. Int. Electron. Devices Meet. 2004, 369 (2004)
90. S.A. DiBenedetto, A. Facchetti, M.A. Ratner, T.J. Marks, Adv. Mater. 21, 1407 (2009)
91. T. Miyadera, T. Minari, S.D. Wang, K. Tsukagoshi, Appl. Phys. Lett. 93, 213302 (2008)
92. L. Wang, M.-H. Yoon, G. Lu, Y. Yang, A. Facchetti, T.J. Marks, Nat. Mater. 5, 893 (2006)
93. M.H. Yoon, H. Yan, A. Facchetti, T.J. Marks, J. Am. Chem. Soc. 127, 10388 (2005)
94. G.W. Hyung, J. Park, J.R. Koo, K.M. Choi, S.J. Kwon, E.S. Cho, Y.S. Kim, Y.K. Kim, Solid State Electron. 69, 27 (2012)
95. K. Lee, G. Ko, G.H. Lee, G. Han, M.M. Sung, T.W. Ha, J.H. Kim, S. Im, Appl. Phys. Lett. 97, 082110 (2010)

96. P.K. Nayak, J. Kim, C. Lee, Y. Hong, Phys. Status Solidi A 207, 1664 (2010)
97. J. Robertson, J. Vac. Sci. Technol. B. 18, 1785 (2000)
98. R.K. Nahar, V. Singh, A. Sharma, J. Mater. Sci. Mater. Electron. 18, 615 (2007)
99. G.D. Wilk, R.M. Wallace, J.M. Anthony, J. Appl. Phys. 87, 484 (2000)
100. H. Takahashi, S. Toyoda, J. Okabayashi, H. Kumigashira, M. Oshima, Y. Sugita, G.L. Liu, Z. Liu, K. Usuda, Appl. Phys. Lett. 87, 012903 (2005)
101. S.A. Campbell, D.C. Gilmer, X.C. Wang, M.T. Hsieh, H.S. Kim, W.L. Gladfelter, J.H. Yan, IEEE Trans. Electron. Devices 44, 104 (1997)
102. J.P. Chang, Y.-S. Lin, K. Chu, J. Vac. Sci. Technol. B 19, 1782 (2001)
103. A. Bahari, A. Anasari, Z. Rahmani, J. Eng. Technol. Res. 3, 203 (2001)
104. I. Levin, D. Brandon, J. Am. Ceram. Soc. 81, 1995 (1998)
105. J.H. Choi, Y. Mao, J.P. Chang, Mater. Sci. Eng. R 72, 97 (2011)
106. M. Kadoshima, M. Hiratani, Y. Shimamoto, K. Torii, H. Miki, S. Kimura, T. Nabatame, Thin Solid Films 424, 224 (2003)
107. D.K. Smith, C.F. Cline, J. Am. Ceram. Soc. 45, 249 (1962)
108. C.T. Lynch, F.W. Vahldiek, L.B. Robinson, J. Am. Ceram. Soc. 44, 147 (1961)
109. M.-H. Cho, D.-H. Ko, K. Jeong, S.W. Whangbo, C.N. Whang, S.C. Choi, S.J. Cho, Thin Solid Films 349, 266 (1999)
110. J. Zabicky, The Chemistry of Metal Enolates Part 1, (John Wiley and Sons, New York 2009), ISBN 9780–470061688
111. Bansal, Narottam P., Low Temperature Synthesis of Monolithic Transparent Ta2O5 Gels from Hydrolysis of Metal Alkoxide, (National Aeronautics and Space Administration (NASA), Washington, DC, 1993), ISBN 9781234352011
112. A.P. Huang, Z.C. Yang, Paul K. Chu, in Hafnium-based High-k Gate Dielectrics, Advances in Solid State Circuit Technologies, ed. Paul K. Chu, (InTech, Shanghai, China, 2010) ISBN: 978-953-307- 086-5, Available from: http://www.intechopen.com/books/ advances-in-solid-state-circuit-technologies/hafnium-based-high-kgatedielectrics
113. U. Diebold, Surf. Sci. Rep. 48, 53 (2003)
114. M.-H. Cho, D.-H. Ko, Y.G. Choi, K. Jeong, I.W. Lyo, D.Y. Noh, H.J. Kim, C.N. Whang, J. Vac. Sci. Technol. A 19, 192 (2001)
115. J. McPherson, J.Y. Kim, A. Shanware, H. Mogul, Appl. Phys. Lett. 82, 2121 (2003)
116. S.C. Choi, M.H. Cho, S.W. Whangbo, C.N. Whang, J. Korean Phys. Soc. 31, 144 (1997)
117. M. Gurvitch, L. Manchanda, J.M. Gibson, Appl. Phys. Lett. 51, 919 (1987)
118. D.-H. Son, D.-H. Kim, J.H. Kim, S.-J. Sung, E.A. Jung, J.-K. Kang, Electrochem. Solid State Lett. 13, H274 (2010)
119. J.-S. Park, J.K. Jeong, Y.-G. Mo, S. Kim, Appl. Phys. Lett. 94, 042105 (2009)
120. J.S. Lee, S. Chang, S.-M. Koo, S.Y. Lee, IEEE Electron. Device Lett. 31, 225 (2010)
121. Y.K. Moon, S. Lee, J.-W. Park, J. Korean Phys. Soc. 55, 1906 (2009)
122. L. Zhang, J. Li, X.W. Zhang, X.Y. Jiang, Z.L. Zhang, Appl. Phys. Lett. 95, 072112 (2009)
123. Z.L. Pei, L. Pereira, G. Gonçalves, P. Barquinha, N. Franco, E. Alves, A.M.B. Reqo, R. Martins, E. Fortunato, Electrochem. Solid State Lett. 12, G65 (2009)
124. D.H. Kim, H. Seo, K.-B. Chung, N.G. Cho, H.-G. Kim, I.-D. Kim, J. Electrochem. Soc. 157, H964 (2010)
125. J.-Y. Kwon, J.S. Jung, K.S. Son, K.-H. Lee, J.S. Park, T.S. Kim, J.-S. Park, R. Choi, J.K. Jeong, B. Koo, S.Y. Lee, Appl. Phys. Lett. 97, 183503 (2010)
126. X. Zou, G. Fang, L. Yuan, X. Tong, X. Zhao, Semicond. Sci. Technol. 25, 055006 (2010)
127. D. Yi, J. Yuan, H. Liu, Y. Shen, Y.-H. Lin, C.-W. Nan, X. Xi, J. He, J. Electroceram. 29, 95 (2012)
128. M. Zhang, H. Wang, H. Yang, W. Liu, H. Zhou, X. Yao, J. Electroceram. 26, 99 (2011)
129. X. Chou, J. Wang, Z. Zhao, W. Geng, W. Zhang, J. Zhai, J. Electroceram. 26, 185 (2011)
130. H.-S. Kim, H.-G. Kim, I.-D. Kim, K.-B. Kim, J.-C. Lee, Appl. Phys. Lett. 87, 212903 (2005)
131. H.-S. Kim, T.-S. Hyun, H.-G. Kim, I.-D. Kim, T.-S. Yun, J.-C. Lee, Appl. Phys. Lett. 89, 052902 (2006)
132. Y.-S. Ham, S.W. Yun, J.-H. Koh, J. Electroceram. 26, 32 (2011)
133. J.B. Kim, C. Fuentes-Hernandez, B. Kippelen, Appl. Phys. Lett. 93, 242111 (2008)
134. K. Kang, M.-H. Lim, H.-G. Kim, I.-D. Kim, J.-M. Hong, Appl. Phys. Lett. 90, 043502 (2007)
135. K. Kang, I.-D. Kim, M.-H. Lim, H.-G. Kim, J.-M. Hong, Thin Solid Films 516, 1218 (2008)
136. Y.-B. Kim, J.-U. Kim, D.-K. Choi, J.-M. Hong, I.-D. Kim, J. Electroceram. 23, 76 (2009)
137. D.-H. Kim, N.G. Cho, H.-G. Kim, H.-S. Kim, J.-M. Hong, I.-D. Kim, Appl. Phys. Lett. 93, 032901 (2008)
138. D.-H. Kim, N.G. Cho, H.-G. Kim, I.-D. Kim, Electrochem. Solid State Lett. 13, H370 (2010)
139. W. Li-Ping, L. Ai-Xia, D. Wei, W. Qing, Chin. Phys. Lett. 27, 078502 (2010)
140. W. Fu, Z. Xu, K. Liu, W. Wang, X. Bai, E. Wang, Appl. Phys. Lett. 93, 213107 (2008)
141. W. Liu, H. Wang, J. Electroceram. 27, 209 (2011)
142. A.F. Qasrawi, B.H. Kmail, E.M. Nazzal, A. Mergen, J. Electroceram. (2013). doi:10.1007/s10832-013-9809-9
143. W. Liu, H. Wang, J. Electroceram. 29, 183 (2012)
144. J. Lu, S. Stemmer, Appl. Phys. Lett. 83, 2411 (2003)
145. I.-D. Kim, Y.W. Choi, H.L. Tuller, Appl. Phys. Lett. 87,

043509 (2005)

146. N.G. Cho, D.H. Kim, H.-G. Kim, J.-M. Hong, I.-D. Kim, Thin Solid Films 518, 2843 (2010)

147. G. Vasta, T.J. Jackson, A. Fronmmhold, J. Bowen, E.J. Tarte, J. Electroceram. 27, 176 (2011)

148. M. Søgaard, A. Bieberle-Hütter, P.V. Hendriksen, M. Mogensen, H.L. Tuller, J. Electroceram. 27, 134 (2011)

149. A.A. Al-Ghamdi, F. Al-Hazmi, F. Alnowaiser, R.M. Al-Tuwirqi, A.A. Al-Ghamdi, O.A. Alhartomy, F. El-Tantawy, F. Yakuphanoglu, J. Electroceram. 29, 198 (2012)

150. T. Arai, T. Sasaoka, SID Symp. Dig. Tech. Pap. 42, 710 (2011)

151. K. Ebata, S. Tomai, Y. Tsuruma, T. Litsuka, S. Matsuzaki, K. Yano, Appl. Phys. Express 5, 011102 (2012)

152. S. Oh, B.S. Yang, Y.J. Kim, M.S. Oh, M. Jang, H. Yang, J.K. Jeong, C.S. Hwang, H.J. Kim, Appl. Phys. Lett. 101, 092107 (2012)

153. S.Y. Park, K.H. Ji, H.Y. Jung, J.-I. Kim, R. Choi, K.S. Son, M.K. Ryu, S. Lee, J.K. Jeong, Appl. Phys. Lett. 100, 162108 (2012)

154. S.-I. Kim, C.J. Kim, J.C. Park, I. Song, S.W. Kim, H. Yin, E. Lee, J.C. Lee, Y. Park, IEDM Tech. Dig. 73 (2008)

155. J.K. Jeong, H.W. Yang, J.H. Jeong, Y.G. Mo, H.D. Kim, Appl. Phys. Lett. 93, 123508 (2008)

156. J.H. Shin, J.S. Lee, C.S. Hwang, S.H.K. Park, W.S. Cheong, C.W.Byun, J.I. Lee, H.Y. Chu, J. ETRI 31, 62 (2009)

157. K.-H. Lee, J.S. Jung, K.S. Son, J.S. Park, T.S. Kim, R. Choi, J.K. Jeong, J.-Y. Kwon, B. Koo, S. Lee, Appl. Phys. Lett. 95, 232106 (2009)

158. J.S. Jung, K.S. Son, K.-H. Lee, J.S. Park, T.S. Kim, J.-Y. Kwon,

159. K.-B. Chung, J.-S. Park, B. Koo, S. Lee, Appl. Phys. Lett. 96, 193506 (2010)

160. S.-I. Kim, S.W. Kim, C.J. Kim, J.-S. Park, J. Electrochem. Soc. 158, H115 (2011)

161. H.-S. Kim, K.-B. Park, K.S. Son, J.S. Park, W.-J. Maeng, T.S. Kim, K.-H. Lee, E.S. Kim, J. Lee, J. Suh, J.-B. Seon, M.K. Ryu, S.Y. Lee, K. Lee, S. Im, Appl. Phys. Lett. 97, 102103 (2010)

162. K.H. Ji, J.-I. Kim, H.Y. Jung, S.Y. Park, R. Choi, U.K. Kim, C.S. Hwang, D. Lee, H. Hwang, J.K. Jeong, Appl. Phys. Lett. 98, 103509 (2011)

CHAPTER 2

IMPACTS OF TI ON ELECTRICAL PROPERTIES OF GE METAL–OXIDE–SEMICONDUCTOR CAPACITORS WITH ULTRATHIN HIGH-K LATION GATE DIELECTRIC

H.X. Xu, J.P. Xu, C.X. Li, C.L. Chan, P.T. Lai

INTRODUCTION

As the Si-based metal-oxide-semiconductor field-effect transistor (MOSFET) is approaching its physical limit, germanium MOSFETs with high-permittivity (high-k) gate dielectric (e.g. ZrO_2 [1], HfO_2 [2–4], and Al_2O_3 [5]) are extensively discussed as alternative candidates for future complementary MOS technology due to the high carrier mobilities of germanium [6–8]. However, the growth of unstable and water-soluble germanium oxide during deposition and post-deposition annealing hinders the fabrication of competitive G_e-based MOS devices [9, 10]. To overcome this problem, various techniques including NH3 surface treatment [2–5] and Si interlayer [11, 12] have been applied in Ge-based MOS devices to enhance device performances. Recently, it was reported that rare-earth metal oxides such as CeO_2 [13, 14], Gd_2O_3 [15] and Dy_2O_3 [16] produced promising electrical properties when directly in contact with Ge. Especially, La_2O_3 thin film shows many advantages, including low interface-states density, very small frequency dispersion and hysteresis due to formation of stable lanthanum germanate (La–G_e–O) [16–18]. However, a relatively low permittivity due to Ge diffusion into the highk dielectric prevents further decrease in equivalent oxide thickness [17]. Ti incorporation has been studied by several groups to increase the permittivity of H_f-based oxides because of the extremely high relative permittivity (k) of Ti-based oxides (~80) [19, 20]. Higher permittivity has been obtained with Ti incorporation while leading to a gate-leakage increase [21]. On the other hand, N incorporation in HfO_2 results in beneficial characteristics including reduction of gate-leakage current, improved thermal stability of the dielectric material, due to suppression of the onset of dielectric crystallization [22, 23]. In view of the above two effects on H_f-based oxides, it is expected that better dielectric properties could be achieved by simultaneously incorporating Ti and N into La_2O_3 thin film. Therefore, in this work, capacitors with LaTiON gate dielectric are fabricated and the impact of Ti content on the electrical properties of the capacitors is carefully examined. It is found that some electrical properties, e.g. k value, interface-state density, gate-leakage current and device reliability, strongly depend on the Ti content incorporated in the La_2O_3 gate dielectric. When Ti content is set at an optimal value, a good trade-off among the electrical properties can be obtained.

EXPERIMENTS

The used G_e substrate was (100) S_b-doped n-type G_e wafers with a resistivity of 0.10 ~ 0.11 cm. The wafers were cleaned in organic solvents, and lastly with de-ionized water rinsing followed by 30-s diluted HF (1:50) dipping for several cycles to remove the native oxide. After drying in N_2, LaON thin film was deposited by sputtering of La_2O_3 target at a power of 30 W in A_r and N2 (Ar:N2 = 24:6) ambient at room temperature (denoted as LON). In order to investigate the impact of Ti content on the electrical properties of the capacitors, Ti was added into the LaON thin film by reactive co-sputtering of

Citation: H. X. Xu, J. P. Xu, C. X. Li, C. L. Chan, P. T. Lai, Impacts of Ti on electrical properties of Ge metal–oxide–semiconductor capacitors with ultrathin high-k LaTiON gate dielectric, DOI 10.1007/s00339-010-5665-5.

Ti and La_2O_3 targets with different Ti-target powers (11.7 W, 17.4 W, 26.1 W and 33.3 W) as shown in Table 1 (denoted as LTON1, LTON2, LTON3 and LTON4, respectively). All LaTiON films have roughly the same thickness (7.5 ~ 7.8 nm).

Table 1 Preparation conditions and physical thickness of the gate dielectrics. The Ti/La_2O_3 ratio is calculated from the ratio of the deposition rates of Ti (at different sputtering powers) and La_2O_3 (at a sputtering power of 30 W) in the same Ar:N_2 = 24:6 ambient

Sample	Ti-target power (W)	Physical thickness (nm)	Ti/La_2O_3 (%)
LON	-	8.05	-
LTON1	11.7	7.47	7.3
LTON2	17.4	7.55	14.7
LTON3	26.1	7.60	23.8
LTON4	33.3	7.83	34.0

A post-deposition annealing (PDA) was carried out in N_2 ambient with a flow rate of 500 ml/min at 500°C for 5 min. Subsequently, Al was evaporated and patterned by lithography as gate electrodes with an area of A = 7.85 × 10^{-5} cm^{-2}, followed by a forminggas annealing at 300°C for 20 min. High-frequency (HF, 1-MHz) capacitance–voltage (C–V) curves were measured by HP4284A precision LCR meter. Gate-leakage current was measured by HP4156A precision semiconductor parameter analyzer. Physical thickness of the gate dielectrics was determined by a multi-wavelength ellipsometer and TEM. High-field stress (10 MV cm^{-1} for 3600 s), with the capacitors biased in accumulation by the HP 4156A, was used to examine the device reliability in terms of gate-leakage increase and flatband-voltage shift. All electrical measurements were carried out under a lighttight and electrically-shielded condition at room temperature.

RESULTS AND DISCUSSION

The Ti content of the dielectric films is evaluated by the Ti/La_2O_3 ratio, which is calculated from the ratio of the deposition rates of Ti (at different sputtering powers) and La_2O_3 (at a sputtering power of 30 W) in the same Ar:N2 = 24:6 ambient, to represent the relative amount of Ti in the dielectric films. Figures 1(a) and (b) show the TEM pictures of the LON and LTON2 samples, respectively. It can be seen that the interlayer between the high-k dielectric and G_e substrate can be hardly observed in the two samples. This is ascribed to the N incorporation, which inhibits O diffusion and thus suppresses the formation of interlayer [24]. Comparing the two TEM pictures, it can be found that the interface is rougher for the LTON2 sample than the LON sample. A possible reason is that the incorporated Ti in LaTiON thin film strongly reacts with G_e at the surface of the G_e substrate [21]. Typical HF C–V curves of the samples are depicted in Fig. 2. The four Ti-incorporated samples have larger accumulation capacitances (Cox) than the LON sample, with larger Cox for higher Ti-target power (thus higher Ti content), implying larger k value and thus smaller capacitance equivalent thicknesses (CET) can be obtained for higher Ti content (LTON4 sample has the largest k value of 30.8 and smallest CET of 1.0 nm), as shown in Table 2. This should be attributed to an increase of Ti-based oxide in the LaTi-based oxide for higher Ti content, because Ti-based oxide has larger k value than La-based oxide. However, the

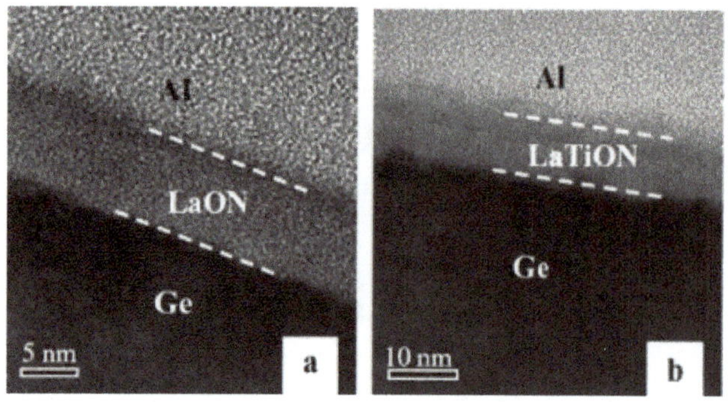

Figure. 1 TEM images of (a) LaON/G_e (LON), (b) LaTiON/G_e (LTON2) capacitors

Table 2 Electrical parameters of the n-Ge MOS capacitors without or with Ti incorporation, extracted from 1-MHz C–V curves

Sample	C_{ox} (pF)	*CET* (nm)	V_{fb} (V)	Q_{ox} (cm^{-2})	D_{it} (eV^{-1} cm^{-2})	*k*
LON	129	2.1	0.19	-8.5×10^{11}	1.2×10^{11}	14.9
LTON1	173	1.6	0.30	-2.7×10^{12}	2.4×10^{11}	18.6
LTON2	227	1.2	0.34	-4.3×10^{12}	3.1×10^{11}	24.6
LTON3	252	1.1	0.40	-6.0×10^{12}	1.2×10^{12}	27.5
LTON4	274	1.0	0.46	-7.2×10^{12}	4.1×10^{12}	30.8

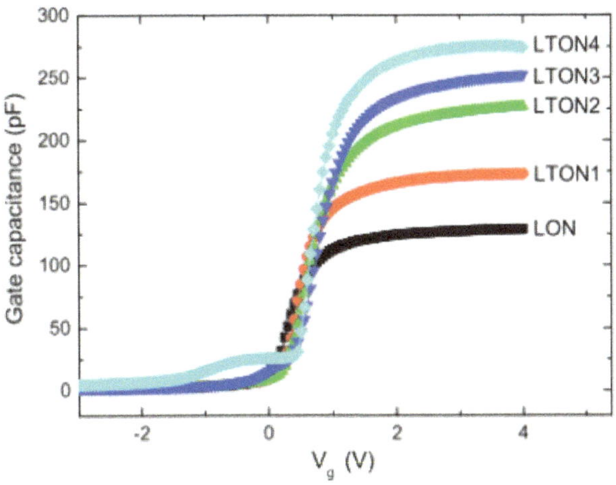

Figure. 2 High-frequency (1-MHz) C–V curves for the n-G_e MOS capacitors without or with Ti incorporation

interface-state density (D_{it}) increases with Ti content, which is obvious from the distorted C–V curves of the LTON3 and LTON4 samples. This is confirmed by the extracted D_{it} near midgap using Terman's method [25] for the purpose of comparison, with the largest D_{it} for the LTON4 sample (4.1 × 10^{12} eV−1 cm^{-2}), as shown in Table 2. This is probably associated with the fact that stronger reaction between the Ge substrate and LaTiON film could happen as the Ti content increases, similar to the situation of HfTiO in contact with G_e [21]. The higher equivalent oxide-charge density (Q_{ox}) of the Ti-incorporated samples should be mainly due to the high oxide charges near the interface and interface trap charges. The negative Q_{ox} indicates that these traps could be acceptor-like interface and near-interface traps due to the G_e diffusion from the substrate into the high-k layer and the reaction between Ge and Ti near the interface. So, a trade-off consideration between k and D_{it} is necessary when Ti is added into LaON, e.g. for our studied range of Ti-target power, a sputtering power of 17.4 W (LTON2 sample) seems to be suitable, producing a relatively high-k value (24.6) and relatively low D_{it} (3.1 × 10^{11} eV−1 cm^{-2}). Further optimization of the Ti content is necessary for obtaining better trade-off among the device properties.

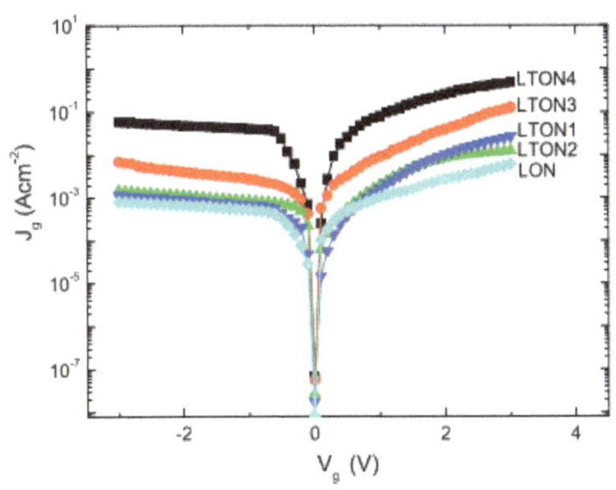

Figure. 3 Gate-leakage current density (J_g) vs. gate voltage (Vg) for the n-G_e MOS capacitors without or with Ti incorporation

The gate-leakage properties of the samples are illustrated in Fig. 3. Obviously, the Ti-incorporated samples exhibit larger gate-leakage current than the LON sample, and the higher the Ti content, the larger is the gate-leakage current. The gate-leakage current density (J_g) of the LTON4 sample is 8.1 × 10^{-2} A cm−2 at Vg = 1 V, which is nearly 80 times higher than that of the LON sample (1.0 × 10^{-3} A cm^{-2}) and 40 times higher than that of the LTON1 sample (1.7 × 10^{-3} A cm^{-2}), although they have almost the same gatedielectric thickness. This should be closely related to the gradually deteriorated quality of the dielectric/G_e interface as the Ti content increases (see Table 2 and Fig. 2). For the LTON2 sample with relatively few interface defects, an acceptably small J_g of 2.0 × 10^{-3} A cm−2 can be obtained at Vg = 1 V. In order to investigate the device reliability, a high-field stress at 10 MV cm−1 for 3600 s is performed on the capacitors, with the samples biased in accumulation. Their HF (1-MHz) C–V curves and gate-leakage currents before and after the stress are measured. The shift of flatband voltage (V_f b) extracted from the C–V curve and the increase of gate-leakage current density (J_g) at Vg = 1 V are summarized in Fig. 4. Obviously, the four Ti-incorporated samples

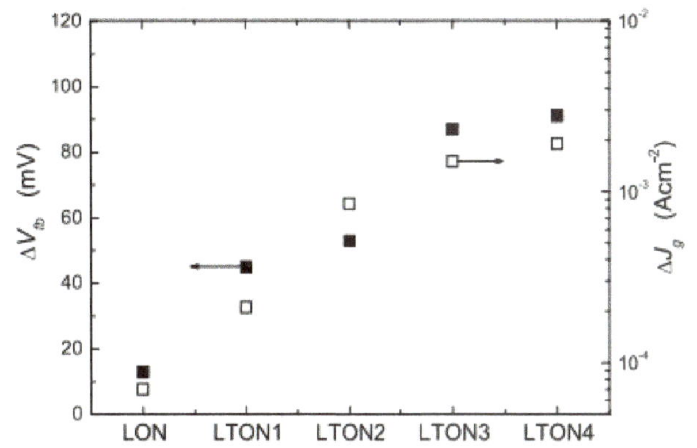

Figure. 4 Flatband-voltage shift (V_f b) and gate-leakage increase (J_g) at Vg = 1 V for the n-G_e MOS capacitors without or with Ti incorporation after a high-field stress (at 10 MV/cm for 3600 s)

have larger V_fb and J_g than the LON sample, which should be attributed to the poor interface quality associated with Ti-induced defects [26]. So, the Ti-target power and thus Ti-content control is the key for excellent electrical properties of the LaTi-based gate-dielectric MOS devices.

CONCLUSIONS

The effects of Ti incorporation into LaON gate dielectric on the electrical properties of MOS capacitors have been investigated. Results indicate that Ti addition can signifi- cantly increase the k value and decrease CET due to the extremely high permittivity of Ti-based oxides, but deteriorate the dielectric/ Ge interface quality, and thus gate leakage properties and device reliability due to Ti-induced defects. Therefore, the Ti content incorporated into LaON has to be carefully chosen to achieve a good trade-off between the k value and Dit. For our Ti-incorporated samples, a suitable Ti/La_2O_3 ratio of 14.7% leads to a high k value of 24.6, low interface-state density of 3.1 × 10^{11} eV−1 cm^{-2}, relatively low gate-leakage current density of 2.0 × 10^{-3} A cm−2 at Vg = 1 V and good device reliability

ACKNOWLEDGEMENTS

This work is financially supported by the National Natural Science Foundation of China (Grant No. 60776016), the

RGC of HKSAR, China (Project No. HKU 713308E), and the University Development Fund (Nanotechnology Research Institute, 00600009) of the University of Hong Kong.

REFERENCES

1. C.O. Chui, H. Kim, D. Chi, B.B. Triplett, P.C. McIntyre, K.C. Saraswat, in IEEE IEDM Tech. Dig. (2002), p. 437
2. W.P. Bai, N. Lu, J. Liu, A. Ramirez, D.L. Kwong, D. Wristers, A. Ritenour, L. Lee, D. Antoniadis, in VLSI Symp. Tech. Dig. (2003), p. 121
3. A. Ritenour, S. Yu, M.L. Lee, N. Lu, W. Bai, A. Pitera, E.A. Fitzgerald, D.L. Kwong, D.A. Antoniadis, in IEEE IEDM Tech. Dig., Washington, DC, USA, December 8–10, 2003, p. 433
4. C.O. Chui, H. Kim, P.C. Mcintyre, K.C. Saraswat, in IEEE IEDM Tech. Dig., Washington, DC, USA, December 8–10, 2003, p. 437
5. J.J. Chen, N.A. Bojarczuk, H. Shang, M. Copel, J.B. Hannon et al., IEEE Trans. Electron. Dev. 51, 1441 (2004)
6. A. Ritenour, S. Yu, M.L. Lee, N. Lu, Wu. Bai, A. Pitera, E.A. Fitzgerald, D.L. Kwong, and D.A. Antoniadis, in IEEE IEDM Tech. Dig., vol. 433, Washington, DC, USA, December 8–10, 2003, p. 18.2.1
7. P. Zimmerman, G. Nicholas, B. De Jaeger, B. Kaczer, A. Stesmans, L.A. Ragnarsson, D.P. Brunco, F.E. Leys, M. Caymax, G. Winderickx, K. Opsomer, M. Meuris, M.M. Heyns, in IEEE IEDM Tech. Dig., San Francisco, USA, December 10–13, 2006, p. 655
8. S. Joshi, C. Krug, D. Heh, H.J. Na, H.R. Harris, J.W. Oh, P.D. Kirsch, P. Majhi, B.H. Lee, H.-H. Tseng, R. Jammy, J.C. Lee, S.K. Banerjee, IEEE Electron. Dev. Lett. 28, 308 (2007)
9. N. Wu, Q. Zhang, C. Zhu, C.C. Yeo, S.J. Whang, D.S.H. Chan, M.F. Li, B.J. Cho, A. Chin, D.-L. Kwong, A.Y. Du, C.H. Tung, N. Balasubramanian, Appl. Phys. Lett. 84, 3741 (2004)
10. Y.A. Hu, J.-T. Zettler, S. Chongsawangvirod, Y.Q. Wang, E.A. Irene, Appl. Phys. Lett. 61, 1098 (1992)
11. W.P. Bai, N. Lu, D.-L. Kwong, IEEE Electron. Dev. Lett. 26, 378 (2005)
12. N. Wu, Q. Zhang, C. Zhu, D.S.H. Chan, A. Du, N. Balasubramanian, M.F. Li, A. Chin, J.K.O. Sin, D.-L. Kwong, IEEE Electron. Dev. Lett. 25, 631 (2004)
13. D.P. Brunco, A. Dimoulas, N. Boukos, M. Houssa, T. Conard, K. Martens, C. Zhao, F. Bellenger, M. Caymax, M. Meuris, M.M. Heyns, J. Appl. Phys. 102, 24104 (2007)
14. G. Nicholas, D.P. Brunco, A. Dimoulas, J. van Steenbergen, F. Bellenger, M. Houssa, M. Caymax, M. Meuris, Y. Panayiotatos, A. Sotiropoulos, IEEE Trans. Electron. Dev. 54, 1425 (2007)
15. C.H. Lee, T.D. Lin, L.T. Tung, M.L. Huang, M. Hong, J. Kwo, J. Vac. Sci. Technol. 26, 1128 (2008)
16. A. Dimoulas, D.P. Brunco, S. Ferrari, J.W. Seo, Y. Panayiotatos, Thin Solid Films 515, 6337 (2007)
17. G. Mavrou, S. Galata, P. Tsipas, A. Sotiropoulos, Y. Panayiotatos, A. Dimoulas, E.K. Evangelou, J.W. Seo, Ch. Dieker, J. Appl. Phys. 103, 14506 (2008)
18. J. Song, K. Kakushima, P. Ahmet, K. Tsutsui, N. Sugii, T. Hattori, H. Iwai, Microelectron. Eng. 84, 2336 (2007)
19. N. Lu, H.J. Li, J.J. Peterson, D.L. Kwong, Appl. Phys. Lett. 90, 82911 (2007)
20. V.V. Afanasev, A. Stesmans, F. Chen, M. Li, S.A. Campbell, J. Appl. Phys. 95, 5583 (2004)
21. C.X. Li, X. Zou, P.T. Lai, J.P. Xu, C.L. Chan, Microelectron. Reliab. 48, 526 (2008)
22. M.R. Visokay, J.J. Chambers, A.L.P. Rotondaro, A. Shanware, L. Colombo, Appl. Phys. Lett. 80, 3183 (2002)
23. C.H. Choi, T.S. Jeon, R. Clark, D.L. Kwong, IEEE Electron Dev. Lett. 24, 215 (2003)
24. P.S. Lysaght, J. Barnett, G.I. Bersuker, J.C. Woicik, D.A. Fischer, B. Foran, H.H. Tseng, R. Jammy, J. Appl. Phys. 101, 24105 (2007)
25. L.M. Terman, Solid State Electron. 5, 285 (1962)
26. H. Arimura, N. Kitano, Y. Naitou, Y. Oku, T. Minami, Appl. Phys. Lett 92, 212902 (2008)

CHAPTER 3 CHARACTERISTICS OF A 4H-SIC METAL-SEMICONDUCTOR-METAL ULTRAVIOLET PHOTODETECTOR

Chen Bin, Yang YinTang, XIE XuanRong, Wang Ning, MA ZhenYang, Song Kun & Zhang XianJun

Keywords: Silicon carbide, ultraviolet photodetector, temperature dependence

We investigate the temperature dependence of current-voltage and spectral response characteristics of a 4H-SiC metal-semiconductor-metal (MSM) ultraviolet photodetector in the temperature range from room temperature to 800 K with two-dimensional (2D) numerical simulator ISE-DESSIS. It is found that the dark current and photocurrent increase with the increasing temperature. For the range of 500–800 K, the dark current increases by nearly a factor 3.5 every 150 K larger than that of photocurrent, leading to a negative effect on photodetector current ratio (PDCR). Nevertheless, the PDCR is still greater than 200 even at 800 K, which exhibits the excellent thermal stability. In addition, the responsivity has an unsymmetrical trend. As temperature rises, it is clear that a remarkable red-shift of 12 nm occurs and overall responsivity is enhanced for longer wavelength. While the short-wavelength response remains relatively independent of temperature. The mechanism of indirect and direct band absorption transition is responsible for temperature-dependent spectrum distribution. These findings provide a significant insight on the design of the MSM detector operated at elevated temperature.

In despite of having an indirect bandgap, silicon carbide (SiC) has emerged as an appropriate material for the fabrication of ultraviolet (UV)/visible optoelectronic devices over the last few years due to the available substrates [1], the lower defect density and a more mature process technology [2]. Typically, 4H-SiC based UV photodetectors achieve photosensitivity spectrum of 220–380 nm and peak responsivity at around 290 nm. Photodetectors based on Schottky contacts have an enhanced sensitivity to shorter wavelength radiation as well as a reduced response time as no minority carriers are involved [3]. Especially, the metalsemiconductor-metal (MSM) photodetectors possess potential merits of large device bandwidth, small intrinsic capacitance, and sub-nanosecond response time. Whilst the planar process of MSM detectors may be accomplished with as little as one photolithography step and allows their monolithic integration with other optical and electronic circuits [4]. Recent achievements include nano-structured MSM photodetector with high peak voltage [5], MSM solar-blind photodetector with fast rise and decay times of 10 ns and 150 ns [6], high responsivity MSM photodetector with 26000 A/W at 8 V bias via a carrier-trapping process [7]. More recently, MSM UV detectors exhibit UV-photo generated current to dark current ratio up to 1.34×108 with colloidal nanoparticles [8], realize a photocurrent gain around 40 at 2 V bias using CaF2 as the insulator [9], and achieve a very low dark current density of 3.84 nA/cm2 at 5 V bias based on sol-gel-derived TiO2 films [10]. Most reported devices meet the requirement of optical detection in terms of low dark current, high responsivity and fast response. Up to now, several groups have reported the temperature dependence for photodetectors, including Ge-on-Si p-i-n detector [11], nonreach-through 4H-SiC separate absorption avalanche detectors [12], terahertz quantum-well detectors [13] and Schottky detectors [14]. Lately, the dependence of dark current and response characteristics for GaN-based avalanche photodiodes can be found in [15]. However, a thorough description of temperature dependence of the dark/illuminated current and spectral response of 4H-SiC MSM detectors, particularly quantitative modeling analysis, is still lacking, whereas it can determine the temperature stability and identify the best operating regions. One of the considerable puzzles is the deficiency of temperaturedependent absorption coefficients. In this work, we use a self-consistent calculation method to establish a numerical model of a 4H-SiC MSM photodetector. With temperature-dependent physical model, an investigation of dark current variation at elevated temperature is carried out via

Citation: Chen B, Yang Y T, Xie X R, et al. Characteristics of a 4H-SiC metal-semiconductor-metal ultraviolet photodetector. Chin Sci Bull, 2012, 57: 4427-4433, doi: 10.1007/s11434-012-5494-3.

a two-dimensional (2D) device simulator ISE-DESSIS. For the optical features at wide range of temperature, we address an exploratory method to extrapolate the UV absorption coefficients at different temperatures with consideration of the effects of variant phonon population, bandgap narrowing and direct band transition [16]. Finally, the calculated results are verified with relevant experiment and the electrical and optical characteristics of the MSM detector above room temperature (RT) are discussed in detail to evaluate thermal stability for practical application.

METHODOLOGY

Device Structure and Characterization

The cross section of a 4H-SiC MSM photodetector is illustrated in Figure 1. The layer structure consists, from bottom to top, of three parts: an ideal 400-μm n^+ 4H-SiC substrate with doping concentration of 1020 cm^{-3} , an n- epilayer with a thickness of 5 μm and a doping level lower than 1016 cm^{-3} , and Au Schottky contacts formed planar interdigital fingers with a height of 80 nm. The fingers of the contact electrodes are 3 μm wide and 500 μm long with a spacing of 3 μm. The optical sensitive area of the UV photodetector is 200 μm× 500 μm

Numerical Modeling

The analysis of the device behavior at only room temperature does not give detail information about the nature of device reliability and their conduction process. Different from the previous calculation models at room temperature (RT), we must determine the governing physics for the device operation at elevated temperature with consideration of

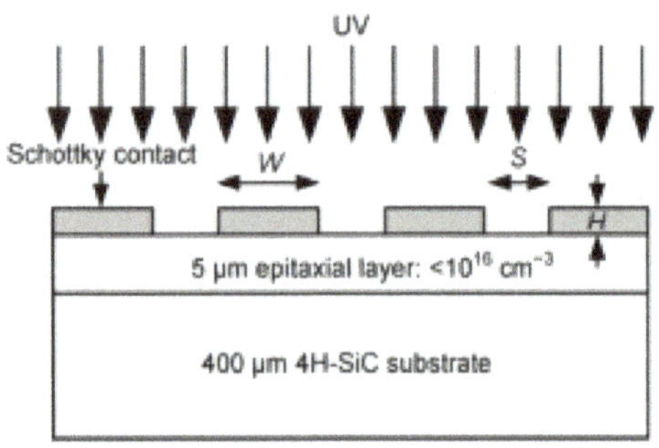

Figure 1 Schematic cross section of a 4H-SiC MSM UV detector.

the temperature effect on dark/illuminated conduct mechanism. In this study, a series of physical models utilized in ISE-DESSIS are the Shockley-Read-Hall (SRH) generation/ recombination model including the band-to-band tunneling and the trap-assisted tunneling items, the thermionic emission model, the barrier lowering model and the incomplete ionization model. A more detailed description of the models can be specified in [17] and similar modeling and simulation methods are presented elsewhere [18]. Here, we elaborate on the appropriate models employed for high temperature operation in combination with experiments and published data. Parts of the model parameters used in our calculation are summarized in Table 1. (i) Dark current simulating model above RT. An accurate temperature model is indispensable for describing the device performance at elevated temperatures. Hence, hydrodynamic transport model is used to properly model electron transport, energy relaxation and calculate a spatial electron and lattice temperatures. Unlike the drift-diffusion model, a temperature-dependent hydrodynamic Canali model is selected for the high field saturation of the mobility,

Table 1 Temperature-dependent model parameters used in this study [12,16,17]

Parameters	Unit	Value
β_0		1.23
β_{exp}		0.17
$E_{g,0}$	eV	3.359
α	eV/K	3.3×10^{-4}
β	K	0
a	cm	$1.98\times10^{6}-4000\times(T-300)$
b	V/cm	9.46×10^{6}
c		1.42
$E_{ad,0}$ (Γ-M transition)	eV	3.25
θ (Γ-M transition)	K	638
A_{ad} (Γ-M transition)		6.2×10^{4}
$E_{ad,0}$ (Γ-L transition)	eV	4
θ (Γ-L transition)	K	3480
A_{ad} (Γ-L transition)		1.8×10^{6}
$E_{ad,0}$ (Γ-Γ transition)	eV	5
A_{ad} (Γ-Γ transition)		3.2×10^{6}

which can be coupled with thermal differential equations and boundary conditions:

$$\mu = \frac{\mu_{low}}{[\sqrt{1+\alpha^2[3k_B/2(T_C-T_L)]^{\beta}}+\alpha[3k_B/2(T_C-T_L)]^{\beta/2}]^{2/\beta}}, \quad (1)$$

$$\beta = \beta_0\left(\frac{T}{T_0}\right)^{\beta_{exp}}, \quad (2)$$

$$\alpha = \frac{1}{2}\left(\frac{\mu_{low}}{q\tau_{e,c}v_{sat}^2}\right)^{\beta/2}, \quad (3)$$

where μ_{low} denotes the low field mobility. $3k_BT_C/2$ is the average carrier thermal energy; $3k_BT_L/2$ gives the equilibrium thermal energy. $\tau_{e,c}$ denotes the energy relaxation time, T_C is the carrier temperature, and T_L denotes the lattice temperature. In addition, we model temperature-dependent band gap [17] as

$$E_g(t) = E_{g,0} + \delta E_{g,0} - \frac{\alpha t^2}{t+\beta}, \quad (4)$$

where $E_{g,0}$ is an adjustable parameter, α and β are material parameters. It is important to utilize an appropriate impact ionization model since the MSM detectors are operated with the variation of temperature. The temperature-dependent impact ionization coefficients for electrons are given by [16]

$$\alpha(T,E) = a\exp\left[-\left(\frac{b}{E}\right)^{c}\right]. \quad (5)$$

It is reported that the coefficient a in 4H-SiC varies linearly with temperature and the exponential term is fairly constant [19]. Thermal generation-recombination is important in calculation of electrical characteristics of the MSM photodetector above RT. It is noted that the dominating recombination mechanism in 4H-SiC is a non-radiative capture of carriers by mid-gap defects [20]. We consider the temperature-dependent carriers life times for electron in the course of Shockley-Read-Hall (SRH) recombination, which can be described by [16]

$$\tau_{n,p} = \frac{\tau_0 (T/300)^{\alpha}}{1 + N/N_{ref}}, \tag{6}$$

where $\alpha = 1.7$, $\tau_0 = 5\times10^{-9}$ s, and $N_{ref} = 6\times10^{15}$ cm^{-3} . (ii) UV absorption coefficient model above RT. In order to determine the optical features of the detector, we adopt an accurate optical model to compute the net photon flux of the input UV optical signal via multiple vertical photon beams illuminating on the surface of the device. The key input parameters including optical power, wavelength and incidence window are 2.5×10^{-4} W/cm^2 , $2.1\times10^{-5} - 4\times10^{-5}$ cm and 2×10^{-2} cm, respectively. The specification of the optical generation rate model can be found in our previous work [21]. To the best of our knowledge, the absorption coefficient related to temperature is vital for calculating the characteristics of the photodetector at elevated temperature and general properties of the carrier recombination and transport. Unfortunately, the optical absorption coefficient for 4H-SiC has been mainly studied only at RT. Inspiringly, Galeckas et al. [22] investigated the temperature dependence of 4H-SiC absorption coefficient at 355 nm. A complete wavelength- and temperature-dependent absorption coefficient in UV region is still lacking. Due to an indirecttransition material, for lower absorption coefficients, we calculate the values as a function of doping, temperature and incident photon energy with the Matlab package. The numerical formula is expressed by [23]

$$\alpha_{indirect} = A_{id}\left[\frac{(\frac{h\upsilon - E_{id}(N,T) - k\theta}{h\upsilon})^2}{1-\exp(-\theta/T)} + \frac{(\frac{h\upsilon - E_{id}(N,T) + k\theta}{h\upsilon})^2}{\exp(\theta/T)-1}\right], \tag{7}$$

where Aid is a fitting coefficient with a unit of cm^{-1} , h_{υ} is the incident photon energy, $E_{id}(N,T)$ is energy spacing from the valence band to the conduction band, k is the Boltzmann constant, k_{θ} denotes the momentum conserving phonon energy. The first term in eq. (7) represents the photon absorption with phonon emission, and the second term represents the photon absorption with phonon absorption. For photon energies higher than band-gap spacing of 4H-SiC, the direct band transition becomes predominant and the absorption coefficient is related to the following expression [24]:

$$\alpha_{direct} \propto [h\upsilon - E_{id}(N,T)]^2. \tag{8}$$

The energy spacing $E_{id}(N,T)$ and $E_d(N,T)$ are given by [16]

$$E_{id,d}(N,T) = E_{id,d}(N) - 3.3\times10^{-4}\cdot(T-300), \tag{9}$$

$$E_{id,d}(N) = E_{id,d(0)} - 3.4\times10^{-8}\cdot N^{1/3} - 1.17\times10^{-11}\cdot N^{1/2}. \tag{10}$$

RESULTS AND DISCUSSION

Verification of Numerical Model

Due to the lack of relevant experimental results of temperature-dependent dark current characteristics of 4H-SiC MSM photodetector, based on the above models, we calculate the dark current characteristics at 293 K and 321 K of a GaAs MSM detector, whose structure is extremely similar to the measured one with the dark current features at different temperatures reported in [25]. The results are illustrated in Figure 2. As is shown, the calculated dark current is consistent with the measured data especially when the reverse voltage is larger than -4 V. To verify the photocurrent model, the calculated and experimental results with 310 nm

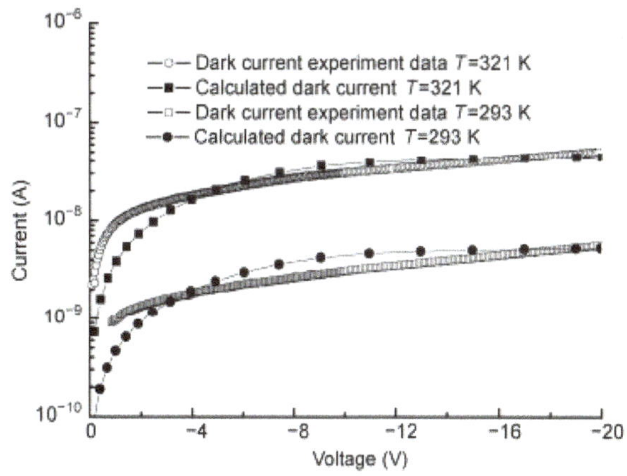

Figure 2 Comparison of calculated dark current of a GaAs MSM detector at different temperature with experimental data [25].

UV illumination are plotted in Figure 3 for comparison [26]. It is found that the calculated photocurrent is somewhat larger than that of measured one with applied bias lower than 5 V. The discrepancy is probably due to the difference between the calculated absorption coefficient based on model and the experimental data. The schematic comparison of absorption model calculated result and measured data is visible in Figure 4, which can account for the phenomenon. However, the approximate agreement between the simulated and experimental results of dark current and photocurrent is achieved, which proves the present numerical models are reasonable.

The Influences of Temperature on Dark Current and Photocurrent Characteristics of the Detector

The dark current characteristics are calculated with temperature ranging from 300 to 800 K. In Figure 5, it is clear that the dark current increases with the bias and temperature. For a voltage lower than 5 V, the every curve achieves a fast increase and then tends to saturation with larger bias. For the variation of temperature, the changes of dark current are

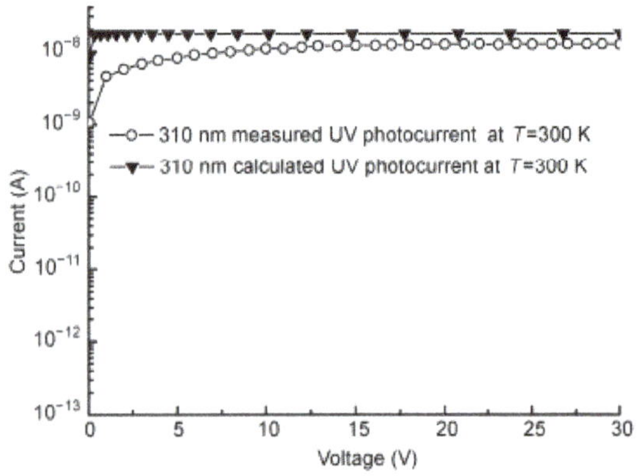

Figure 3 Comparison between the calculated photocurrent of the 4H-SiC MSM detector under 310 nm illumination with experiment [26].

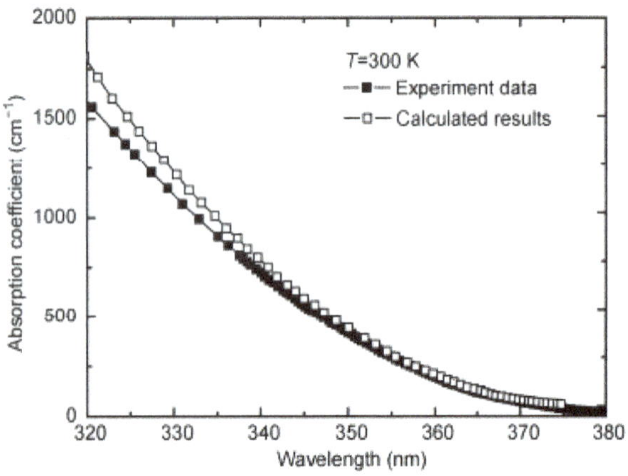

Figure 4 Schematic comparison between the model calculated 4H-SiC absorption coefficient and the measured result [27].

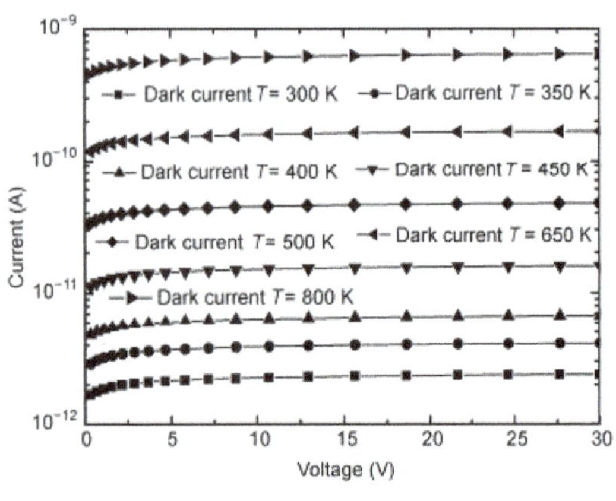

Figure 5 The dark current of 4H-SiC MSM photodetectors versus applied bias with various temperatures between RT and 800 K.

not unified with slightly different dependencies. Specifically, the dark current increases by a factor about 1.7 every 50 K with range of 300–500 K. However, a larger factor about 3.5 is achieved when the detector at 500–800 K. The maximum and minimum dark currents at 800 K and 300 K are 0.646 nA and 2.39 pA. It is clear that the dark current at 800 K is two orders of magnitude higher than that at RT. The band gap narrowing at high temperature and relevant thermionic emission mechanisms are responsible for the phenomenon. The temperature-dependent photocurrent properties are important characteristics for UV photodetector. As is clearly visible in Figure 6, the variations of photocurrent of MSM UV detector are calculated with the temperature ranging from 300 to 800 K under 330 nm UV illumination. A remarkable increase can be observed at elevated temperatures. With 30 V applied bias, it is found that the photocurrent are 10.8, 12.3, 13.7, 15.2, 16.6, 17.9 and 19.2 nA when the temperature increases from 300 to 800 K. In addition, we can see that the increscent trend of photocurrent is asymmetry. With the range of 300–450 K, the photocurrent rises

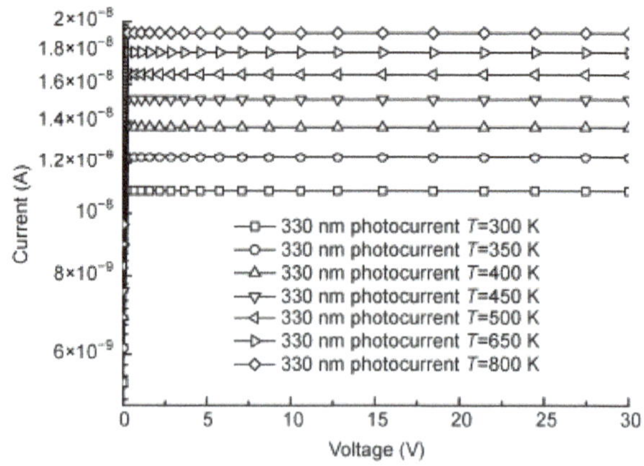

Figure 6 The calculated photocurrent-voltage curves of 4H-SiC MSM detector under 330 nm illumination from RT to 800 K.

very fast and their increment is almost equidistant. Above 450 K, it is obvious that the increase rate of photocurrent slows down. However, the photocurrent at 800 K increases near twice than that of at RT. As the temperature rises, the UV absorption coefficient increases as well. On the other hand, the changes of temperature also have influence on SRH recombination. For the absorption of longer wavelength, the former factor is prevailing as compared with the latter. Therefore, the enhanced absorption effect for the photodetector indicates that more photo-generated carriers can be produced and swept rapidly between electrode contacts to form photocurrent. Moreover, device performance at elevated temperature can be further estimated by calculation of photodetector current ratio (PDCR), which is defined as [28]

$$\mathrm{PDCR} = (I_p - I_d) / I_d, \tag{11}$$

where I_p denotes photocurrent and I_d dark current. Figure 7 shows the results of PDCR as a function of temperature. The photocurrent is obtained under 300 nm illumination and 30 V applied voltage. It can be seen that the PDCR decreases with the increase of temperature. The PDCR values are larger than 8400 at 300 K and above 5300 at 350 K. Clearly, a steep decrease in PDCR can be observed with further increase in temperature to 450 K. While the temperature is larger than 500 K, the PDCR values tend to saturation. We can see from the Figure 7 that the values decrease gradually from 578 to 249 with temperature up to 800 K. The temperature-dependent current-voltage (I-V) characteristics of the detector may account for the decay of PDCR. In contrast to photocurrent, temperature has a relatively direct impact on dark current and the considerable increment in dark current at higher temperature is responsible for the even worse PDCR. However, a comparable value of two orders of magnitude is achieved even at 800 K, which distinctly indicates that the 4H-SiC MSM UV detectors have excellent thermal stability and can be applied under high temperature and harsh condition.

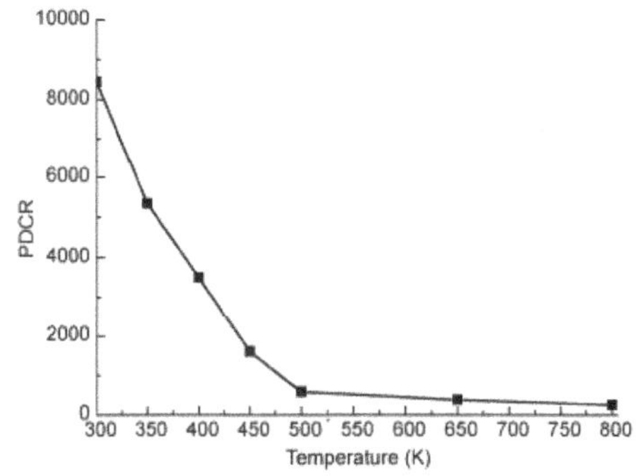

Figure 7 PDCR of 4H-SiC MSM detector obtained at various temperatures under a bias of 30 V and 300 nm UV illumination.

The Influences of Temperature on Spectral Response and Quantum Efficiency of the Detector

Figure 8 shows the responsivity curves of 4H-SiC MSM UV detector at different temperature under 10 V applied voltage. At 300 K, it is found that the responsivity rises with the increasing wavelength. When wavelength is around 290 nm, the responsivity achieves the maximum of 85.2 mA/W. Then, the spectral response decreases gradually and reaches the minimum of 0.048 mA/W at 400 nm. The cutoff of short wavelength (near 240 nm) and long wavelength (near 380 nm) are caused by strong recombination of photon-generated carriers and the decay of optical absorption, respectively. A point to note is that no abrupt spectrum cutoff can be observed with the wavelength ranging from 380 to 400 nm. The reason mainly relates to the fact that the 4H-SiC is an indirect material and does not possess a sharp cutoff edge near the band edge. As is shown in Figure 8, it is found that the peak responsivity increases from 85.2 to 120.3 mA/W, when the temperature increases from 300 to 800 K. A similar behavior of responsivity versus temperature can be found in [29]. Also, it is concluded that the spectral response curves

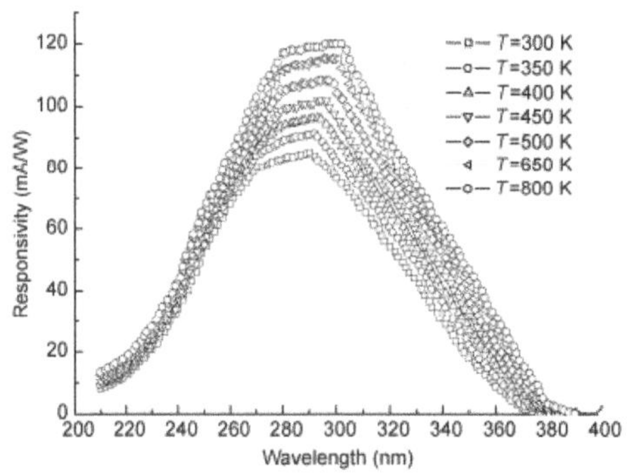

Figure 8 Spectral response of the 4H-SiC MSM photodetector for temperature ranging from RT to 800 K under a bias of 10 V.

become more extensive at the elevated temperature but with discrepant variations. Specifically, a distinct boundary around 270 nm exhibits a nonuniform tendency. When the wavelengths are lower than 270 nm, all curves arranges compactly with the increasing temperature and slight increment of responsivity can be observed, indicating that at short wavelength the spectral response seems to be independent of the temperature. While the wavelength is larger than 270 nm, the variation of responsivity with elevated temperature is more remarkable than the former. Not only does the overall responsivity is enhanced but also the peak increment of responsivity at 800 K is over 41% as compared with RT spectral response. The peak responsivities of the top two curves at 650 K and 800 K are near 299 and 302 nm. Consequently, it is clear that a visible red-shift about 12 nm occurs to longer wavelengths as the temperature increases. The fact that the responsivity shifts to longer wavelength at higher temperature is ascribed to temperature-dependent changes of band gap and optical absorption. Due to thermal expansion, the periodic lattice vibration becomes more violent, which has a profound impact on band structure. When the temperature increases, the band gap narrowing effect is more salient, resulting in the spectrum curves shift towards longer wavelength. Additionally, the elevated temperature varies the phonon distribution and increases the phonon flux. On the other hand, the absorption coefficient is a function of temperature and photon energy. For the wavelength near the band gap, the lower photon energy can be represented by Γ-M indirect band transition in 4H-SiC. The higher the temperature is, the greater absorption will be and, in turn, enhanced responsivity feature is achieved. However, for the high photon energies obtained at short wavelength, the direct band transition is predominant. While the temperature increases, an explosive increment of absorption coefficient effectively reduces the penetration depth of photon owing to the fact that the former is reversely proportional to the latter. Thus, photon-generated carriers cannot form a photocurrent due to intense recombination occurring at device surface. Therefore, the high recombination rate offsets the advantage of higher absorption so that the response curves at low wavelength region remain independent of temperature. Under a bias of 10 V, the quantum efficiencies versus diverse temperatures are sketched in Figure 9. It is noted that all curves, no matter short or long wavelength region, shift upwards as temperature rises. The peak zones of quantum efficiency of the curves locate at 280 nm. The maximum quantum efficiency of 51.82% and the minimum of 36.94% are obtained with corresponding temperature of 800 K and RT, respectively. In comparison with the both, we can find that the peak value of quantum efficiency increases approximately about 41%, indicating that temperature intensely affects the photoelectric conversion process of the detector.

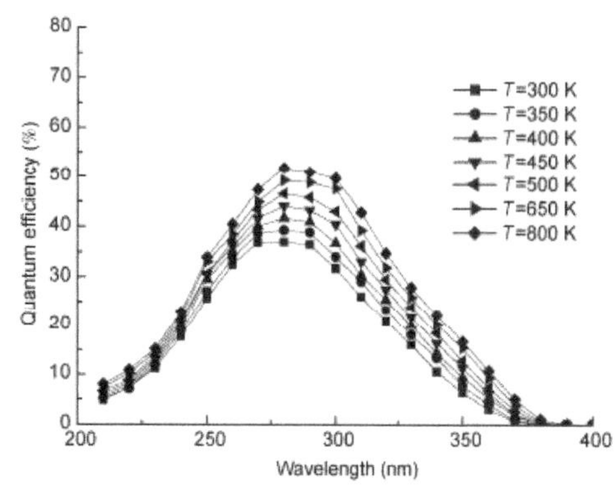

Figure 9 Quantum efficiency of the 4H-SiC MSM detector as a function of wavelength and temperature at 10 V bias.

CONCLUSIONS

With consideration of temperature-dependent physical model, we investigate the temperature dependence of a 4HSiC MSM photodetector in terms of dark/illuminated I-V features and spectral characteristics in the range of 300–800 K by utilizing a 2D device simulator ISE-DESSIS. The calculations are in good agreement with the measured characteristics, thereby, validating the correctness of our numerical modeling. Up to 30 V bias, a distinct increase of the dark current and photocurrent can be observed as the temperature rises from RT to 800 K. The impact of temperature on dark current becomes predominant as compared with the counterpart, leading to the decrease of PDCR. However, the PDCR value obtained even at 800 K is still greater than 200, which reveals the excellent thermal stability of the detector. Furthermore, it is found that the red shift of responsivity shifts to longer wavelength. For spectral response and quantum efficiency, a noticeable increment with elevated temperature can be found. At 800 K, the maximum responsivity of 120.3 mA/W and peak quantum efficiency of 51.82% can be obtained at 302 and 280 nm, respectively. In brief, the systematic investigations, providing valuable information about temperature-dependent properties of the 4H-SiC MSM UV detector, are of paramount essentiality in the assessment of the device stability and reliability under harsh and high-temperature UV detection condition.

This work was supported by the National Defense Pre-Research Foundation of China (51323040118, 513080302).

REFERENCES

1. 1 Powell A R, Rowland L B. SiC materials-progress, status, and potential roadblocks. Proc IEEE, 2002, 90: 942–955
2. 2 Razeghi M. Short-wavelength solar-blind detectors-status, prospects, and markets. Proc IEEE, 2002, 90: 1006–1014
3. 3 Wright N G, Horsfall A B. SiC sensors: A review. J Phys D, 2007, 40: 6345–6354
4. 4 Ito M, Wada O, Nakai K, et al. Monolithic integration of a metal-semiconductor-metal photodiode and a GaAs preamplifier. IEEE Electron Device Lett, 1984, 5: 531–532
5. 5 Borsa T, Williams D F, Hale P D, et al. Novel nano-structured metal-semiconductor-metal photodetector with high peak voltage. Jpn J Appl Phys, 2009, 48: 06FD031
6. 6 Wang L K, Ju Z G, Shan C X, et al. MgZnO metal-semiconductormetal structured solar-blind photodetector with fast response. Solid State Commun, 2009, 149: 2021–2023
7. 7 Liu J S, Shan C X, Li B H, et al. High responsivity ultraviolet photodetector realized via a carrier-trapping process. Appl Phys Lett, 2010, 97: 251102
8. 8 Qin L Q, Shing C, Sawyer S. Metal-semiconductor-metal ultraviolet photodetectors based on zinc-oxide colloidal nanoparticles. IEEE Electron Device Lett, 2011, 32: 51–53
9. 9 Sang L W, Liao M Y, Koide Y, et al. High-performance metal-semiconductor-metal InGaN photodetectors using CaF2 as the insulator. Appl Phys Lett, 2011, 98: 103502
10. 10 Xie Y N, Huang H L, Yang W F, et al. Low dark current metal-semiconductor-metal ultraviolet photodetectors based on sol-gel-derived TiO2 films. J Appl Phys, 2011, 109: 023114
11. 11 Balbi M, Sorianello V, Colace L, et al. Analysis of temperature dependence of Ge-on-Si p-i-n photodetectors. Physica E, 2009, 41: 1086–1089
12. 12 Cha H Y, Soloviev S, Zelakiewicz S, et al. Temperature dependent characteristics of nonreach-through 4H-SiC separate absorption and multiplication APDs for UV detection. IEEE Sensors J, 2008, 8: 233–237
13. 13 Tan Z Y, Guo X G, Cao J C, et al. Temperature dependence of current-voltage characteristics of terahertz quantum-well photodetectors. Semicond Sci Technol, 2009, 24: 115014
14. 14 Blank T V, Goldberg Yu A, Kalinina E V, et al. Temperature dependence of the photoelectric conversion quantum efficiency of 4H-SiC Schottky UV photodetectors. Semicond Sci Technol, 2005, 20: 710–715
15. 15 Wang X D, Hu W D, Chen X S, et al. Dependence of dark current and photoresponse characteristics on polarization charge density for GaN-based avalanche photodiodes. J Phys D, 2011, 44: 405102
16. 16 Cha H Y, Sandvik P M. Electrical and optical modeling of 4H-SiC avalanche photodiodes. Jpn J Appl Phys, 2008, 47: 5423–5425
17. 17 DESSIS. 2D Semiconductor Device Simulator. Version 10.0 2005, Integrated Systems Engineering, Zurich
18. 18 Chen B, Yang Y T, Chai C C, et al. Optical coupling optimization in a novel metal-semiconductor-metal ultraviolet photodetector based on semicircular Schottky electrodes. J Semicond, 2012, 33: 054010
19. 19 Raghunathan R, Baliga B J. Temperature dependence of hole impact ionization coefficients in 4H and 6H-SiC. Solid State Electron, 1999, 43: 199–211
20. 20 Bergman J P, Kordina O, Janzen E. Time resolved spectroscopy of defects in SiC. Phys Status Solidi A, 1997, 162: 65–77
21. 21 Chen B, Yang Y T, Chai C C, et al. Two-dimensional numerical computation of the structure-dependent spectral response in a 4H-SiC metal-semiconductor-metal ultraviolet photodetector with consideration of reflection and absorption on contact electrodes. J Semicond, 2011, 32: 084001
22. 22 Galeckas A, Grivickas P, Grivickas V, et al. Temperature dependence of the absorption coefficient in 4H- and 6H-silicon carbide at 355 nm laser pumping wavelength. Phys Status Solidi A, 2002, 191: 613–620
23. 23 Nallet F, Planson D, Isoird K, et al. Comparison of static, switching and thermal behavior between a 1500 V silicon and silicon carbide bipolar diodes. Proc Int Semicond Conf, 1999, 1: 195–198
24. 24 Lindefelt U. Doping-induced band edge displacements and band gap narrowing in 3C-, 4H-, 6H-SiC, and Si. J Appl Phys, 1998, 84: 2628

25. 25 Averine S, Chan Y C, Lam Y L. Evaluation of Schottky contact parameters in metal-semiconductor-metal photodiode structures. Appl Phys Lett, 2000, 77: 274–276

26. 26 Yang W F, Zhang F, Liu Z G, et al. Effects of annealing on the performance of 4H-SiC metal-semiconductor-metal ultraviolet photodetectors. Mater Sci Semicond Proc, 2008, 11: 59–62

27. 27 Sridhara S G, Devaty R P, Choyke W J. Absorption coefficient of 4H silicon carbide from 3900 to 3250 angstrom. J Appl Phys, 1998, 84: 2963

28. 28 Chang W R, Fang Y K, Ting S F, et al. The hetero-epitaxial SiCN/Si MSM photodetector for high-temperature deep-UV detecting applications. IEEE Electron Device Lett, 2003, 24: 565–567

29. 29 Sou I K, Ma Z H, Zhang Z Q, et al. Temperature dependence of the responsivity of ZnS-based UV detectors. J Cryst Growth, 2000, 214: 1125–1129

CHAPTER 4

GAINNAS-BASED HELLISH-VERTICAL CAVITY SEMICONDUCTOR OPTICAL AMPLIFIER FOR 1.3 μM OPERATION

Faten Adel Ismail Chaqmaqchee, Simone Mazzucato, Murat Oduncuoglu, Naci Balkan, Yun Sun, Mustafa Gunes, Maxime Hugues, Mark Hopkinson

INTRODUCTION

III-V semiconductors are indispensable for today's optoelectronic devices, such as lasers modulators, photodetectors and optical amplifiers in optical fibre communication systems. One potentially important material for such applications is the quaternary alloy GaInNAs [1, 2]. In the 1.3-μm optical communications window, GaInNAs may be grown pseudomorphically on GaAs, allowing the use of high quality AlAs/GaAs distributed Bragg reflectors (DBRs), with potential cost advantages compared to InP-based approaches. It can be used to fabricate several devices, among which vertical cavity semiconductor optical amplifiers (VCSOAs) are important components in optical fibre networks. They have improved performance over SOAs as they have inherent polarization insensitivity, lower noise figures, high-fibre coupling, easy chip testing and potential for integration into high-density two-dimensional arrays. Furthermore the narrower bandwidth of vertical cavity structures makes these devices good for filtering applications [3, 4, 5, 6].

A VCSOA can be simply described as a vertical cavity surface emitting laser (VCSEL) operating in the linear regime below threshold, with a reduced number of top DBR layers. However, in this article, a novel VCSOA based on the Hellish structure as an alternative to conventional VCSOAs is investigated [7]. Hellish devices utilise the transport of non-equilibrium carriers parallel to the layers. Spontaneous emission of ultra bright Hellish structures has been demonstrated [8, 9]. VCSEL operation was achieved by addiction of DBR layers [10, 11, 12, 13]. That design is adapted in this study to make a GaInNAs-based Hellish-VCSOA structure, which differs from the conventional VCSEL by the reduced number of top DBR layers [14]. The structure is designed to operate in the 1.3-μm wavelength region via electrical pumping.

The authors demonstrate for the first time the operation of a Hellish VCSOA with a multiple quantum well (MQW) GaInNAs/GaAs active region, at temperatures between 77 and 300 K. Optical and electrical pumping (photoluminescence—PL, electroluminescence—EL) were used, and a 1.28-μm emission at room temperature was observed. By combining the two measurements, an electro-photoluminescence (EPL) technique was performed, from which light amplification is demonstrated. The authors also present the results of the reflectivity spectrum and cavity resonance calculations using the matrix formulation for multi-layer structures [15], and compare these with experimental results.

Experimental Results and Discussion

The structure of the investigated device, shown in Figure 1a, contains 11 layers of 6 nm-thick $Ga_{0.35}In_{0.65}N_{0.02}As_{0.08}$ quantum wells separated by 10 nm GaAs barriers. The use of MQWs, placed at the electric field antinode of 3λ/2 cavity length, is done in order to provide optical gain (Figure 1b). The active region is enclosed between two 150 nm-thick doped cladding layers Si-doped ($n = 1 \times 10^{17}$ cm^{-3}) on the bottom side, and C-doped ($p = 1 \times 10^{17}$ cm^{-3}) on the top side. The structure is sandwiched between two DBRs. The bottom DBR has 20.5 pairs of AlAs/GaAs quarter-wave stacks and provides a reflectivity in excess of 99% at 1.3-μm. On the other side, the top DBR has six pairs of AlAs/GaAs quarter-wave stacks giving around 60% reflectivity. This is lower than the bottom DBR, thus allowing light emission from the top surface. Both DBRs are undoped except for the first bottom AlAs/GaAs period which is 1×10^{17} cm^{-3} doped.

Citation: Faten Adel Ismail ChaqmaqcheeEmail author, Simone Mazzucato, Murat Oduncuoglu, Naci Balkan, Yun Sun, Mustafa Gunes, Maxime Hugues and Mark Hopkinson, GaInNAs-based Hellish-vertical cavity semiconductor optical amplifier for 1.3 μm operation, DOI: 10.1186/1556-276X-6-104.

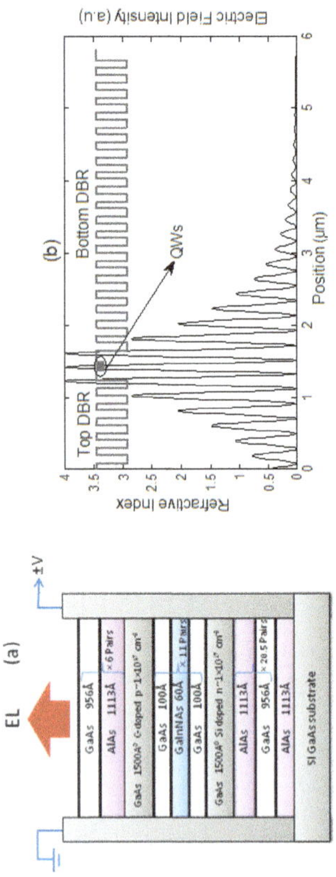

Figure 1 Schematic diagram illustrates (a) the layer structure of the simple bar Hellish-VCSOA and (b) the refractive index profile and distribution of the electric filed intensity across the sample, in which the QWs are situated at the antinode of the electric field, i.e. where it reaches its maximum intensity.

Ohmic contacts are formed by diffusing Au/GeAu/Ni/Au through all the layers and into the substrate, defining a simple bar-shaped sample, with 1-mm contact separation and 4.5-mm width. This is done by annealing the contacts for 60 s at 430°C. Once fabricated, the device is electrically biased with positive voltage pulses 390-ns duration and a 3-ms repetition rate. The duty cycle is small enough to prevent damage by excessive Joule heating. The applied electric field is varied from 0.01 to 1 kV/cm. Figure 2 shows the current-voltage (I-V) characteristics at 77 and 300 K. The sample exhibits ohmic behaviour at electric fields below 600 and 900 V/cm at 77 and 300 K, respectively. The small deviation from ohmic behaviour is an indication of carrier heating [16, 17].

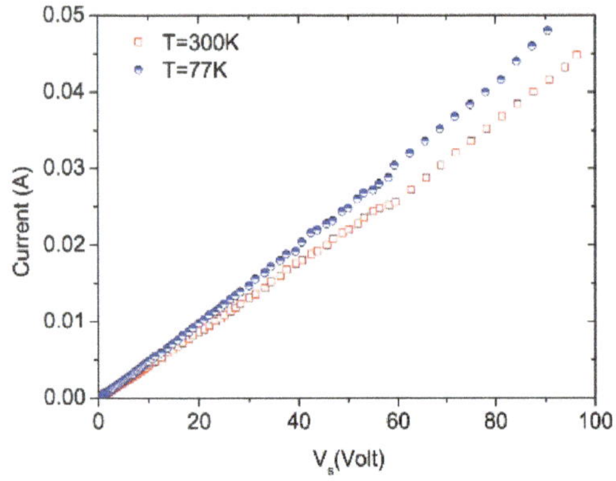

Figure 2 I - V characteristics of simple bar Hellish-VCSOA at 77 and 300 K .

The operation of Hellish device is based on the longitudinal injection of electron and hole pairs in their respective channels, due to the diffusion of both top contacts through all layers. Without the applied electric field, if the sample is illuminated, photogenerated carriers will eventually recombine radiatively in the QW without drifting laterally in the longitudinal channels. On the other hand, when the device is biased, the energy bands tilt up, with the degree of tilting being proportional to the applied voltage. At low bias, a quasi-flat region is established by the tilted energy bands, and a small number of carriers are then able to drift diagonally into the p-n junction. This is illustrated in Figure 3. With an increase in the electric field, the energy bands will tilt up more, so that more carriers will flow into the active region, enhancing the emitted light. In view of the operational diagram depicted in Figure 3, the application of a negative bias results in a tilting and the diffusion of the holes to the region where electrons are injected, and recombination occurs in the vicinity of the cathode. This allows for spatial confinement and control of the light emission area. Luminescence from the opposite site (anode) appears by inverting the bias polarity [16].

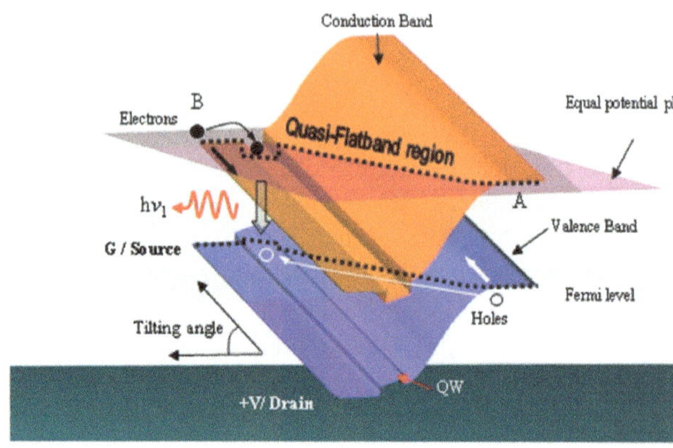

Figure 3 Schematic diagram to illustrate the emission of light under quasi-flat band region condition [16].

Experiments have been carried out using PL, EL and EPL techniques at different temperatures between liquid nitrogen and room temperature. The experimental arrangement for these techniques is illustrated in Figure 4.

In PL and EPL, the optical excitation source is a CW Argon laser operating at 488-nm wavelength with 20-mW output power. The laser beam is chopped using a mechanical chopper and directed to the sample surface. The emitted light is dispersed by a Bentham M300 1/3 m monochromator and collected with a cooled InGaAs photomultiplier. The outcoming electrical signal is sent to a Gated Integrator & Boxcar Averager Module (Stanford Research Systems, model SR250) or a lock-in amplifier (Stanford Research Systems, model SR830) according to the experiment performed.

Figure 5 shows the integrated emission light from the device as a function of applied electric field. The threshold light emission varies between 110 and 290 V/cm according to the sample temperature. Above the threshold, the integrated EL increases linearly with applied electric field.

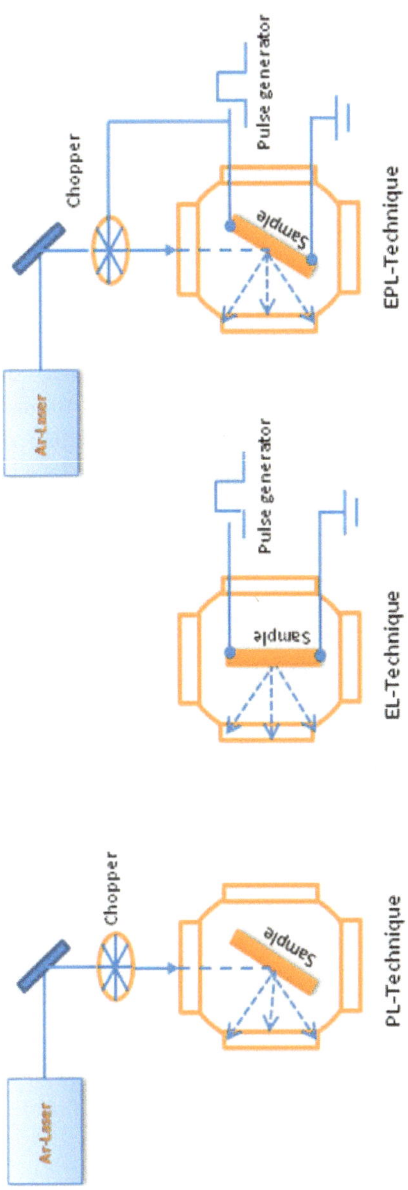

Figure 4 PL, EL and EPL experimental arrangement.

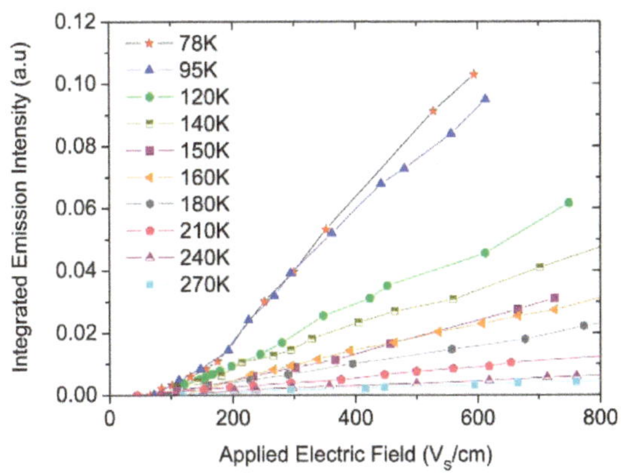

Figure 5 Integrated EL intensity versus applied electric field at various temperatures.

Figure 6 shows the PL spectra measured at different temperatures. The PL peak red-shifts from 1245 nm at 77 K to 1270 nm at 300 K.

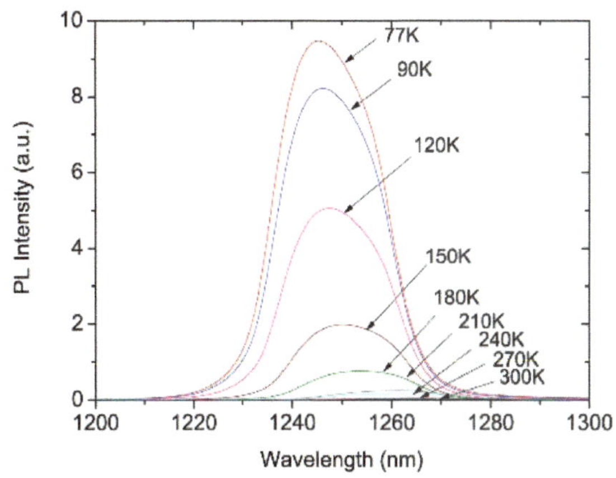

Figure 6 PL spectra measured at different temperatures.

Spectral EL is also measured with applied voltage pulses of amplitude between 0.3 and 100 V, where the pulse duration is kept at about 390 ns. The EL spectra are obtained at different temperatures between 80 and 300 K, and according to Figure 7, it shows a broad spectra. Approximately, the EL spectrum shifts in wavelength from 1239 nm at 80 K to 1281 nm at 300 K.

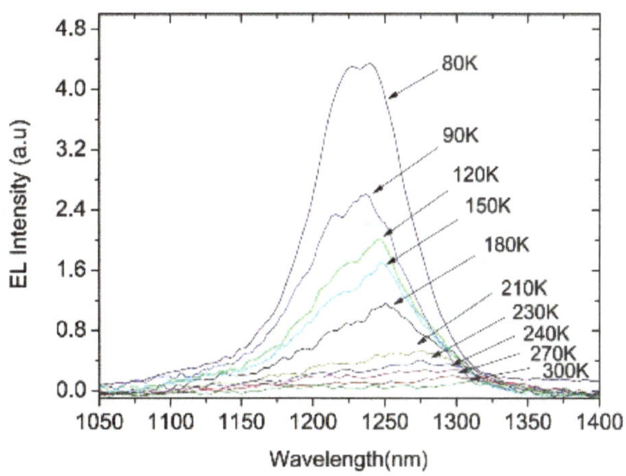

Figure 7 EL spectra measured at a fixed bias voltage of 97 V corresponding to an electric field of 0.97 kV/cm.

There is good agreement between the EL and PL peak positions. However, the EL emission is considerably broader than the PL. This observation is attributed to growth non-uniformities and material fluctuations. PL is measured from a small spot (excitation spot size 0.5 mm^2), while the EL is collected from the whole sample surface. Therefore, the EL may be expected to be broader if the QWs and/or DBRs width have monolayer fluctuations. In order to prove this, the PL at different spots on the sample (Figure 8) was measured and the reflectivity spectrum for small fluctuations in the thickness of the layers in the cavity of around 2 nm (Figure 9) was calculated. The effect of layer fluctuations is clear.

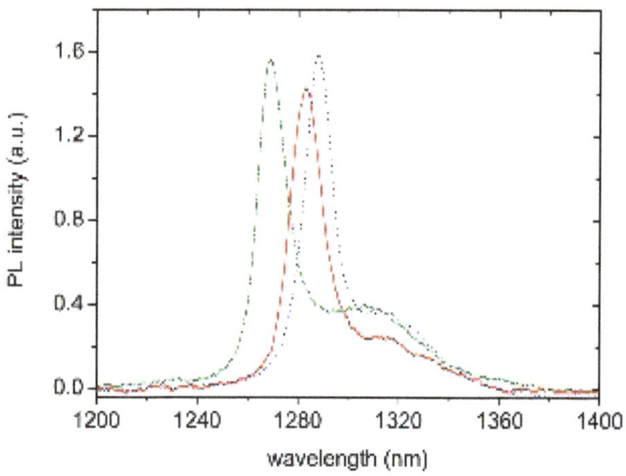

Figure 8 Room temperature PL spectra taken at different laser spot positions across the sample, showing an approximate 20-nm uncertainty in the peak position.

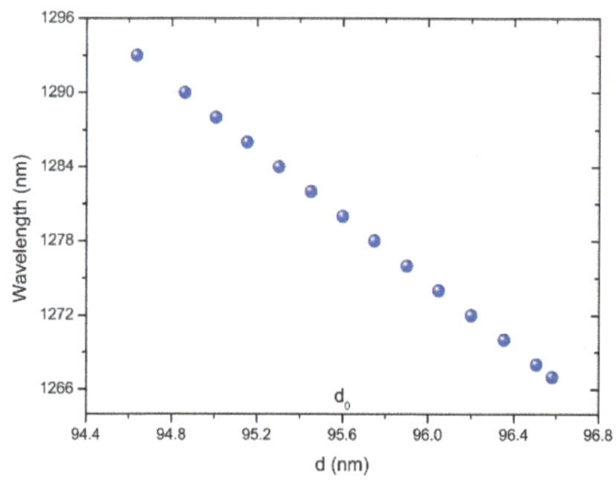

Figure 9 Dependence of the cavity resonance position with ±1 nm fluctuations of the GaAs thickness in the DBR (d_0 is the nominal GaAs DBR layer thickness) but weaker behaviour takes place when fluctuations occur in the AlAs layers.

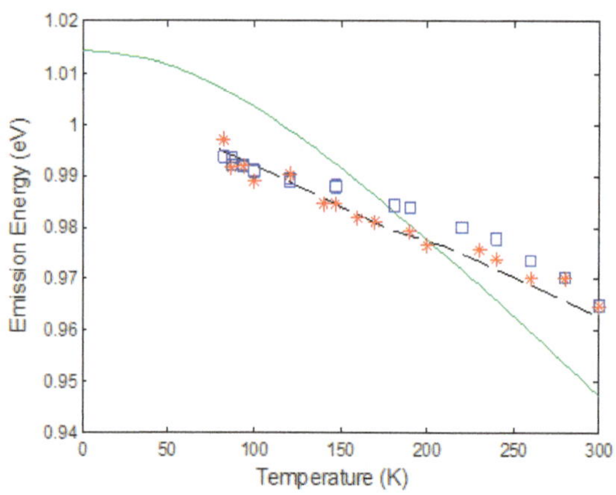

Figure 10 Continuous line represents the calculated temperature dependence of bandgap energy for the device active area (GaInNAs/GaAs QW) using the BAC model, while the expected cavity resonance position is plotted with a dashed line and finally the scattered points represent the experimental data for PL (asterisks) and EL (squares).

The temperature dependence of EL and PL peaks and the cavity resonance are plotted in Figure 10, together with the active material bandgap energy curve [18]. The behaviour differs extremely from the change of the GaInNAs/GaAs bandgap energy with temperature. Theoretically, a red shift of the active material peak wavelength at a rate of 0.38 nm/K was predicted, while the resonance cavity moves with temperature at 0.18 nm/K. At these rates, the optimum operating temperature for this device will be at around 220 K, where the maximum peak material gain coincides with the DBR resonance cavity position.

The EPL technique was performed by combining the two experimental techniques, namely PL and EL. In order to synchronise optical and electrical pulses, the pulse generator is triggered by a mechanical chopper. PL, EL and EPL spectra for Hellish-VCSOA are measured as a function of temperature. In both EL and EPL, the electric field was kept constant at 0.7 kV/cm.

In Figure 11, the T = 87 K EL, PL, EPL spectra and the sum of EL and PL are plotted. The PL spectrum presents a broad peak at around 1250-nm wavelength and a full-width-at-half-maximum of 13 meV. As stated before, it corresponds to the overlap of the active region gain spectrum and the cavity resonance reflectivity that filters and narrows the emission. Variations in the peak position are ascribable to fluctuations in the cavity resonance. The EL spectrum measured at the same temperature shows the emission peak at around 1.03 eV and by comparing the SUM (EL + PL) and EPL spectrum, the presence of optical gain was clearly visible. Signal amplification occurs when both electrical and optical inputs are applied.

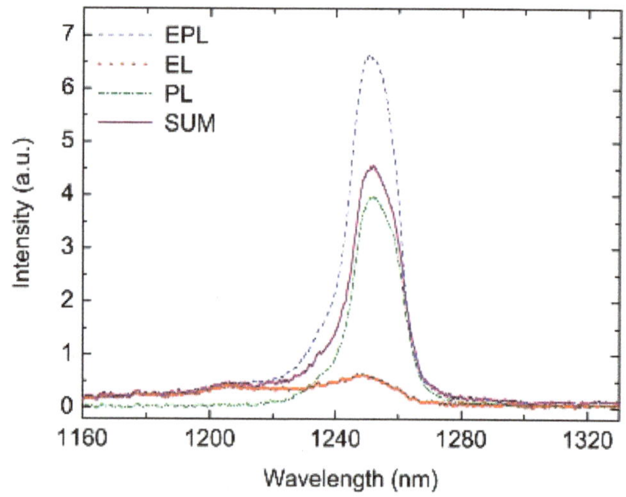

Figure 11 EL, PL, EPL and SUM (EL + PL) spectra at T = 87 K.

This investigation was focussed on the gain at room temperature. The integrated intensities of PL, EL and EPL, together with the calculated SUM (EL + PL) and gain are plotted in Figure 12, as function of applied voltage, up to 800 V/cm, with laser excitation power of 10 mW. Finally, Figure 13 displays the evolution of the gain with applied voltage, which reaches its maximum at around 50 V. It should be noted that the wavelength of the laser (λ = 488 nm) is very different from the cavity resonance position shown in Figure 10. Therefore, most of the excitation is lost through absorption. In order to give a quantitative value to the VCSOA gain, the PL gain is defined as the ratio of the PL peak when the device is electrically pumped to that when the device is not biased. This gain should not be confused with conventional VCSOA gain as ratio of output power to input power.

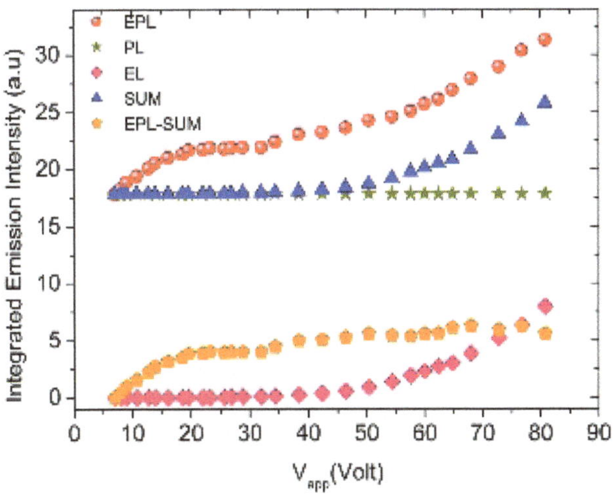

Figure 12 Integrated EPL, PL, EL, SUM (EL + PL) and EPL-SUM (EL + PL) measured as a function of applied voltages at T= 300 K.

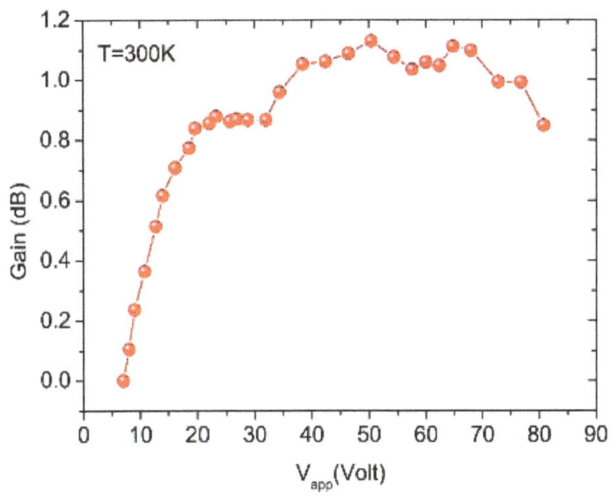

Figure 13 Gain characteristics are measured as a function of applied voltages at T = 300 K.

Further improvements in gain characteristic and device performance will be expected by optimising the Hellish-VCSOA structure for 1.3-μm application via electrically pumping, and by reducing the device length so that the operating voltage will be much lower than the one used here.

CONCLUSIONS

Optical gain at $\lambda \sim 1.28$ μm is demonstrated in a Hellish-VCSOA device consisting of $Ga_{0.35}In_{0.65}N_{0.02}As_{0.08}/GaAs$ QWs and AlAs/GaAs DBRs. The advantage of using such device is that longitudinal electric fields are applied parallel to active layer so that the current flows along the p and n layers without passing through the DBRs. The operation of the device is independent of the polarity of the applied electric field. The emission and amplification characteristics are investigated as a function of temperature and applied voltage. Thus, the Hellish-VCSOA is a good candidate for electrically pumped optical amplifier operating at around 1.3 μm.

ABBREVIATIONS

DBRs: distributed Bragg reflectors

EL: electroluminescence

EPL: electro-photoluminescence

MQWs: multiple quantum wells

PL: photoluminescence

QWs: quantum wells

VCSEL: vertical cavity surface emitting laser

VCSOA: vertical cavity semiconductor optical amplifier.

DECLARATIONS

Acknowledgements

F.AI. Chaqmaqchee is grateful to the Ministry of Higher Education and Scientific Research of IRAQ for their financial support. M. Oduncuoglu is grateful to Kilis 7 Aralik University/Turkey research fund and the financial support provided by TUBITAK. The authors also acknowledge A. Boland-Thoms for technical assistance. Finally we are grateful to the COST Action MP0805 for providing the scientific platform for collaborative research.

Competing interests

The authors declare that they have no competing interests.

Authors' contributions

FAI, YS and MO designed the structure. MHu and MHo grew the sample according to the specifications. FAI fabricated the devices, carried out the experiments and the theoretical calculations, in collaboration with MO, YS, SM, and MG. FAI and SM wrote up the article. NB, is the inventor of the original device and the overall supervisor of the project.

All authors read and approved the final manuscript.

REFERENCES

1. Kondow M, Uomi K, Niwa A, Kitatani T, Watahiki S, Yazawa Y: GaInNAs: A novel material for long-wavelength-range laser diodes with excellent high-temperature performance. Jpn J Appl Phys 1996, 35: 1273. 10.1143/JJAP.35.1273
2. Buyanova IA, Chen WM, Monemar B: Electronic properties of Ga(In)NAs alloys. MRS Internet J Nitride Semicond Res 2001, 6: 2.
3. Piprek J, Bjorlin ES, Bowers JE: Optical gain-bandwidth product of vertical-cavity laser amplifiers. Electron Lett 2001, 37: 298. 10.1049/el:20010201
4. Karim A, Bjorlin S, Piprek J, Bowers JE: Long-wavelength vertical-cavity lasers and amplifiers. IEEE J Sel Top Quantum Electron 2000, 6: 1244. 10.1109/2944.902174
5. Bjorlin S, Riou B, Keating A, Abraham P, Chiu Yi-J, Piprek J, Bowers JE: 1.3-μm vertical-cavity amplifier. IEEE Photonics Technol Lett 2000, 12: 951. 10.1109/68.867971
6. Bjorlin S, Abraham P, Pasquariello D, Piprek J, Chiu Yi-

J, Bowers JE: High gain, high efficiency vertical-cavity semiconductor optical amplifiers. Indium Phosphide and Related Materials Conference 2002., 307:

7. Balkan N, Serpenguzel A, O'Brien-Davies A, Sokmen I, Hepburn C, Potter R, Adams MJ, Roberts JS: VCSEL structure hot electron light emitter. Mater Sci Eng 2000, B74: 96. 10.1016/S0921-5107(99)00542-5

8. Straw A, Balkan N, O'Brien A, da Cunha A, Gupta R: Hot electron light-emitting and lasing semiconductor heterostructures--type 1. Superlatt Microstruct 1995, 18: 33. 10.1006/spmi.1995.1085

9. O'Brien A, Balkan N, Boland-Thoms A, Adams M, Bek A, Serpenguzel A, Aydinli A, Roberts J: Super-radiant surface emission from a quasi-cavity hot electron light emitter. Opt Quantum Electron 1999, 31: 183.

10. Balkan N, Sokmen I, O'Brien A, Potter R, Hepburn C, Boland-Thoms A, Adams MJ, Roberts J: Hot electron VCSEL. Proc SPIE 1999, 3625: 336.

11. Balkan N, O'Brien A, Boland-Thoms A, Potter R, Poolton N, Adams M, Masum J, Bek A, Serpenguzel A, Aydinli A, Roberts J: The operation of a novel hot electron Vertical Cavity Surface Emitting Laser. Proc SPIE 1998, 3228: 162.

12. Erol A, Balkan N, Arıkan MC, Serpenguzel A, Roberts J: Temperature Dependence of the Threshold Electric Field in Hot Electron VCSELs. IEE Proc Optoelectron 2003, 150: 535. 10.1049/ip-opt:20030433

13. Sceats R, Balkan N: Hot Electron Light Emission at 1.3 μm from a GaInAsP/InP structure with distributed Bragg reflectors. Phys Status Solidi 2003, 198: 495. 10.1002/pssa.200306600

14. Wah JY, Loubet N, Mazzucato S, Balkan N: Bi-directional field effect light emitting and absorbing heterojunction with Ga 0.8 In 0.2 N 0.015 As 0.985 at 1250 nm. IEE Proc Optoelectron 2003, 150: 72. 10.1049/ip-opt:20030051

15. Yeh P: Chapt. 5 of Optical Waves and Layered Media. New York: John Wiley & Sons; 1991.

16. Wah JY, Balkan N: Low field operation of hot electron light emitting devices: quasi-flat-band model. IEE Proc Optoelectron 2004, 151: 482. 10.1049/ip-opt:20040783

17. Wah JY, Balkan N, Potter RJ, Roberts JS: The operation of a Wavelength Converter based on Field Effect Light emitter and absorber heterojunction. Phys Status Solidi A 2003, 196: 503.

18. Potter RJ, Balkan N: Optical Properties of GaInNAs and GaNAs QWs. J Phys Condens Matter 2004, 16: 3387. 10.1088/0953-8984/16/31/026

CHAPTER 5 EMERGING NONVOLATILE MEMORY TECHNOLOGIES

Jagan Singh Meena, Simon Min Sze, Umesh Chand, Tseung-Yuen Tseng

Keywords: Emerging nonvolatile memory technologies, Magnetic storage, Market memory technologies, Memristors, Phase change memories, Random-access storage, Flash memory technologies, Three-dimensional memory, Transparent memory, Unified memory

BACKGROUND

General overview

The idea of using a floating gate (FG) device to obtain a nonvolatile memory device was suggested for the first time in 1967 by Kahng D and Sze SM at Bell Labs [1]. This was also the first time that the possibility of nonvolatile MOS memory device was recognized. From that day, semiconductor memory has made tremendous contributions to the revolutionary growth of digital electronics since a 64-bit bipolar RAM chip to be used in the cache memory of an IBM computer was reported in 1969 [2]. Semiconductor memory has always been an indispensable component and backbone of modern electronic systems. All familiar computing platforms ranging from handheld devices to large supercomputers use storage systems for storing data temporarily or permanently [3]. Beginning with punch card which stores a few bytes of data, storage systems have reached to multiterabytes of capacities in comparatively less space and power consumption. Regarding application aspects, the speed of storage systems needs to be as fast as possible [4]. Since Flash memory has become a common component of solid-state disks (SSDs), the falling prices and increased densities have made it more cost-effective for many other applications [5]. Memory devices and most SSDs that use Flash memory are likely to serve very different markets and purposes. Each has a number of different attributes which are optimized and adjusted to best meet the needs of particular users. Because of natural inherent limitations, the long-established memory devices have been shorted out according to their inventions to match with portable electronic data storage systems. Today, the most prominent one is the limited capacity for continued scaling of the electronic device structure. Research is moving along the following paths for embedded Flash devices: (i) scaling down the cell size of device memory, (ii) lowering voltage operation, and (iii) increasing the density of state per memory cell by using a multilevel cell. To sustain the continuous scaling, conventional Flash devices may have to undergo revolutionary changes. Basically, it is expected that an entire DVD collection be in the palm of a hand. Novel device concepts with new physical operationing principles are needed. It is worthwhile to take a look at semiconductor memories against the background of digital systems. The way semiconductor devices are used in a systems environment determines what is required of them in terms of density, speed/power, and functions. It is also worthwhile to look into the economic significance of semiconductor memories and the relative importance of their various types. For the past three and a half decades in existence, the family of semiconductor memories has expanded greatly and achieved higher densities, higher speeds, lower power, more functionality, and lower costs [3, 6, 7]. At the same time, some of the limitations within each type of memory are also becoming more realized. As such, there are several emerging technologies aiming to go beyond those limitations and potentially replace all or most of the existing semiconductor memory technologies to become a universal semiconductor memory (USM). In addition, the rewards for achieving such a device would be to gain control of an enormous market, which has expanded from computer applications to all of consumer electronic products. Looking forward to the future, there are wide ranges of emerging memory applications for automation and information technology to health care. The specification of nonvolatile memory (NVM) is based on the floating gate configuration, which is the feature of an erased gate put into many cells to facilitate block erasure. Among them, designed Flash memories such as NOR and NAND Flash have been

Citation: Jagan Singh Meena, Simon Min Sze, Umesh Chand, Tseung-Yuen Tseng, emerging nonvolatile memory technologies, DOI: 10.1186/1556-276X-9-526.

developed and then proposed as commercial products into bulk market. They have been considered as the most important products. NOR has high operation speed for both code and data storage applications; on the other hand, NAND has high density for large data storage applications [8]. Since the inception of Flash memory, there has been an exponential growth in its market driven primarily by cell phones and other types of consumer electronic equipment. While, today, integration of a silicon chip is not economical, toys, cards, labels, badges, value paper, and medical disposables could be imagined to be equipped with flexible electronics and memory. With growing demands for high-density digital information storage, memory density with arriving technology has been increased dramatically from the past couple of years. The main drive to develop organic nonvolatile memory is currently for applications of thin-film, flexible, or even printed electronics. One needs a technology to tag everything to electronic functionality which can be foreseen in a very large quantity and at a very low cost on substrates such as plastic and paper. Accessible popularization of roll-to-roll memory commercialization is a way to make an encounter interesting and challenging to have charge storage devices of choice for applications with enormous flexibility and strength. Recently, polymer (plastic memory) and organic memory devices have significant consideration because of their simple processes, fast operating speed, and excellent switching ability [9, 10]. One significant advantage polymer memory has over conventional memory designs is that it can be stacked vertically, yielding a three-dimensional (3-D) use of space [11]. This means that in terabyte solid-state devices with extremely low transistor counts such as drives about the size of a matchbook, the data persists even after power is removed. The NAND Flash market is continually growing by the successive introduction of innovative devices and applications. To meet the market trend, 3-D NVMs are expected to replace the planar ones, especially for 10-nm nodes and beyond. Moreover, simple-structure organic bistable memory exhibiting superior memory features has been realized by employing various nanoparticles (NPs) blended into a single-layered organic material sandwiched between two metal electrodes [12, 13]. The NPs act as traps that can be charged and discharged by suitable voltage pulses. NP blends show promising data retention times, switching speed, and cycling endurance, but the on-state current is too low to permit scaling to nanometer dimensions [10, 14]. A lot of these great ideas tend to die before reaching this point of development, but that is not to say that we will be seeing plastic memory on store shelves next year. There are still many hurdles to get over; software alone is a big task, as is the manufacturing process, but it does bring this technology one step closer to reality [15]. It is not an exaggeration to say that the equivalent of 400,000 CDs, 60,000 DVDs, or 126 years of MPG music may be stored on a polymer memory chip the size of a credit card.

The Vision of This Review

In this review, we focus on electrically programmable nonvolatile memory changes from silicon nanocrystal memory scaling to organic and metallic NP memory devices. Further, the scaling trend move towards the emerging NVM to flexible and transparent redox-based resistive switching memory technologies. This review is intended to give an overview to the reader of storage systems and components from conventional memory devices that have been proposed in the past years of recent progress in current NVM devices based on nanostructured materials to redox-based resistive random-access memory (RRAM) to 3-D and transparent memory devices. We describe the basics of Flash memory and then highlight the present problems with the issue of scaling tunnel dielectric in these devices. We briefly describe a historical change, how the conventional FG nonvolatile memory suffers from a charge loss problem as the feature size of the device continues to shrink. A discrete polysilicon-oxide-nitride-oxide-silicon (SONOS) memory is then proposed as a replacement of the conventional FG memory. The NC memory is expected to efficiently preserve the trapped charge due to the discrete charge storage node while also demonstrating excellent features such as fast program/erase speeds, low programming potentials, and high endurance. We also discuss current ongoing research in this field and the solutions proposed to solve the scaling problems by discussing a specific solution in detail which would be the centerpiece in recent memory work progress. Moreover, this review makes distinct emerging memory concepts with more recent molecular and quantum dot programmable nonvolatile memory concepts, specifically using charge trapping in conjugated polymers and metal NPs. We classify several possible devices, according to their operating principle, and critically review the role of π-conjugated materials in the data storage device operation. We describe specifications for applications of emerging NVM devices as well as already existing NAND memory and review the state of the art with respect to these target specifications in the future. Conclusions are drawn regarding further work on materials and upcoming memory devices and architectures.

CLASSIFICATION OF SOLID-STATE MEMORY TECHNOLOGIES

Data storage devices can be classified based on many functional criteria. Of them, silicon-based semiconductor memories are categorized into two: volatile and nonvolatile [3, 16]. In volatile memories, the information eventually fades while power supply is turned off unless the devices used to store data will be periodically refreshed. On the other hand, nonvolatile memories retain the stored information even when the power supply is turned off. Volatile memories, such as static random-access memory (SRAM) and dynamic random-access memory (DRAM), need voltage supply to hold their information while nonvolatile memories, namely Flash memories, hold their information without one. DRAM (dynamic stands for the periodical refresh) is needed for data integrity in contrast to SRAM. The basic circuit structures of DRAM, SRAM, and Flash memories are shown in Figure 1. DRAM, SRAM, and Flash are today's dominant solid-state memory technologies, which have been around for a long time, with Flash the youngest, at 25 years. DRAM is built using only one transistor and one capacitor component, and SRAM is usually built in CMOS technology with six transistors. Two cross-coupled inverters are used to store the information like in a flip-flop. For the access control, two further transistors are needed. If the write line is enabled, then data can be read and set with the bit lines. The Flash memory circuit works with the FG component. The FG is between the gate and the source-drain area and isolated by an oxide layer. If the FG is uncharged, then the gate can control the source-drain current. The FG gets filled (tunnel effect) with electrons when a high voltage at the gate is supplied, and the negative potential on the FG works against the gate and no current is possible. The FG can be erased with a high voltage in reverse direction of the gate. DRAM has an advantage over SRAM and Flash of only needing one MOSFET with a capacitor. It also has the advantage of cheap production as well as lower power consumption as compared to SRAM but slower than SRAM. On the other hand, SRAM is usually built in CMOS technology with six transistors and two cross-coupled inverters, and for the access control, two further transistors are needed. SRAM has the advantage of being quick, easy to control, integrated in the chip, as well as fast because no bus is needed like in DRAM. But SRAM has the disadvantages of needing many transistors and hence expensive, higher power consumption than DRAM. In comparison to DRAM and SRAM, Flash memory has FG between the gate and the source-drain area and isolated with an oxide layer. Flash memory does not require power to store information but is slower than SRAM and DRAM.

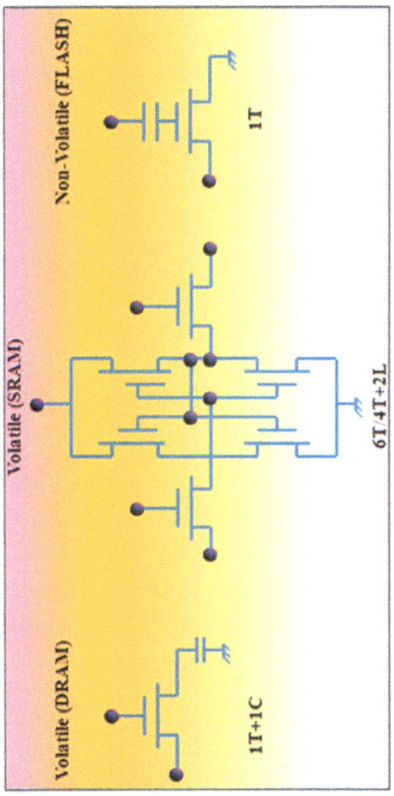

Figure 1: The circuitry structures of DRAM, SRAM, and Flash memories.

Both types of memories can be further classified based on the memory technology that they use and based on data volatility as shown in the classification flow chart depicted in Figure 2. Volatile memories consist mostly of DRAM [17], which can be further classified into SDRAM and mobile RAM which only retain information when current is constantly supplied to the device [18]. Another small but very important memory device is SRAM. The market for DRAM devices far exceeds the market for SRAM devices, although a small amount of SRAM devices is used in almost all logic and memory chips. However, DRAM uses only one transistor and one capacitor per bit, allowing it to reach much higher densities and, with more bits on a memory chip, be much cheaper per bit. SRAM is not worthwhile for desktop system memory, where DRAM dominates, but is used for its cache memories. SRAM is commonplace in small embedded systems, which might only need tens of kilobytes or less. Forthcoming volatile memory technologies that hope to replace or compete with SRAM and DRAM include Z-RAM, TTRAM, A-RAM, and ETA RAM. In the industry, new universal and stable memory technologies will appear as real contenders to displace either or both NAND Flash and DRAM. Flash memory is presently the most suitable choice for nonvolatile applications for the following reasons: Semiconductor nonvolatile memories consist mostly of the so-called 'Flash' devices and retain their information even when the power is turned off. Other nonvolatile semiconductor memories include mask read-only memory (MROM), antifuse-based one-time programmable (OTP) memory, and electrically erasable read-only memory (EEPROM). Flash is further divided into two categories: NOR, characterized by a direct write and a large cell size, and NAND, characterized by a page write and small cell size. Nonvolatile memory is a computer memory that can retain the stored information even when not powered [3, 19, 20]. Nonvolatile semiconductor memories are generally classified according to their functional properties with respect to the programming and erasing operations, as shown in the flow chart described in Figure 2. These are floating gate, nitride, ROM and fuse, Flash, emerging, and other new next-generation memory technologies. Today, these nonvolatile memories are highly reliable and can be programmed using a simple microcomputer and virtually in every modern electronic equipment, which are expected to replace existing memories.

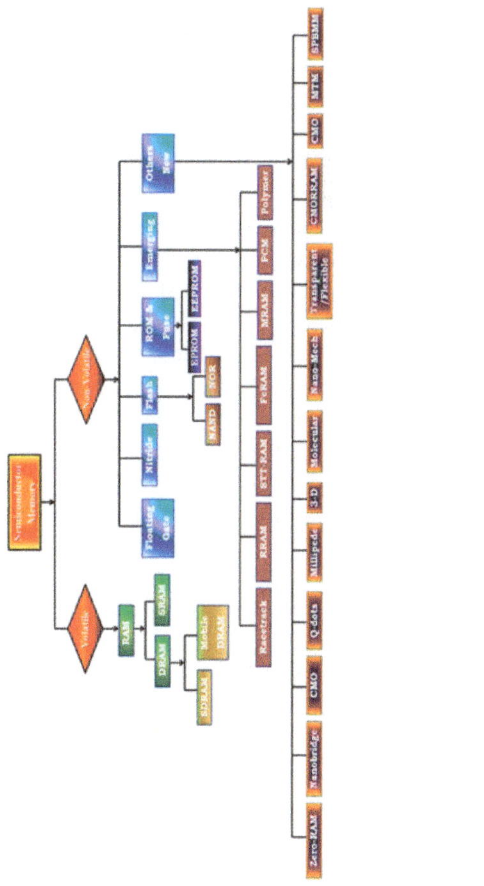

Figure 2 Flow chart for the semiconductor memory classification according to their functional criteria.

Among them, emerging nonvolatile memories are now very captivating. The next-generation memory market will cover up these emerging memory technologies [21]. There are mainly five types of nonvolatile memory technology: Flash memory, ferroelectric random-access memory (FeRAM), magnetic random-access memory (MRAM), phase-change memory (PCM), and RRAM. Nonvolatile memory, specifically 'Flash' memory, which is characterized by a large-block (or 'sector') erasing mechanism, has been the fastest growing segment of the semiconductor business for the last 10 years. Some of these newer emerging technologies include MRAM, FeRAM, PCM, spin-transfer torque random-access memory (STT-RAM), RRAM and memristor. MRAM is a nonvolatile memory [10, 22]. Unlike DRAM, the data is not stored in an electric charge flow, but by magnetic storage elements. The storage elements are formed by two ferromagnetic plates, each of which can hold a magnetic field, separated by a thin insulating layer. One of the two plates is a permanent magnet set to a particular polarity; the other's field can be changed to match that of an external field to store memory. STT-RAM is an MRAM (nonvolatile) but with better scalability over traditional MRAM. The STT is an effect in which the orientation of a magnetic layer in a magnetic tunnel junction or spin valve can be modified using a spin-polarized current. Spin-transfer torque technology has the potential to make MRAM devices combining low current requirements and reduced cost possible; however, the amount of current needed to reorient the magnetization is at present too high for most commercial applications. PCM is a nonvolatile random-access memory, which is also called ovonic unified memory (OUM), based on reversible phase conversion between the amorphous and the crystalline state of a chalcogenide glass, which is accomplished by heating and cooling of the glass. It utilizes the unique behavior of chalcogenide (a material that has been used to manufacture CDs), whereby the heat produced by the passage of an electric current switches this material between two states. The different states have different electrical resistance which can be used to store data. The ideal memory device or the so-called unified memory would satisfy simultaneously three requirements: high speed, high density, and nonvolatility (retention). At the present time, such memory has not been developed. The floating gate nonvolatile semiconductor memory (NVSM) has high density and retention, but its program/erase speed is low. DRAM has high speed (approximately 10 ns) and high density, but it is volatile. On the other hand, SRAM has very high speed (approximately 5 ns) but limited from very low density and volatility. It is expected that PCM will have better scalability than other emerging technologies. RRAM is a nonvolatile memory that is similar to PCM. The technology concept is that a dielectric, which is normally insulating, can be made to conduct through a filament or conduction path formed after application of a sufficiently high voltage. Arguably, this is a memristor technology and should be considered as potentially a strong candidate to challenge NAND Flash. Currently, FRAM, MRAM, and PCM are in commercial production but still, relative to DRAM and NAND Flash, remain limited to niche applications. There is a view that MRAM, STT-RAM, and RRAM are the most promising emerging technologies, but they are still many years away from competing for industry adoption [23]. Any new technology must be able to deliver most, if not all, of the following attributes in order to drive industry adoption on a mass scale: scalability of the technology, speed of the device, and power consumption to be better than existing memories. The NVSM is in inspiring search of novel nonvolatile memories, which will successfully lead to the realization and commercialization of the unified memory.

In progress, another new class of nonvolatile memory technologies will offer a large increase in flexibility compared to disks, particularly in their ability to perform fast, random accesses. Unlike Flash memory, these new technologies will support in-place updates, avoiding the extra overhead of a translation layer. Further, these new nonvolatile memory devices based on deoxyribonucleic acid (DNA) biopolymer and organic and polymer materials are one of the key devices for the next-generation memory technology with low cost. Nonvolatile memory based on metallic NPs embedded in a polymer host has been suggested as one of these new cross-point memory structures. In this system, trap levels situated within the bandgap of the polymer are introduced by the NPs [24, 25]. Memory devices play a massive role in all emerging technologies; as such, efforts to fabricate new organic memories to be utilized in flexible electronics are essential. Flexibility is particularly important for future electronic applications such as affordable and wearable electronics. Much research has been done to apply the flexible electronics technology to practical device areas such as solar cells, thin-film transistors, photodiodes, light-emitting diodes, and displays [26–28]. Research on flexible memory was also initiated for these future electronic applications. In particular, organic-based flexible memories have merits such as a simple, low-temperature, and low-cost manufacturing process. Several fabrication results of organic resistive memory devices on flexible substrates have been reported [29, 30]. In addition, with growing demand for high-density digital information storage, NAND Flash memory density has been increased dramatically for the past couple of decades. On the other hand, device dimension scaling to increase memory density is expected to be more and more difficult in a bit-cost scalable manner due to various physical and electrical limitations. As a solution to the problems, NAND Flash memories having stacked layers are under developing extensions [31, 32]. In 3-D memories, cost can be reduced by building multiple stacked cells in vertical direction without device size scaling. As a breakthrough for the scaling limitations, various 3-D stacked memory architectures are under development and expecting the huge market of 3-D memories in the near future. With lots of expectation, future-generation memories have potential to replace most of the existing memory technologies. The new and emerging memory technologies are also named to be a universal memory; this may give rise to a huge market for computer applications to all the consumer electronic products.

MARKET MEMORY TECHNOLOGIES BY APPLICATIONS

The semiconductor industry has experienced many changes since Flash memory first appeared in the early 1980s. The growth of consumer electronics market urges the demand of Flash memory and helps to make it a prominent segment within the semiconductor industry. The Flash memories were commercially introduced in the early 1990s, and since that time, they have been able to follow Moore's law and the scaling rules imposed by the market. There are expected massive changes in the memory market over the next couple of years, with more density and reliable technologies challenging the dominant NAND Flash memory now used in SSDs and embedded in mobile products. Server, storage, and application vendors are now working on new specifications to optimize the way their products interact with NVM - moves that could lead to the replacement of DRAM and hard drives alike for many applications, according to a storage networking industry association (SNIA) technical working group [33,34]. The Flash memory marketplace is one of the most vibrant and exciting in the semiconductor industry, not to mention one of the most competitive. The continuous invention of new memory technologies and their applications in the memory market also increase performance demands. These new classes of memories with the latest technology increase the vertical demand in the future memory market. In the next coming years, cumulative price reductions should become disruptive to DVDs and hard disk drives (HDDs), stimulate huge demand, and create new Flash markets.

The nonvolatile memories offer the system a different opportunity and cover a wide range of applications, from consumer and automotive to computer and communication. Figure 3 shows NVSM memory consumption by various applications in the electronics industry by market in 2010 extending upwards from computers and communication to consumer products [22]. It is noticed that there is a faster growth rate of the digital cellular phone since 1990; the volume of production has increased by 300 times, e.g., from 5 million units per year to about 1.5 billion units per year. Nowadays, flexibility and transparency are particularly of great significance for future electronic applications such as affordable and wearable electronics. Many advanced research technologies are applied to flexible technology to be used in a real electronics area [35]. Although silicon-based semiconductor memories have played significant roles in memory storage applications and communication in consumer electronics, now, the recent focus is turning from rigid silicon-based memory technology into a soft nonvolatile memory technology for low-cost, large-area, and low-power flexible electronic applications. Further, the memory market for the long term is continuously growing, even if with some ups and downs, and this is expected to continue in the coming years [36]. Since innovation drives the semiconductor industry, a new trend with transparency as well as flexibility and 3-D technologies will be attractive and move towards continuous growth in the near future.

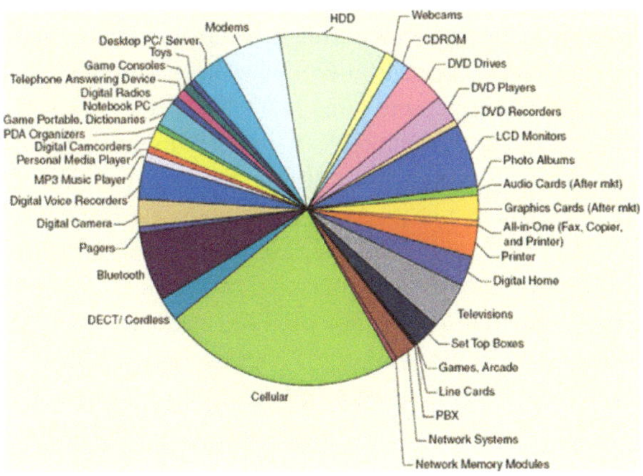

Figure 3 Various NVSM applications in the electronics industry by market size in 2010. Reprinted from ref. [22].

Successive creation of new mobile devices leads to the continual growth of NAND products as shown in Figure 4. To meet this market demand, early this year, 30-nm node technologies are in ramping-up phase, 20-nm node technologies are in the phase of transition to mass production, and a 10-nm node technology is under development. In addition, the future market requires high-speed operation even up to approximately 1,500 MB/s in order to satisfy a large amount of data correspondence [37]. However, high-speed operations cause high power consumption and chip temperature increase, which can deteriorate NAND reliability. Hence, reduction of operating voltage is inevitable to achieve the future NAND. Opportunities for the use of 3-D as well as polymer memory design in modern electronic circuits are rapidly expanding, based on the very high performance and unique functionality. However, their practical implementation in electronic applications will ultimately be decided by the ability to produce devices and circuits at a cost that is significantly below that needed to manufacture conventional electronic circuits based on, for example, silicon. If successful, these low-cost fabrication processes will ultimately result in the printing of large-area organic electronic circuits on a sheet of plastic paper using a roll-to-roll method, where low-temperature deposition of organics is followed by metal deposition and patterning in a continuous, high-speed process analogous, perhaps, to processes used in the printing of documents or fabrics.

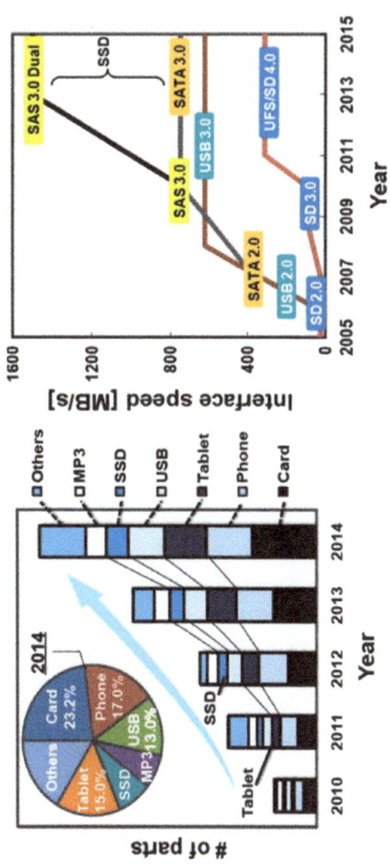

Figure 4 Growth of NAND Flash market up to 2014 (iSuppli) and the interface speed of various NAND applications. Reproduced from ref. [37].

In recent years, IDTechEx finds that the total market for printed, flexible, and organic electronics will grow from $16.04 billion in 2013 to $76.79 billion in 2023 and this growing trend is expected to continue in the coming years (see Figure 5a).

The majority of that is OLEDs (only organic, not printed) and conductive ink used for a wide range of applications. On the other hand, stretchable electronics, logic and memory, and thin-film sensors are much smaller ingredients but having huge growth potential as they emerge from R&D [38]. The report specifically addresses the big picture that over 3,000 organizations are pursuing printed, organic, flexible electronics, including printing, electronics, materials, and packaging companies. While some of these technologies are in use now - indeed there are main sectors of business which have created billion-dollar markets - others are commercially embryonic.

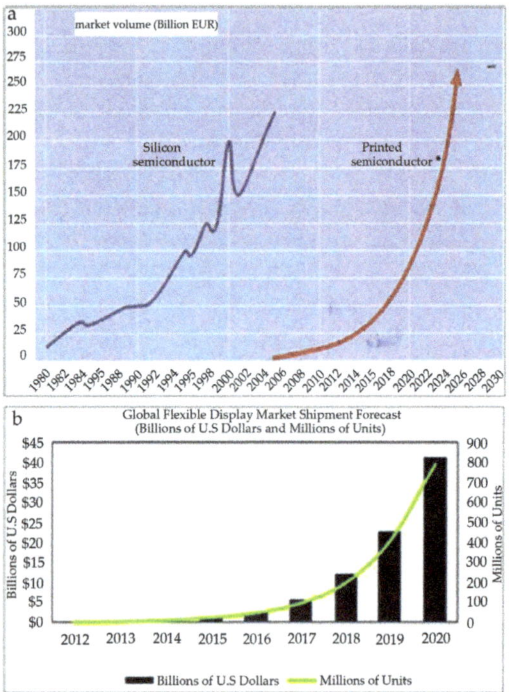

Figure 5 Market volume (a) and global flexible display market shipment forecast (b).Reproduced from refs. [38, 39].

Another key potential market for printed/flexible electronics is next-generation transparent conductive film to replace brittle and expensive indium tin oxide (ITO) in touch screens and displays, lighting, and photovoltaics. Touch display research says that the market for non-ITO transparent conductors will be about \$206 million this year and grow to some \$4 billion by 2020 as shown in Figure 5b. 'High demand for touchscreens for notebook and PC size displays has created a shortage of ITO touch sensors since the end of last year to drive more interest in these technologies, and the more flexible and potentially cheaper replacement technologies are getting more mature, notes Jennifer Colegrove, president and analyst, who will speak at the FlexTech workshop on transparent conductors. She notes that Atmel, Fujifilm, Unipixel and Cambrios are all in some phase of production' [39]. A large amount of the semiconductor market (approximately 20%) is given by the semiconductor memories; thus, the market for chips will develop in the next few years. This study reports that there is an analysis of the production process and the subsequent value chain, which comprises a benchmark analysis of the main segments of the semiconductor industry.

Recently, the 3-D nonvolatile memory structure has also attracted considerable attention due to its potential to replace conventional Flash memory in next-generation NVM applications [37,40]. 3-D memories are gathering increasing attention as future ultra-high-density memory technologies to keep a trend of increasing bit density and reducing bit cost. The NAND Flash market is continuously growing by the successive introduction of innovative devices and applications. To meet the market trend, 3-D NVMs are expected to replace the planar one, especially for 10-nm nodes and beyond. Therefore, the fundamentals and current status of the 3-D NAND Flash memory are reviewed and future directions are discussed [41]. 3-D integration promises to be an excellent replacement of current technologies for the development of NAND Flash memory. Time is running out for planar NAND technology. It will not be long that planar NAND will be completely replaced by 3-D NAND. 3-D NAND promises to satisfy the growing need of NAND memory [37].

Finally, NVM technologies have a bright future since every end-use application needs to store some parameters or some amount of an application program in the on-board NVM to enable it to function. The upcoming NVMs are the big hope for a semiconductor memory market, which provides memories for systems to run with flexibility, reliability, high performance, and low power consumption in a tiny footprint in nearly every electronic application. Recent market trends have indicated that commercialized or near-commercialized circuits are optimized across speed, density, power efficiency, and manufacturability. Flash memory is not suited to all applications, having its own problems with random-access time, bit alterability, and write cycles. With the increasing need to lower power consumption with zero-power standby systems, observers are predicting that the time has come for alternative technologies to capture at least some share in specific markets such as automotive smart airbags, high-end mobile phones, and RFID tags. An embedded nonvolatile memory with superior performance to Flash could see widespread adoption in system-on-chip (SoC) applications such as smart cards and microcontrollers.

EMERGING NVM TECHNOLOGIES FOR APPLICATIONS

The new emerging nonvolatile random-access memory products address the urgent need in some specific and small-form devices. Therefore, *iRAP* felt a need to do a detailed technology update and market analysis in this industry [42]. Recently, Yole Dévelopement reports describe that emerging memory technologies have great potential to improve future memory devices to be increasingly used in various markets of industry and transportation, enterprise storage, mobile phones, mass storage, and smart cards [43]. Emerging NVM applications in various markets are shown in Figure 6. But there are numerous opportunities existing for novel architectures and applications that these emerging memory technologies can enable. These new emerging NVM products address the urgent need in some specific and small-form devices. Therefore, emerging nonvolatile memory products provide market data about the size of growth of the application segments and the developments of business opportunities. Until now, only FeRAM, PCM, and MRAM were industrially produced and available in low-density chips to only a few players. Thus, the market was quite limited and considerably smaller than the volatile DRAM- and nonvolatile Flash NAND-dominant markets (which enjoyed combined revenues of \$50+ billion in 2012). However, in the next 5 years, the scalability and chip density of those memories will be greatly improved and will spark many new applications with NVM market drivers explained in more detail.

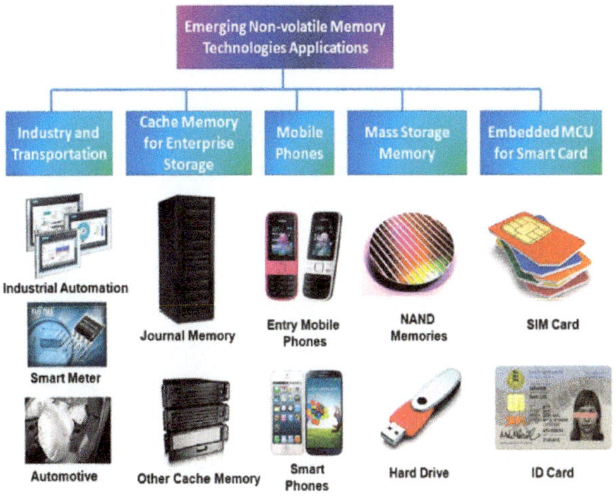

Figure 6 Emerging NVM applications in various markets.

Accompanied by the adoption of STT-MRAM and PCM cache memory, enterprise storage will be the largest emerging NVM market. NVM will greatly improve the input/output performance of enterprise storage systems whose requirements will intensify with the growing need for web-based data supported by floating mass servers. In addition, mobile phones will increase their adoption of PCM as a substitute to Flash NOR memory in MCP packages to 1-gigabyte (GB) chips made available by Micron in 2012. Higher-density chips, expected in 2015, will allow access to smart phone applications that are quickly replacing entry-level phones. STT-MRAM is expected to replace SRAM in SoC applications, thanks to lower power consumption and better scalability. Smart cards and microcontrollers (MCU) will likely adopt MRAM/STT-MRAM and PCM as a substitute to embed Flash. Indeed, Flash memory cell size reduction is limited in the future. The NVM could reduce the cell size by 50% and thus be more cost-competitive. Additional features like increased security, lower power consumption, and higher endurance are also appealing NVM attributes. The mass storage markets served by Flash NAND could begin using 3-D RRAM in 2017 to 2018, when 3-D NAND will slow down its scalability as predicted by all of the main memory players. If this happens, then a massive RRAM ramp-up will commence in the next decade that will replace NAND; conditional 3-D RRAM cost-competitiveness and chip density are available. It is expected surely that the emerging NVM business will be very dynamic over the next 5 years, thanks to improvements in scalability/cost and density of emerging NVM chips [44].According to a recently published report from Yole Développement, Emerging Non-volatile Memory Technologies, Industry Trends and Market Analysis, the global market for emerging nonvolatile random-access memory products was projected to have reached \$200 million in 2012. This market is expected to increase to \$2,500 million by 2018 at an average annual growth at a CAGR of +46% through the forecast period with mobile phones, smart cards, and enterprise storage as main growth drivers (Figure 7). Market adoption of memory is strongly dependent on its scalability. This Yole Développement report provides a precise memory roadmap in terms of technological nodes, cell size, and chip density for each emerging NVM such as FeRAM, MRAM/STT-MRAM, PCM, and RRAM. A market forecast is provided for each technology by application, units, revenues, and also market growth as given a detailed account of emerging NVM market forecast (Figure 7). PCM devices, the densest NVM in 2012 at 1 GB, will reach 8 GB by 2018, which are expected to replace NOR Flash memory in mobile phones and will also be used as a storage class memory in enterprise storage. MRAM/STT-MRAM chips will reach 8 to 16 GB in 2018. They will be widely sold as a storage class memory and possibly as a DRAM successor in enterprise storage after 2018. By 2018, MRAM/STT-MRAM and PCM will surely be the top two NVM on the market. Combined, they will represent a \$1.6 billion business by 2018, and their sales will almost double each year, with double-density chips launched every 2 years. FeRAM will be more stable in terms of scalability, with 8- to 16-MB chips available by 2018; the development of a new FRAM material could raise scalability, but we do not expect it to be widely industrialized and commercialized before 2018. FeRAM will grow at a steady growth rate (10% per year) and will focus on industrial and transportation applications because of the low-density availability, whereas RRAM revenues would not really surge by 2018, with the availability of high-density chips of several tens of gigabytes that could replace NAND technology. Meanwhile, it has also been considered by memory technologist experts that for large-volume markets like mass storage NAND, only one technology will be adopted in order to reduce production cost and RRAM seems to be the best candidate. But the real massive adoption of emerging NVM as a replacement for NAND and DRAM will happen after 2020.

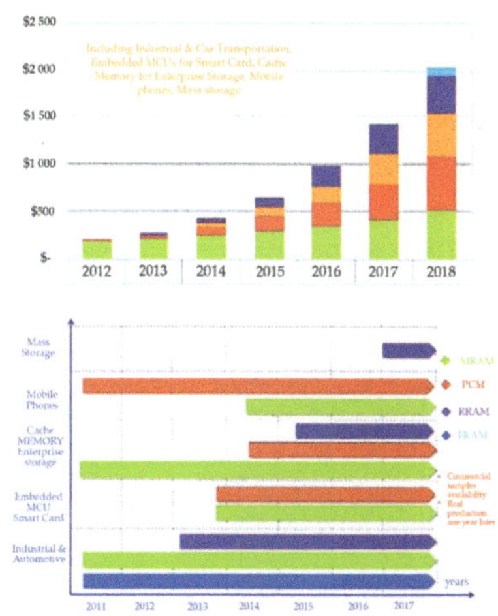

Figure 7 Emerging NVM market forecast by applications from 2012 to 2018 (in M\$).Reproduced from ref. [43].

ADVANCES IN FLASH MEMORY TECHNOLOGIES

Flash memory is basically a MOSFET nonvolatile device that can be electrically erased and reprogrammed [3, 45]. It is a technology that is primarily used in memory cards and Flash drives for general storage and transfer of data between computers and other digital products. Since the invention of the transistor, NVSM had been the most important invention in the electron device field. The floating gate memory was used to store the information and a tunneling current for programming and erasing operations. The charge is injected into or removed from the floating gate and the floating gate remains in that state, even after power is removed, which means that Flash memory is nonvolatile. The invention of NVSM further gave rise to a new class of memory devices and hence broadened its applications to become ubiquitous. There are a large number of products in the market now which use Flash devices exclusively as secondary storage. Few examples of their applications include medical diagnostic systems, notebook computers, digital audio players, digital cameras, mobile phones, personal digital assistants, digital televisions, universal serial bus (USB) Flash personal disks, Global Positioning Systems, and many more. Semiconductor storage devices store data in tiny memory cells made of very small transistors and capacitors made of semiconductor materials such as silicon. Each cell can hold 1 bit of information and an array of cells stores a large chunk of information. Flash devices are gaining popularity over conventional secondary storage devices like hard disks. The Flash memory fabrication process is compatible with the current CMOS process and is a suitable solution for embedded memory applications. A Flash memory cell is simply a MOSFET cell, except that a polysilicon floating gate [46] (or a silicon nitride charge trap layer) is sandwiched between a tunnel oxide and an inter-polyoxide to form a charge storage layer [47]. Although Flash memory is likely the standard charge storage device for the next generation, scaling may eventually be limited by the tunnel oxide limit [8]. In terms of the operation speed of program and erase, Flash memory requires a thin tunnel oxide to enhance the carrier transport between the floating gate and the silicon substrate. However, the very thin tunnel oxide suffers from many reliability issues like reduction in operation voltage, and after a considerable number of program and erase cycles, the tunnel oxide undergoes deterioration loss [48]. Thus, researchers have focused on possible solutions and proposed alternate technologies, including nitride-based memory, nanocrystal memory, and switching memory. All other nonvolatile memories require integration of new materials that are not as compatible as the conventional CMOS process.

NOR and NAND Flash memory technologies

NOR and NAND Flash, two major Flash types, are dominant in the memory market. NOR Flash has lower density but a random-access interface, while NAND Flash has higher density and interface access through a command sequence [49]. Their corresponding structures are shown in Figure 8. NOR and NAND Flash come from the structure used for the interconnections between memory cells. Intel is the first company to introduce a commercial (NOR type) Flash chip in 1988, and Toshiba released the world's first NAND Flash in 1989 [50]. Depending on how the cells are organized in the matrix, it is possible to distinguish between NAND Flash memories and NOR Flash memories. In NOR Flash, cells are connected in parallel to the bit lines, which notably allow the cells to be read and programmed individually. The parallel connection of NOR Flash cells resemble the parallel connection of transistors in a CMOS NOR gate architecture. On the other hand, in NAND Flash, the cells are connected in series, resembling a NAND gate. The series connections consume less space than the parallel ones, reducing the cost of NAND Flash. It does not, by itself, prevent NAND cells from being read and programmed individually. Most of the engineers and scientists are not so familiar with the differences between these two technologies. Generally, they usually refer to the NOR architecture as 'Flash' and are unaware of the NAND Flash technology and its many benefits over NOR [51]. This could be due to the fact that most Flash devices are used to store and run codes (usually small), for which NOR Flash is the default choice, although we are providing some major differences between NOR and NAND Flash technologies by their architecture and the internal characteristic features of the individual Flash.

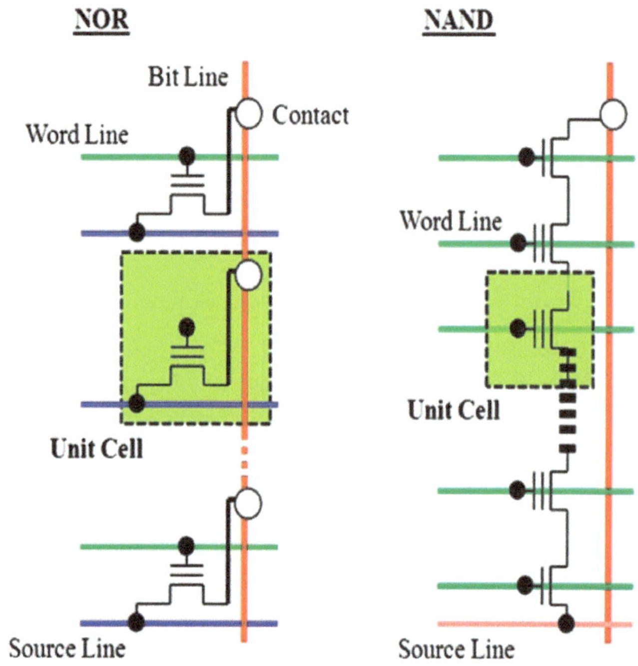

Figure 8 Comparison of NOR Flash array and NAND Flash array architectures.

NOR Flash is slower in erase operation and write operation compared to NAND Flash [52]. This means that NAND Flash has faster erase and write times. Moreover, NAND Flash has smaller erase units, so fewer erases are needed. NOR Flash can read data slightly faster than NAND Flash. NOR Flash offers complete address and data buses to randomly access any of its memory locations (addressable to every byte). This makes it a suitable replacement for older ROM BIOS/firmware chips, which rarely needs to be updated. Its endurance is 10,000 to 1,000,000 erase cycles. NOR Flash is highly suitable for storing codes in embedded systems. Most of today's microcontrollers come with built-in Flash memory [53].

NAND Flash occupies a smaller chip area per cell. This makes NAND Flash available in greater storage densities and at lower costs per bit than NOR Flash. It also has up to ten times the endurance of NOR Flash. NAND is more fit as storage media for large files including video and audio. USB thumb drives, SD cards, and MMC cards are of NAND type [54]. NAND's advantages are fast write (program) and erase operations, while NOR's advantages are random access and byte write capability. NOR's random access ability allows for execute in place (XiP) capability, which is often a requirement in embedded applications. NAND is slow random accessible, while NOR is hampered by having slow write and erase performance. NAND is better suited for filing applications.

However, more processors include a direct NAND interface and can boot directly from NAND (without NOR). However, NAND cannot perform read and write operations simultaneously; it can accomplish these at a system level using a method called shadowing, which has been used on PCs for years by loading the BIOS from the slower ROM into the high-speed RAM.

Table 1 highlights the major differences between NOR and NAND. It shows that NAND is ideal for high-capacity data storage while NOR is best used for code storage and execution, usually in small capacities. There are many other differences between these two technologies which will be further discussed individually. However, those listed in the table are enough to strongly differentiate the types of applications using them: NOR is typically used for code storage and execution. This, mainly in capacities up to 4 MB, is common in applications such as simple consumer appliances, low-end cell phones, and embedded applications, while raw NAND is used for data storage in applications such as MP3 players, digital cameras, and memory cards [55–57]. The codes for raw NAND-based applications are stored in NOR devices.

Table 1 Comparison between NOR and NAND Flash memories[55–57]

Features	NOR	NAND
Memory size	≤512 Mbit	1 to 8 Gbit
Sector size	Approximately 1 Mbit	Approximately 1 Mbit
Program time	9 μs/word	400 μs/page
Erase time	1 s/sector	1 ms/sector
Read access time	<80 ns	20 μs
Write parallelism	8 to 16 words	2 Kbyte
Output parallelism	Byte/word/dword	Byte/word
Read parallelism	8 to 16 words	2 Kbyte
Access method	Random	Sequential
Price	High	Very low
Reliability	Standard	Low

Scaling and Challenges of Flash Memory Technologies

Currently, there have been increasing demands on reducing the feature size in microelectronic products and more interest in the development of Flash memory devices to meet the growing worldwide demand. A conventional FG memory device must have a tunnel oxide layer thickness of 8 nm to prevent charge loss and to make 10 years' data retention certain. This necessity will limit scalability for Flash memory devices [8, 58]. Thus, in order to meet technology scaling in the field of memory and data storage devices, mainstream transistor-based Flash technologies will be developed gradually to incorporate material and structural innovations [59]. Dielectric scaling in nonvolatile memories has been reached near to the point where new approaches will be required to meet the scaling requirements while simultaneously meeting the reliability and performance requirements for future products. High-dielectric-constant materials are being explored as possible candidates to replace both the traditional SiO_2 and oxide/nitride/oxide (ONO) films used in Flash memory cells. Flash cell scaling has been demonstrated to be really possible and to be able to follow Moore's law down to the 90-nm technology generations. The technology development and the consolidated know-how are expected to sustain the scaling trend down to the 50-nm technology node and below as forecasted by the International Technology Roadmap for Semiconductors (ITRS) in Figure 9, which indicates that the silicon MOSFET was already in the nanoscale. The minimum feature size of an individual CMOSFET has shrunk to 15 nm with an equivalent gate oxide thickness (EOT) of 0.8 nm in 2001 [13]. However, semiconductor Flash memory scaling is far behind CMOS logic device scaling. For example, the EOT of the gate stack in semiconductor Flash memory is still more than 10 nm. Moreover, semiconductor Flash memory still requires operation voltages of more than 10 V, which is still far from the operation voltage of CMOS logic devices. It is important to scale the EOT of the gate stack to achieve a small memory cell size and also prolong battery life.

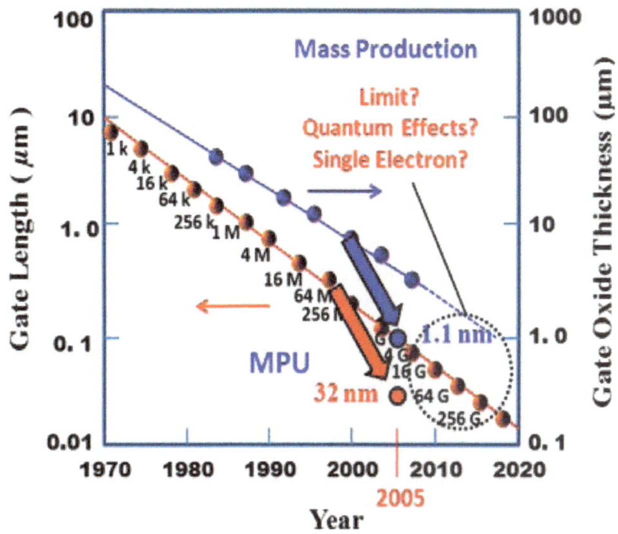

Figure 9 The trend of MOSFET scaling from ITRS. Reproduced from ITRS Corp.

Another limitation of FG technology is that tunnel oxide scaling is limited by stress-induced leakage current (SILC) related to charge transfer problem as indicated in Figure 10[60, 61]. The SILK increases with decreasing oxide thickness. This can be attributed to tunneling assisted by the traps in the bulk of the dielectric. Trap-assisted tunneling can take place at very low electric fields. If the density of traps is increased, the leakage will also increase. Electrical stress can increase the number of these traps. So it becomes an important limitation of scaling down the memory device [62]. For EOT<8 nm, a single oxide trap will cause to complete the charge loss in the FG Flash cell. The scaling of the gate stacks and operation voltages are often related to each other. A tunnel oxide thickness of more than 8 nm is currently used in the commercial Flash memory chip to meet the 10 years› data retention time requirement. If the tunnel oxide were to be scaled below 2 nm, the operation voltage could be reduced from more than 10 V to below 4 V [63]. Unfortunately, the retention time would also be reduced, from 10 years to several seconds. This physical damage to the tunnel oxide during the cycling process causes data retention problems, program disturbance, read disturbance, and erratic characteristic behavior of the FG memory cell. Such problems severely limit the reliability and multilevel cell operation. This basic limitation of the tunnel oxide thickness becomes increasingly important with scaling. New storage node concepts are also becoming attractive as an alternative approach to address some of the dielectric scaling limitations. Flash memory adopts a charge stored in a silicon nitride as the trapping layer, which exhibits significantly reduced defect-related leakage current and very low SILC as compared to SiO_2 with a similar EOT [64]. Such a relentless reduction of device dimensions has many challenges like retention, endurance, reduction in the number of electrons in the FG, dielectric leakage, cell-to-cell cross talk, threshold voltage shift, and reduction in memory window margins [65, 66]. The key concept of real scaling issues such as material and structural changes in Flash memory technologies is provided in detail in the next distinct part.

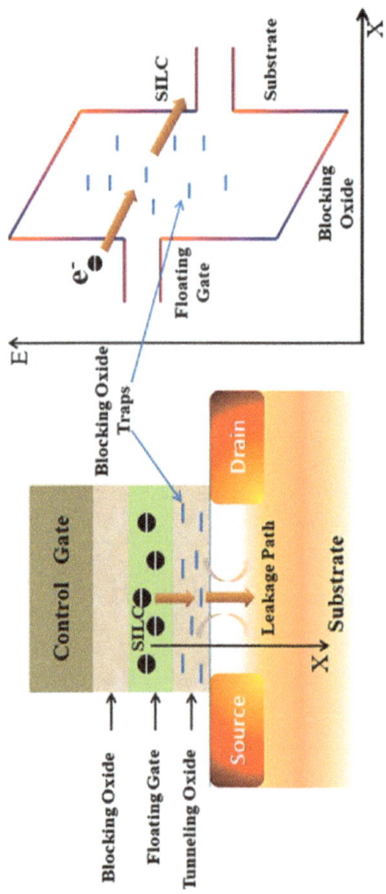

Figure 10 Schematic plots of a Flash memory cell and the degradation of its tunnel oxide. The degradation leads to the formation of percolation paths responsible for the FG charge loss, hence the loss of the stored information. The presence of traps in the energy barrier yields the trap-assisted tunneling mechanism and originates the stress-induced leakage current (SILC).

FG Flash memory technology

The FGNV memory is a basic building block of Flash memory, which is based on FG thin-film storage (TFS) memories that have been developed with the addition of an erase gate configuration. The conventional FG memory (Figure 11a) consists of a MOSFET configuration that is modified to include polysilicon as a charge storage layer surrounded by an insulated inner gate (floating gate) and an external gate (control gate). This what makes Flash memory nonvolatile and all floating gate memories to have the same generic cell structure. Charge is transferred to or from the floating gate through a thin (8 to 10 nm) oxide [1, 67]. Because the floating gate is electrically isolated by the oxide layer, any electrons placed on it are trapped there. Flash memory works by adding (charging) or removing (discharging) electrons to and from a floating gate. A bit's 0 or 1 state depends upon whether or not the floating gate is charged or discharged. When electrons are present on the floating gate, current cannot flow through the transistor and the bit state is '0'. This is the normal state for a floating gate. When electrons are removed from the floating gate, current is allowed to flow and the bit state is '1'. The FG memory has achieved high density, good program/erase speed, good reliability, and low operating voltage and promotes endurance for Flash memory application.

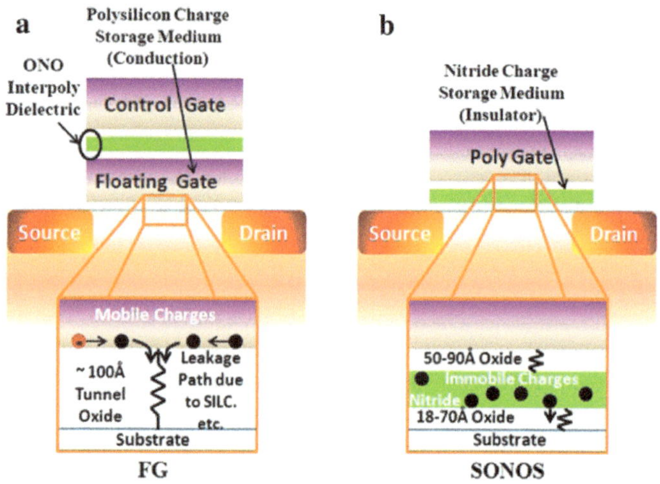

Figure 11 Schematics of the conventional FG memory and SONOS. Schematics of (a)floating gate and thin-film storage-based embedded nonvolatile memory bit cells, depending on the charge stored inside the gate dielectric of a MOSFET, and (b) the nitride traps (SONOS), embedded into the gate oxide of a MOSFET.

SONOS memory technology

In order to solve the scaling issue of the FG memory, the SONOS memory has been proposed as a Flash technology since the 1980s [68, 69]. The acronym SONOS is derived from the structure of the device as shown in Figure 11b. The SONOS device is basically a MOSFET, where the gate has been replaced by an ONO dielectric. The SONOS memory has a better charge retention than the FG memory when the FG bit cell's tunneling oxide layer is below 10 nm [70]. Moreover, the SONOS memory exhibits many advantages, e.g., easy to fabricate, high program/erase (P/E) speed, low programming voltage and power consumption, and better potential for scalability below the 70-nm node, according to the ITRS [71]. The charge, holes or electrons, are injected into the nitride layer using direct tunneling through the tunnel oxide layer. The nitride layer is electrically isolated from the surrounding transistor, although charges stored on the nitride directly affect the conductivity of the underlying transistor channel. Since the SONOS memory possesses spatially isolated deep-level traps, a single defect in the tunneling oxide will not cause discharge of the memory cell. The thickness of the top oxide is important to prevent the Fowler-Nordheim tunneling of electrons from the gate during erase. When the polysilicon control gate is biased positively, electrons from the transistor source and drain regions tunnel through the oxide layer and get trapped in the silicon nitride. This results in an energy barrier between the drain and the source, raising the threshold voltage V_{th} (the gate-source voltage necessary for current to flow through the transistor). Moreover, the nitride layer is electrically isolated from the surrounding transistor, although charges stored on the nitride directly affect the conductivity of the underlying transistor channel. The oxide/nitride sandwich typically consists of a 2-nm-thick oxide lower layer, a 5-nm-thick silicon nitride middle layer, and a 5- to 10-nm-thick oxide upper layer [72, 73]. However, SONOS-type Flash memories have several drawbacks such as shallow trap energy level, erase saturation, and vertical stored charge migration [74]. The programming speed and operating voltage problems can be solved by reducing the tunnel oxide thickness. At low tunnel oxide thickness, the issues that impact SONOS-type memories include erase saturation and vertical charge migration, which seriously degrade the retention capability of the memory [75]. Thus, many concerns still remain for the SONOS type of memories, which will be discussed in the next section.

Limitations of FG and SONOS memory technologies

Scaling demands very thin gate insulators in order to keep short channel effects and control the shrinkage of the device size and maximize the performance. When the tunneling oxide thickness is below 10 nm, the storaged charge in the FG is easy to leak due to a defect in the tunneling oxide formed by repeated write/erase cycles or direct tunneling current.

The tunneling gate oxide thickness in a conventional Flash memory cannot be scaled down to sub-7 nm because of charge retention [76]. The SONOS Flash memory can relieve the problem but still has a relatively thick gate dielectric thickness of about 7 nm. Therefore, conventional SONOS Flash memory also has a scaling-down problem. Many studies have shown that the charge retention characteristics in scaled SONOS nonvolatile memory devices with a low gate oxide thickness and at high temperature are problematic with shallow-level traps [48, 77, 78]. For the conventional SONOS memory, erase saturation and vertical stored charge migration [79, 80] are the two major drawbacks; the most challenging tasks are how to maintain an acceptable charge capability of the discrete storage nodes and how to fabricate nanocrystals with constant size, high density, and uniform distributions [81]. When the trap energy level is shallow, erase saturation and vertical migration occur and the electron charge decay rate increases due to low tunnel oxide thickness, issues that impact SONOS-type memories as shown in Figure 12. This erase saturation makes SONOS erase less as the erase voltage or the tunnel oxide thickness is increased. Since the SONOS memory uses silicon nitride as a charge trapping layer, the electrons in the Si sub-conduction band will tunnel through the tunneling oxide and a portion of the nitride, and this consequently degrades the program speed. Besides this, the conduction band offset of nitride is only 1.05 eV and back-tunneling of the trapped electron may also occur. Although applying a very high electric field may accelerate the de-trapping rate, the gate electron injection current exceeds the de-trapping but resulting in practically an increase in charge and no erasing. Using an ultra-thin (<2 nm) tunnel oxide offers an efficient charge direct tunneling erase and opens a memory window. However, the direct tunneling cannot be turned off at a low electric field, leading to poor retention and read disturb. Thus, the SONOS memory cannot be used for NAND Flash without further innovation of new memory technologies. The main reason for the growth of emerging NVM technologies is that scaling has now become a serious issue for the memory industry. Not only are many of these new technologies inherently more scalable, but also they seem well suited to the next generation of mobile computing and communications that will demand high-capacity memories capable of storing and rapidly accessing video and a large database without overburdening battery power sources.

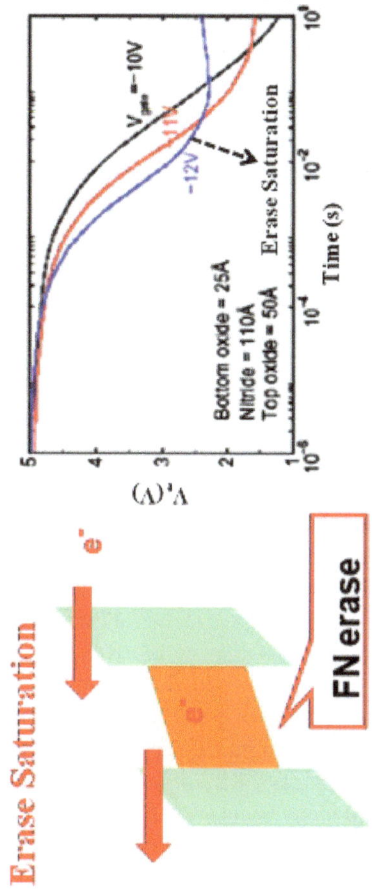

Figure 12 Fowler-Nordheim (FN) tunneling of electrons from the gate during erase and erase saturation in SONOS nonvolatile memory. This indicates the reduced memory window as the erase voltage is increased. Reproduced from ref. [74].

Many alternate device structures are proposed to hopefully circumvent these scaling challenges and to improve the device performance. In an effort to continue Moore›s law and overcome the ultimate limitations of MOS-based memory devices, other storage concepts have been proposed in search of the 'unified memory'. The ideal memory device or the so-called 'unified memory' would satisfy simultaneously three requirements: high speed, high density, and nonvolatility. At the present time, such an ideal memory has not been developed. FGNVSM has high density and nonvolatility, but its P/E speed is low. DRAM has high speed (approximately 10 ns) and relatively high density, but it is volatile. SRAM has very high speed (approximately 5 ns), but it suffers from very low density and volatility. Many nonvolatile memory devices have been proposed on the basis of changing charge storage materials and new device concepts for the 'unified memory'. These structures will be considered in the next sections. In light of such issues, emerging memory solutions seem to be a key technology.

CURRENT EMERGING MEMORY TECHNOLOGIES

Recent studies have revealed that there is a close correlation among existing and emerging memory technologies in view of scalability. The scaling trend of memory transition leads to smaller and smaller memory devices, which have been routinely observed. To further support this assertion, another set of current progress in memory technology is described to the increasing importance of memory to users' experience and the importance of memory to system performance. There are many emerging memory technologies which are trying to replace existing memory technologies in the market. These new memory devices such as RRAM, PCM, and STT-RAM have read/write/retention/endurance characteristics different from those of conventional SRAM, DRAM, and Flash [82]. But the ideal characteristics of new emerging memory technologies have to be meeting the performance of SRAM and the density of NAND Flash in terms of stability, scalability, and switching speed. Thus, going beyond the traditional bistable memory, the possibilities of multilevel, high-performance memory devices suitable for market must be explored. Currently, there are several technologies that show some promise; some of these new emerging technologies are MRAM, FeRAM, PCM, STT-RAM, nano-random-access memory (NRAM), racetrack memory, RRAM and memristor, molecular memory, and many others [10, 83]. Each of these memory technologies will be briefly outlined and discussed in the following sections. In view of the commercial production, currently, MRAM, FeRAM, and PCM are in commercial production but still remain limited to niche applications relative to DRAM and NAND Flash. There is a prospect that among the emerging memory technologies, MRAM, STT-RAM, and RRAM are the most promising ones, but they are still many years away from competing for industry adoption [84]. It is necessary for any new technology to be able to deliver most for industry adoption. For industry adoption on a mass scale, some parameters must be matched with existing memory technologies. In consideration of new technology for industry application, the scalability of the technology, speed of the device, power consumption to be better than existing memories, endurance, densities, better than existing technologies and most importantly the cost; if the emerging technology can only run one or two of these attributes, then, at most desirable, it is likely to be resigned to niche applications.

MRAM

MRAM or magnetic RAM is a nonvolatile RAM technology under development since the 1990s. RRAM methods of storing data bits use magnetic charges instead of the electrical charges used by DRAM and SRAM technologies. MRAM, first developed by IBM in the 1970s [85], is expected to replace DRAM as the memory standard in electronics. MRAM is basically based on memory cells having two magnetic storage elements, one with a fixed magnetic polarity and another with a switchable polarity. These magnetic elements are positioned on top of each other but separated by a thin insulating tunnel barrier as shown in the cell structure in Figure 13. Moreover, scientists define a metal as magnetoresistive if it shows a slight change in electrical resistance when placed in a magnetic field. By combining the high speed of static RAM and the high density of DRAM, proponents say that MRAM could be used to significantly improve electronic products by storing greater amounts of data, enabling it to be accessed faster while consuming less battery power than existing electronic memories. Technically, it works with the state of the cell, which is sensed by measuring the electrical resistance while passing a current through the cell. Because of the magnetic tunnel effect [86], if both magnetic moments are parallel to each other, then the electrons will be able to tunnel and the cell is in the low resistance 'ON' state. However, if the magnetic moments are antiparallel, the cell resistance will be high. The memory characteristics of MRAM of writing and erasing are fulfilled by passing a current through the write line to induce a magnetic field across the cell. MRAM has been slowly getting off the ground but has now entered the market and will become increasingly available for mass production in the couple of years and beyond. Currently, it has reached some level of commercial success in niche applications [87]. Various companies such as Samsung, IBM, Hitachi and Toshiba, and TSMC are actively developing variant technologies of MRAM chips. In view of power consumption and speed, MRAM competes favorably than other existing memories such as DRAM and Flash, with an access time of a few nanoseconds [88–90]. Although it has some limitation during the 'write' operation, the smaller cell size could be limited by the spread of the magnetic field into neighboring cells and need an amendment to compete completely as a universal memory. The price of MRAM is also another issue and considered a limiting factor, with prices far in excess of all the currently established memories at approximately £2 to £3 ($3 to $5) per megabyte [91]. According to this price level, MRAM is in excess of 1,000 times the price of Flash memory and over 10,000 times the price of hard disk drives. It is expected that of the next-generation memory technologies, MRAM, in the future, will have the biggest market, followed by FeRAM, PCRAM, and memristors.

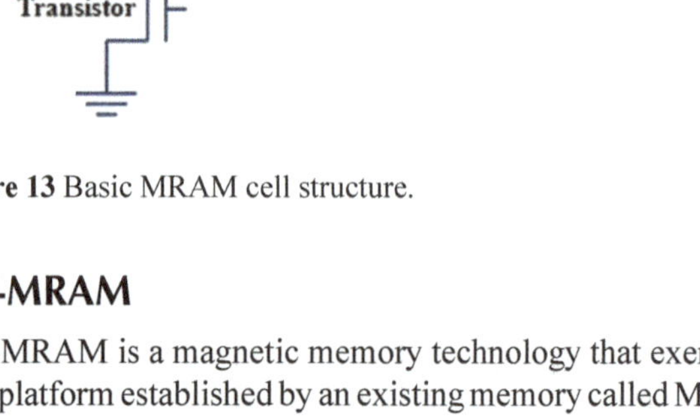

Figure 13 Basic MRAM cell structure.

STT-MRAM

STT-MRAM is a magnetic memory technology that exerts the base platform established by an existing memory called MRAM to enable a scalable nonvolatile memory solution for advanced process nodes [92, 93]. It is a new kind of magnetic RAM with the following features: fast read and write times, small cell sizes, potentially even smaller, and compatibility with existing DRAM and SRAM. As we have discussed in the previous section, MRAM stores data according to the magnetization direction of each bit and the nanoscopic magnetic fields set the bits in conventional MRAM. On the other hand, STT-MRAM uses spin-polarized currents, enabling smaller and less energy-consuming bits. The basic cell structure of STT-RAM is depicted in Figure 14. In addition, STT-RAM writing is a technology in which an electric current is polarized by aligning the spin direction of the electrons flowing through a magnetic tunnel junction (MTJ) element. Data writing is performed by using the spin-polarized current to change the magnetic orientation of the information storage layer in the MTJ element [94]. The resultant resistance difference of the MTJ element is used for information readout. STT-RAM is a more appropriate technology for future MRAM produced using ultra-fine processes and can be efficiently embedded in subsequent generations of such semiconductor devices as FPGAs, microprocessors, microcontrollers, and SoC. A special bonus for embedded designers is the fact that the internal voltage STT-RAM requires is only 1.2 V. The difference between STT-MRAM and a conventional MRAM is only in the writing operation mechanism; the read system is the same. The memory cell of STT-MRAM is composed of a transistor, an MTJ, a word line (WL), a bit line (BL), and a source line (SL) [95]. Currently, STT-RAM is being developed in companies including Everspin, Grandis, Hynix, IBM, Samsung, TDK, and Toshiba. However, for STT-RAM to be adopted as a universal mainstream semiconductor memory, some key challenges should be resolved: the simultaneous achievement of low switching current and high thermal stability. It must be dense (approximately 10 F^2), fast (below 10 ns of read and write speeds), and operating at low power [96].

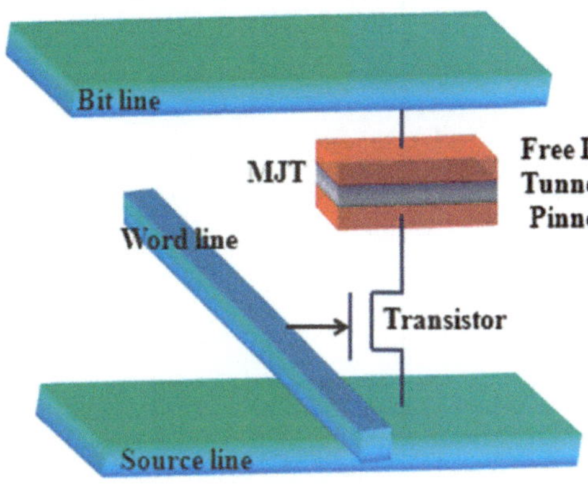

Figure 14 Basic STT-RAM cell structure.

FeRAM

FeRAM is a nonvolatile RAM that combines the fast read and write access of DRAM cells, consisting of a capacitor and transistor structure as shown in Figure 15. The cell is then accessed via the transistor, which enables the ferroelectric state of the capacitor dielectric to be sensed. In spite of its name, FeRAM does not contain iron. The polarization properties of a ferroelectric substance are used as a memory device. Today's FeRAM uses lead zirconate titanate (PZT); other materials are being considered. The main developer of FeRAM is Ramtron International. FeRAM is the most common kind of personal computer memory with the ability to retain data when power is turned off as do other nonvolatile memory devices such as ROM and Flash memory [97]. In a DRAM cell, the data periodically need refreshing due to the discharging of the capacitor, whereas FeRAM maintains the data without any external power supply. It achieves this by using a ferroelectric material in the place of a conventional dielectric material between the plates of the capacitor. When an electric field is applied across dielectric or ferroelectric materials, it will polarize, and while that field is removed, it will depolarize. But the ferroelectric material exhibits hysteresis in a plot of polarization versus electric field, and it will retain its polarization. One disadvantage of FeRAM is that has a destructive *read* cycle. The read method involves writing a bit to each cell; if the state of the cell changes, then a small current pulse is detected by indicating that the cell was in the OFF state. However, it is a fast memory that can endure a high number of cycles (e.g., 10^{14}) [98], meaning that the requirement for a *write* cycle for every *read* cycle will not result in short product lives with a very low power requirement. It is expected to have many applications in small consumer devices such as personal digital assistants (PDAs), handheld phones, power meters, and smart cards, and in security systems. FeRAM is faster than Flash memory. It is also expected to replace EEPROM and SRAM for some applications and to become a key component in future wireless products. Even after FeRAM has achieved a level of commercial success, with the first devices released in 1993 [99, 100], current FeRAM chips offer performance that is either comparable to or exceeding current Flash memories [98, 101], but still slower than DRAM.

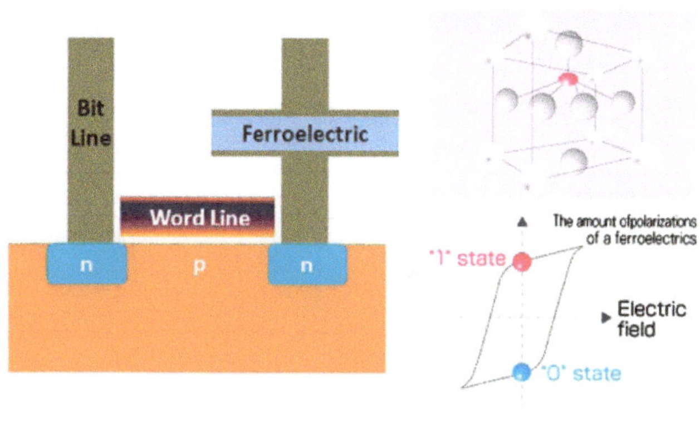

Figure 15 Basic structure of a FeRAM cell. The crystal structure of a ferroelectric and an electric polarization-electric field hysteresis curve are also shown.

PCRAM

PCRAM, also known as PCM, perfect RAM (PRAM), OUM, and chalcogenide RAM (CRAM), is a type of nonvolatile RAM based on a class of material called chalcogenide glasses that can exist in two different phase states (e.g., crystalline and amorphous) [102, 103]. The basic PCRAM cell structure is depicted in Figure 16. Most phase-change materials contain at least one element from group 6 of the periodic table, and the choice of available materials can be further widened by doping these materials [104–107]. In particular, the most promising are the GeSbTe alloys which follow a pseudobinary composition (between GeTe and Sb_2Te_3), referred to as GST. These materials are in fact commonly used as the data layer in rewritable compact disks and digital versatile disks (CD-RW and DVD-RW) where the change in optical properties is exploited to store data. The structure of the material can change rapidly back and forth between amorphous and crystalline on a microscopic scale. The material has low electrical resistance in the crystalline or ordered phase and high electrical resistance in the amorphous or disordered phase. This allows electrical currents to be switched *ON* and *OFF*, representing digital high and low states. This process has been demonstrated to be on the order of a few tens of nanoseconds [108], which potentially makes it compatible with Flash for the *read* operation, but several orders of magnitude faster for the *write* cycle.

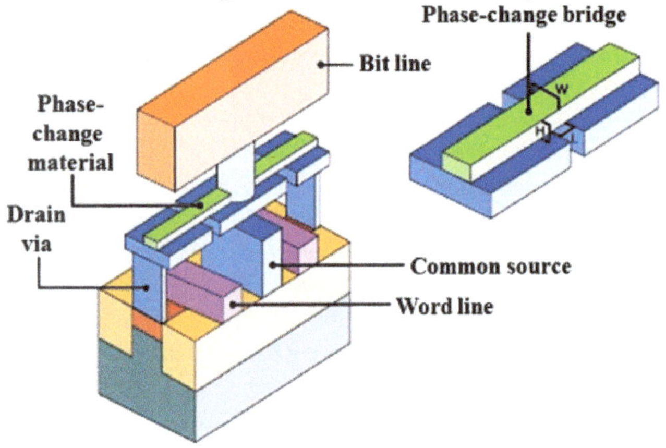

Figure 16 Basic PCRAM cell structure. Reproduced from IBM-Macronix-Qimonda.

This makes it possible for PCM to function many times faster than conventional Flash memory while using less power. In addition, PCM technology has the potential to provide inexpensive, high-speed, high-density, high-volume nonvolatile storage on an unprecedented scale. The physical structure is three-dimensional, maximizing the number of transistors that can exist in a chip of fixed size. PCM is sometimes called perfect RAM because data can be overwritten without having to erase it first. Possible problems facing PCRAM concern the high current density needed to erase the memory; however, as cell sizes decrease, the current needed will also decrease. PCM chips are expected to last several times as long as currently available Flash memory chips and may prove cheaper for mass production. Working prototypes of PCM chips have been tested by IBM, Infineon, Samsung, Macronix, and others. Also, the production of PCM has been announced recently by both collaborations between Intel and STMicroelectronics as well as with Samsung [109, 110].

Comparison of primary contenders for MRAM, STT-RAM, FeRAM, and PCM technologies

Before going to other emerging memories, we herein provide a comparison among MRAM, FeRAM, and PCM. The specific features of these memory devices are provided in Table 2. Relatively mature, new-material memories such as MRAM, STT-RAM, FeRAM, and PCM can offer a variety of features that have potential to be the candidates for next-generation nonvolatile memory devices. Brand-new concepts such as RRAM, molecular, organic/polymer, and other nanowire-based memory technologies have also been proposed. These are discussed in detail in the following section.

Table 2 Summary of primary contenders for MRAM, FeRAM, STT-RAM, and PCM technologies

Features	FeRAM	MRAM	STT-RAM	PCM
Cell size (F^2)	Large, approximately 40 to 20	Large, approximately 25	Small, approximately 6 to 20	Small, approximately 8
Storage mechanism	Permanent polarization of a ferroelectric material (PZT or SBT)	Permanent magnetization of a ferromagnetic material in a MTJ	Spin-polarized current applies torque on the magnetic moment	Amorphous/polycrystal phases of chalcogenide alloy
Read time (ns)	20 to 80	3 to 20	2 to 20	20 to 50
Write/erase time (ns)	50/50	3 to 20	2 to 20	20/30
Endurance	10^{12}	$>10^{15}$	$>10^{16}$	10^{12}
Write power	Mid	Mid to high	Low	Low
Nonvolatility	Yes	Yes	Yes	Yes
Maturity	Limited production	Test chips	Test chips	Test chips
Applications	Low density	Low density	High density	High density

RRAM

RRAM is a disruptive technology that can revolutionize the performance of products in many areas, from consumer electronics and personal computers to automotive, medical, military, and space. Among all the current memory technologies, RRAM is attracting much attention since it is compatible with the conventional semiconductor processes. Memristor-based RRAM is one of the most promising emerging memory technologies and has the potential of being a universal memory technology [111]. It offers the potential for a cheap, simple memory that could compete across the whole spectrum of digital memories, from low-cost, low-performance applications up to universal memories capable of replacing all current market-leading technologies, such as hard disk drives, random-access memories, and Flash memories [112]. RRAM is a simple, two-terminal metal-insulator-metal (MIM) bistable device as shown in the basic configuration in Figure 17. It can exist in two distinct conductivity states, with each state being induced by applying different voltages across the device terminals. RRAM uses materials that can be switched between two or more distinct resistance states. Many companies are investing metal oxide nanolayers switched by voltage pulses. Researchers generally think that the pulses' electric fields produce conducting filaments through the insulating oxide. HP Labs plans to release prototype chips this year based on 'memristors' in which migrating oxygen atoms change resistance [113]. Xu et al. have also defined that among all the technology candidates, RRAM is considered to be the most promising as it operates faster than PCRAM and it has a simpler and smaller cell structure than magnetic memories (e.g., MRAM or STT-RAM) [114]. In contrast to a conventional MOS-accessed memory cell, a memristor-based RRAM has the potential of forming a cross-point structure without using access devices, achieving an ultra-high density. This device is based on the bistable resistance state found for almost any oxide material, including NiO, ZrO_2, HfO_2, $SrZrO_3$, and $BaTiO_3$[115–119]. Currently, Samsung and IBM are actively investigating RRAM.

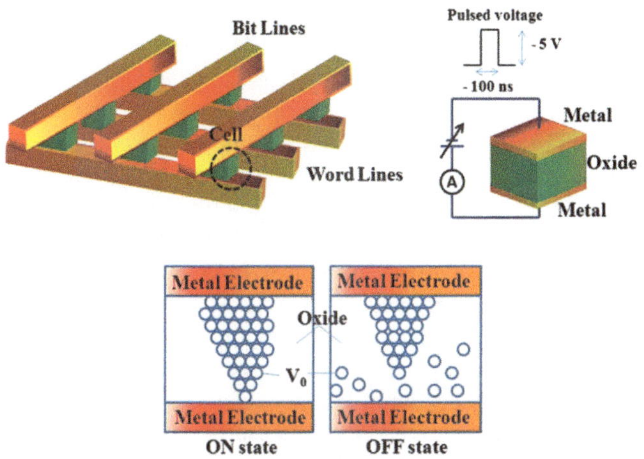

Figure 17 Basic RRAM cell structure. A schematic diagram of the mechanism of the resistive switching in a metal/oxide/metal-structured memory cell is also shown. Reproduced from ref. [123].

Kamiya et al. have revealed by a theoretical mechanism that RRAM shows filamentary-type resistive switching, where the oxygen vacancy is considered to form conductive filaments in the resistive material as shown in Figure 17[120]. The formation and disruption of these filaments are thus the mechanisms responsible for the ON-OFF switching in RRAM devices. The key issue is, therefore, to reveal electronic roles in the formation and disruption of the vacancy filaments. RRAM can be switched between the low resistance state (LRS) and the high resistance state (HRS) of the resistive material by applying voltages to the electrodes. Lee has explained that during the SET process, the current level increases from HRS to LRS as the voltage increases from 0 V to the critical point which is called the set voltage (*V* set), while the current level abruptly decreases from LRS to HRS at the reset voltage (*V* reset) under the RESET process. The SET and RESET processes are repeatedly carried out by sweeping the gate voltage with the binary states LRS and HRS [121]. Wang and Tseng and Lin et al. have indicated that the interface plays an important role in enhancing the performances of RRAM [122, 123]. Recently, Goux et al. have explained that using a stacked RRAM structure has been shown to be one of the most promising methods to improve the memory characteristics [124]. Although being a most promising memory element, critical issues for the future development of RRAM devices are reliable, such as data retention and memory endurance [125]. A data retention time of over 10 years can be extrapolated from retention characteristics measured at high temperatures and a memory endurance of over 10^6 cycles [126]. Therefore, a statistical study of reliability, availability, and maintainability is essential for the future development of RRAM.

Polymer memory

Throughout the last few years, polymers have found growing interest as a result of the rise of a new class of nonvolatile memories. In a polymer memory, a layer consists of molecules and/or nanoparticles in an organic polymer matrix is sandwiched between an array of top and bottom electrodes as illustrated in Figure 18. Moreover, polymer memory has the advantage of a simple fabrication process and good controllability of materials [127]. Polymer memory could be called digital memory with the latest technology. It is not possible for a silicon-based memory to be established in less space, but it is possible for polymer memory. Ling et al. explained that polymer materials have simplicity in structure, free *read* and *write* capability, better scalability, 3-D stacking ability, low-cost potential, and huge capacity of data storage [128]. They revealed that a polymer memory stores information in a manner that is entirely different from that of silicon-based memory devices. Rather than encoding '0' and '1' from the number of charges stored in a cell, a polymer memory stores data on the basis of high and low conductivity while responding to an applied voltage. Among the large number of emerging memory technologies, polymer memory is the leading technology. It is mainly because of its expansion capability in 3-D space [129] since most polymers are organic materials consisting of long chains of single molecules. Prior to polymer memory fabrication, deposition of an organic layer is usually done by the sol-gel spin coating technique. All the other necessary constituent materials are dissolved in a solvent which is then spin-coated over a substrate. When the solvent is evaporated, a thin film of material with 10- to 100-nm thickness is successfully deposited at bottom electrodes. Top electrodes are deposited as the final step. The conductivity of the organic layer is then changed by applying a voltage across the memory cell, allowing bits of data to be stored in the polymer memory cell. When the polymer memory cell becomes electrically conductive, the electrons are introduced and removed. Even the polymer is considered as a 'smart' material to the extent that functionality is built into the material itself of switchability and charge store. This will open up tremendous opportunities in the electronics world, where tailor-made memory materials represent an unknown territory. The nonvolatileness and other features are inbuilt at the molecular level and offers very high advantages in terms of cost. But turning polymer memory into a commercial product would not be easy. Memory technologies compete not only on storage capacity but on speed, energy consumption, and reliability. 'The difficulty is in meeting all the requirements of current silicon memory chips,' says Thomas, the Director of Physical Sciences at IBM›s Watson Research Center in Yorktown Heights, NY. They are likely to be limited to niche applications [130].

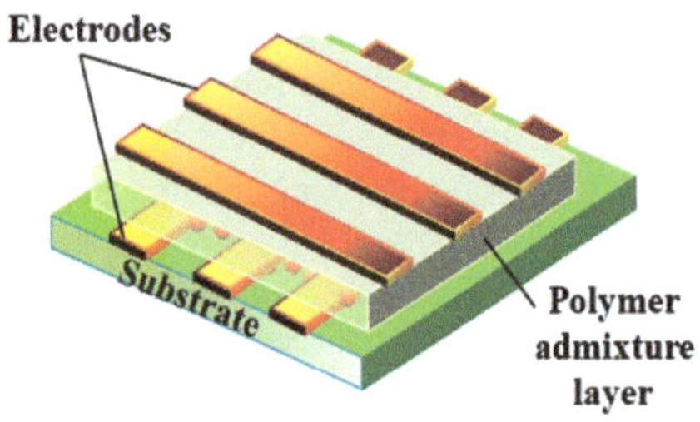

Figure 18 Structure of a polymer memory device.

Racetrack Memory

In a racetrack memory, information is stored on a U-shaped nanowire as a pattern of magnetic regions with different polarities. The U-shaped magnetic nanowire is an array of keys, which are arranged vertically like trees in a forest as shown in Figure 19. Achieving capacities comparable to vertical RM or hard drives would require stacks of these arrays. The nanowires have regions with different magnetic polarities, and the boundaries between the regions represent 1 or 0 s, depending on the polarities of the regions on either side [131, 132]. The magnetic information itself is then pushed along the wire, past the write and read heads by applying voltage pulses to the wire ends. The magnetic pattern to speed along the nanowire, while applying a spin-polarized current, causes the data to be moved in either direction, depending on the direction of the current. A separate nanowire perpendicular to the U-shaped 'racetrack' writes data by changing the polarity of the magnetic regions. A second device at the base of the track reads the data. Data can be written and read in less than a nanosecond. A racetrack memory using hundreds of millions of nanowires would have the potential to store vast amounts of data [133,134]. In this way, the memory requires no mechanical moving of parts and it has a greater reliability and higher performance than HDDs, with theoretical nanosecond operating speeds. For a device configuration where data storage wires are fabricated in rows on the substrate, conventional manufacturing techniques are adequate. However, for the maximum possible memory density, the storage wires are proposed to be configured rising from the substrate in a 'U' shape, giving rise to a 3-D forest of nanowires. While this layout does allow high data storage densities, it also has the disadvantage of complex fabrication methods, with so far, only 3-bit operation of the devices demonstrated [133].

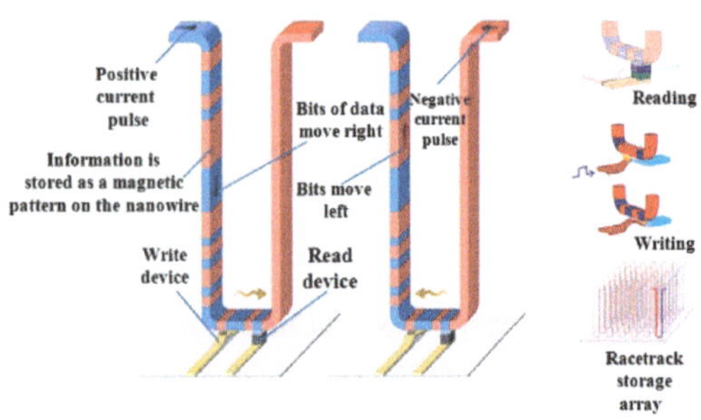

Figure 19 Racetrack memory diagram showing an array of U-shaped magnetic nanowires. The nanowires are arranged vertically like trees in a forest and a pair of tiny devices that read and write the data. Adopted from IBM.

As the access time of the data is also dependent on the position of the data on the wire, these would also be performance losses if long wires are used to increase the storage density further. The speed of operation of the devices has also been an issue during development, with much slower movement of the magnetic domains than originally predicted. This has been attributed to crystal imperfections in the permalloy wire, which inhibit the movement of the magnetic domains. By eliminating these imperfections, a data movement speed of 110 m/s has been demonstrated [133].

OTHER NEW MEMORY TECHNOLOGIES

Researchers are already working hard on several emerging technologies, as discussed in previous sections, to pursue storage-class memories with a more traditional design than that of the racetrack memory, which places the bits in horizontal arrays.

Molecular Memory

A molecular memory is a nonvolatile data storage memory technology that uses molecular species as the data storage element, rather than, e.g., circuits, magnetics, inorganic materials, or physical shapes [135]. In a molecular memory, a monolayer of molecules is sandwiched between a cross-point array of top and bottom electrodes as shown in Figure 20. The molecules are packed in a highly ordered way, with one end of the molecule electrically connected to the bottom electrode and the other end of the molecule connected to the top electrode, and this molecular component is described as a molecular switch [136]. Langmuir-Blodgett (LB) deposition is ideally suited for depositing the molecular layer for the fabrication of molecular memory devices [137, 138]. Then, regarding the molecular memory operation, by applying a voltage between the electrodes, the conductivity of the molecules is altered, enabling data to be stored in a nonvolatile way. This process can then be reversed, and the data can be erased by applying a voltage to the opposite polarity of the memory cell. The increasing demand for nonvolatile electronic memories will grow rapidly in order to keep pace with the requirements for subsystems involved in flight demonstration projects and deep space operations. At the same time, mass, volume, and power must be minimized for mission affordability concerning these requirements; molecular memory could be a very promising candidate to fill this need.

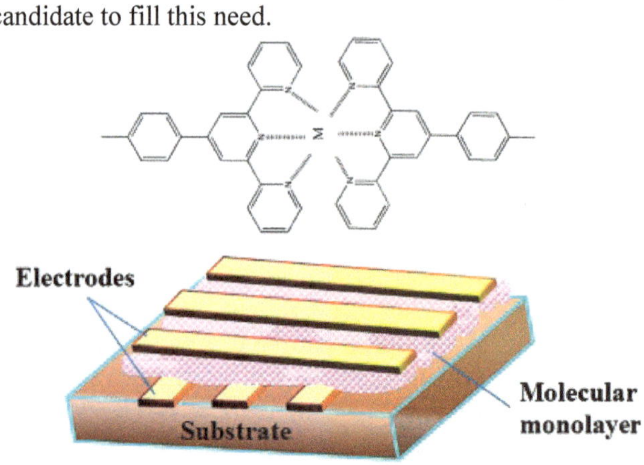

Figure 20 Cell structure of a molecular memory device.

Recently, Plafke has revealed clearly via an article that like most experimental technology that sounds so amazing that we want it right now, the molecular memory cell does not provide enough power for a commercial device [139]. This is currently only able to produce a 20% jump in conductivity. However, the area of molecular switching memory is promising, having eliminated the need for near-absolute zero temperatures and removed some of the constraints of the shape and number of layers of the molecule sheets which intend to convey that two of the biggest barriers are taken away. Thus, molecular memory requires strong attention to work over such issues and needs immediate amendment to see the possibility of a universal memory in the future.

MNW

In the last two decades, an increasing interest is observed for electronics-related devices and the search for a universal memory data storage device that combines rapid *read* and *write* speeds, high storage density, and nonvolatility is driving the investigation of new materials in the nanostructured form [140]. As an alternative to the current Flash memory technology, a novel transistor architecture using molecular-scale nanowire memory cells holds the promise of unprecedently compact data storage. The molecular nanowire array (MNW) memory is fundamentally different from other semiconductor memories; information storage is achieved through the channel of a nanowire transistor that is functionalized with redox-active molecules rather than through manipulation of small amounts of charge. It is relatively slow and lacks the random access capability, wherein data that can be randomly read and written at every byte are being actively pursued. Figure 21 shows the schematic design of a MNW memory cell. Lieber, and Agarwal and Lieber have revealed that the nanowire-based memory technology is a powerful approach to assemble electronic/photonic devices at ultra-small scales owing to their sub-lithographic size, defect-free single-crystalline structure, and unique geometry [141, 142]. Nanowires synthesized by chemical or physical processes are nearly perfect single-crystal structures with a small geometry and perfect surface. The channel of a nanowire transistor is functionalized with redox-active molecules. During programming, control of the voltage acting on the substrate is possible to change the oxidation and reduction states of the active molecules. Finally, by measurement of the conductance of the nanowire with the gate bias fixed at 0 V or a small voltage and from the hysteresis, the two states can be defined as a high-conductance ON state and a low-conductance OFF state. The MNW memory has advantages of low power dissipation, ultra-high density, simple fabrication process, 3-D structure, and multilevel storage, and it functions at the nanoscale with a few electrons but limited by low retention time parameter [143, 144]. Moreover, the deposition of metals onto a monolayer of molecular wires can lead to low device yield, and this problem remains a major challenge [145]. However, mentioning the term emerging class memory, it could be expected that the MNW memory represents an important step towards the creation of molecular computers that are much smaller and could be more powerful than today›s silicon-based computers.

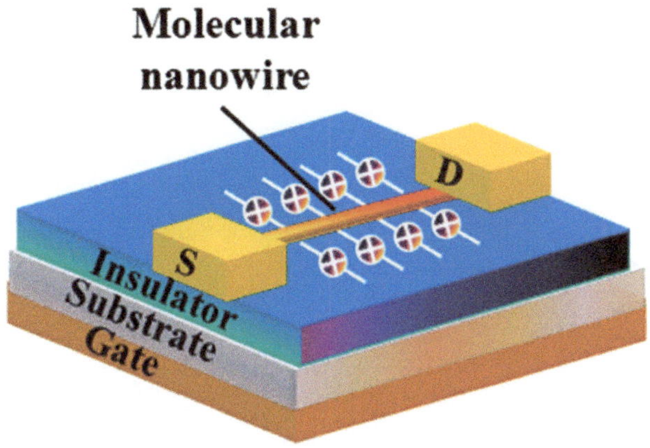

Figure 21 A MNW memory cell structure.

SNW

Semiconductor memory is essential for information processing as a key part of silicon technology; semiconductor memory has been continuously scaled to achieve a higher density and better performance in accordance with Moore's law [146]. Flash memory may reach fundamental scaling limits, however, because a thick tunneling oxide is required to prevent charge leakage and achieve 10 years' retention. As Flash memory approaches its scaling limit, several alternative strategies have been proposed to extend or replace the current Flash memory technology [147]. These approaches are revolutionary, but major challenges must be overcome to achieve small memory size and aggressive technology design architecture. In addition to the engineering of trapping layers, the device performance can also be improved by using innovative nonplanar channel geometries. Among the various nanostructure materials, semiconductor nanowire memory (SNW) has induced great scientific interest as possible building blocks for future nanoelectronic circuitry. In a SNW memory device, nanowires are integrated with SONOS technology. The basic schematic design of SNW is depicted in Figure 22. The SNW memory shows high mobility, less power dissipation, and high performance. Moreover, being 3-D-stacked, the SNW memory enhances cell density and data capacity without relying on advances in process technology. The nanowire-based memory device can store data electrically and is nonvolatile, meaning it retains data when the power is turned off, like the silicon-based Flash memory found in smart phones and memory cards [148], with minimal increase in chip size. In addition, the SNW device exhibits reliable write/read/erase operations with a large memory window and high on-to-off current ratio, which are highly advantageous for applications in nonvolatile memory [149]. The SNW memory cannot hold data as long as the existing Flash, but it is slower and has fewer rewrite cycles and it could potentially be made smaller and packed together more densely. And its main advantage is that it can be made using simple processes at room temperature, which means that it can be deposited even on top of flexible plastic substrates [150]. The SNW could, for instance, be built into a flexible display and could be packed into smaller spaces inside cell phones, MP3 players, plastic RFID tags, and credit cards.

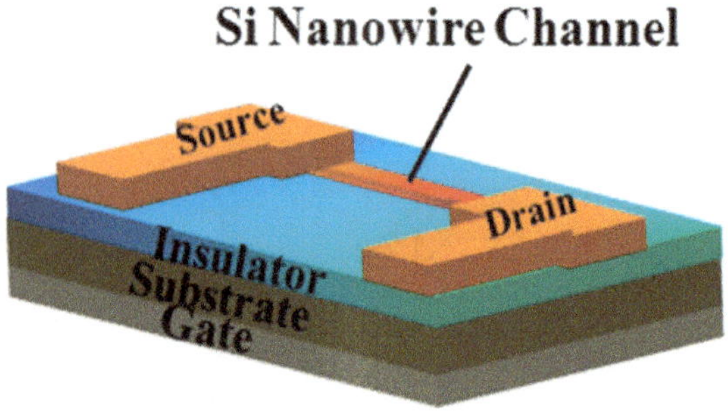

Figure 22 A bottom-gate FET-based nonvolatile SNW memory device.

NRAM

NRAM is a carbon nanotube (CNT)-based memory, which works on a nanomechanical principle, rather than a change in material properties [151]. NRAM uses carbon nanotubes for the bit cells, and the 0 or 1 is determined by the tube's physical state: up with high resistance, or down and grounded. NRAM is expected to be faster and denser than DRAM and also very scalable, able to handle 5-nm bit cells whenever CMOS fabrication advances to that level. It is also very stable in its 0 or 1 state. Produced by Nantero, these memories consist of the structure shown in Figure 23a with an array of bottom electrodes covered by a thin insulating spacer layer [152]. CNTs are then deposited on the spacer layer, leaving them freestanding above the bottom electrodes. Unwanted CNTs are removed from the areas around the electrode, with top contacts and interconnects deposited on top of the patterned CNT layer. During the time that the CNTs are freestanding, there is no conduction path between the bottom and top electrodes and hence the memory cell is in the *OFF* state. However, if a large enough voltage is applied over the cell, the nanotubes are attracted to the bottom electrode where they are held in place by van der Waals forces [153]. Due to the conductive nature of the CNTs, the electrodes are now connected and the cell reads the low conductivity *ON* state as shown in Figure 23b. The *OFF* state can be returned by repelling the nanotubes with the opposite electrode polarity. Nonvolatility is achieved due to the strength of the van der Waals forces overcoming the mechanical strain of the bent nanotubes, hence holding the cell in the *ON* state. NRAM offers the possibility of a simple cell architecture, which could operate at much higher speeds than the conventional Flash and with low power use. Cui et al. reported CNT memory devices exhibiting an extraordinarily high charge storage stability of more than 12 days at room temperature [154]. However, as NRAM is based on CNTs, it suffers from fabrication problems that are inherent in carbon nanotube-based devices. The issues include the cost and fabrication complexity of producing the CNTs, ensuring uniform dispersions of nanotubes, and difficulties in removing nanotubes from the unwanted positions on the substrate.

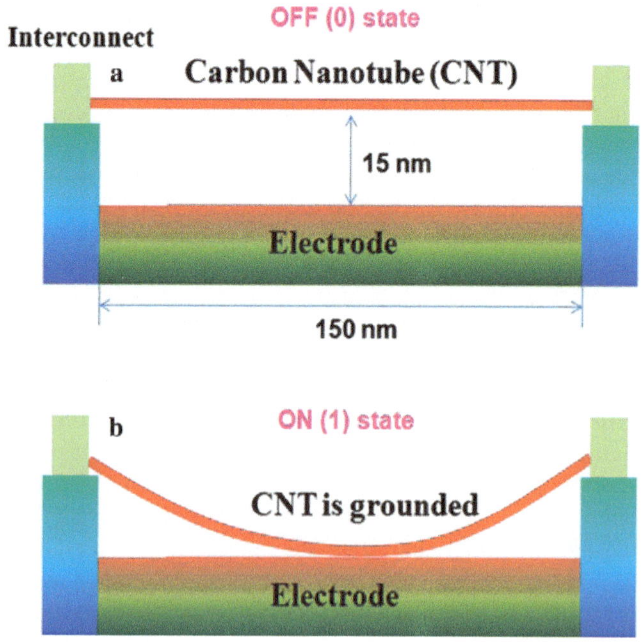

Figure 23 NRAM structure with (a) OFF state and (b) ON state.

Millipede memory

In 2002, IBM developed a punch card system known as Millipede, which is a nonvolatile computer memory stored in a thin polymer sheet with nanoscopic holes to provide a simple way to store binary data [155]. It can store hundreds of gigabytes of data per square centimeter. However, the polymer reverts to its pre-punched form over time, losing data in the process. Millipede storage technology is being pursued as a potential replacement for magnetic recording in hard drives, at the same time reducing the form factor to that of Flash media. The prototype's capacity would enable the storage of 25 DVDs or 25 million pages of text on a postage stamp-sized surface and could enable 10 GB of storage capacity on a cell phone. Millipede uses thousands of tiny sharp points (hence the name) to punch holes in a thin plastic film. Each of the 10-nm holes represents a single bit. The pattern of indentations is a digitized version of the data. The layout of the millipede cantilever/tip in contact with the data storage medium is shown in Figure 24. According to IBM, Millipede can be thought of as a nanotechnology version of the punch card data processing technology developed in the late nineteenth century [156]. However, there are significant differences: Millipede is rewritable, and it may eventually enable storage of over 1.5 GB of data in a space no larger than a single hole in the punch card. Storage devices based on IBM's technology can be made with existing manufacturing techniques, so they will not be expensive to make. According to P. Vettiger, head of the Millipede project, there is not a single step in fabrication that needs to be invented. Vettiger predicts that a nanostorage device based on IBM's technology could be available as early as 2005 [155]. Now, researchers at IBM's Zurich Research Laboratory in Switzerland have clocked the rate of data loss. They have calculated that at 85°C, a temperature often used to assess data retention, it would lose just 10% to 20% of information over a decade, comparable to Flash memory [157].

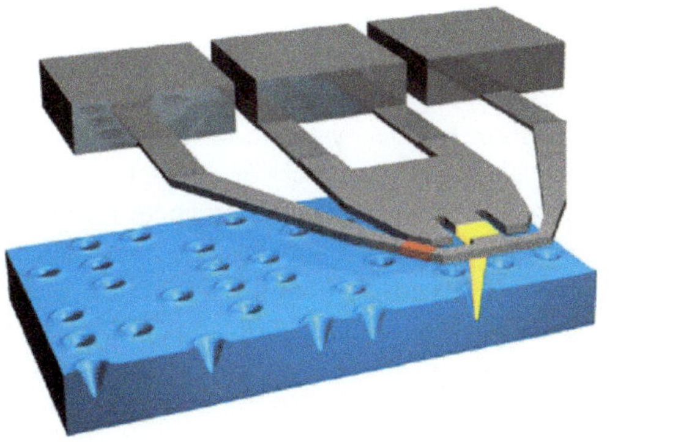

Figure 24 Schematic layout of the millipede cantilever/tip in contact with the data storage medium. Adopted from ref. [157].

WORM memory based on DNA biopolymer Nanocomposite

The use of DNA is well known as a good model for metal NP synthesis due to its affinity to the metal ions [158]. In recent years, DNA has also been shown to be a promising optical material with the material processing fully compatible with conventional polymer for thin-film optoelectronic applications [159, 160]. Researchers from National Tsing Hua University in Taiwan and the Karlsruhe Institute of Technology in Germany have created a DNA-based memory device, that is, write-once-read-many-times (WORM), that uses ultraviolet (UV) light to encode information [161]. The device consists of a single biopolymer layer sandwiched between electrodes, in which electrical bistability is activated by *in situ* formation of silver nanoparticles embedded in a biopolymer upon light irradiation (Figure 25). The device functionally works when shining UV light on the system, which enables a light-triggered synthesis process that causes the silver atoms to cluster into nanosized particles and readies the system for data encoding. For some particular instance, the team has found that using DNA may be less expensive to process into storage devices than using traditional, inorganic materials like silicon, the researchers say [161, 162]. They said that when no voltage or low voltage is applied through the electrodes to the UV-irradiated DNA, only a low current is able to pass through the composite; this corresponds to the 'OFF' state of the device. But the UV irradiation makes the composite unable to hold a charge under a high electric field, so when the applied voltage exceeds a certain threshold, an increased amount of charge is able to pass through. This higher state of conductivity corresponds to the 'ON' state of the device. The team found that this change from low conductivity ('OFF') to high conductivity ('ON') was irreversible: once the system had been turned on, it stayed on, no matter what voltage the team applied to the system. Once information is written, the device appears to retain that information indefinitely. The researchers hope that the technique will be useful in the design of optical storage devices and suggest that it may have plasmonic applications as well. Consequently, WORM memories based on DNA a biopolymer nanocomposite have emerged as an excellent candidate for next-generation information storage media because of their potential application in flexible memory devices. This work combines new advances in DNA nanotechnology with a conventional polymer fabrication platform to realize a new emerging class of DNA-based memory.

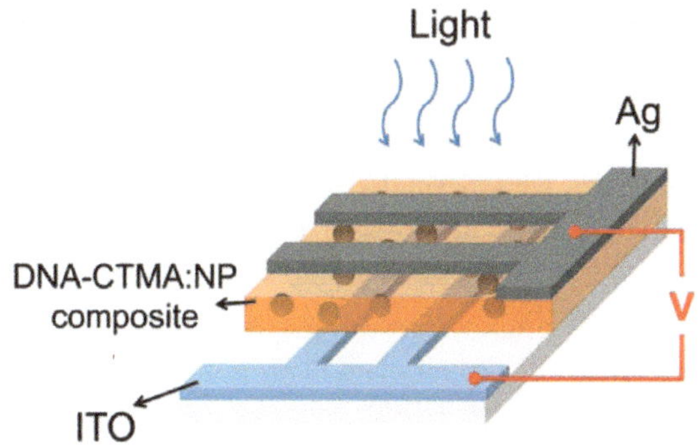

Figure 25 Schematic design of a memory device consisting of a thin DNA biopolymer film sandwiched between electrodes. The memory switching effect is activated upon light irradiation. Adopted from ref. [161].

QD Memory

Memory made from tiny islands of semiconductors - known as quantum dots - could fill a gap left by today's computer memory, allowing storage that is fast as well as long lasting. Researchers have shown that they can write information into quantum dot memory in just nanoseconds. Memory is divided into two forms: DRRAM and Flash [163, 164]. Computers use DRAM, for short-term memory, but data does not persist for long and must be refreshed over 100 times per second to maintain its memory. On the other hand, Flash memory, like that used in memory cards, can store data for years without refreshing but writes information about 1,000 times slower than DRAM. New research shows that memory based on quantum dots can provide the best of both: long-term storage with write speeds nearly as fast as DRAM. A tightly packed array of tiny islands, each around 15 nm across, could store 1 terabyte (1,000 GB) of data per square inch, the researchers say. Dieter Bimberg and colleagues at the Technical University of Berlin, Germany, with collaborators at Istanbul University, Turkey, demonstrated that it is possible to write information to the quantum dots in just 6 ns [165, 166]. The key advantages of quantum dot (QD) NVMs are the high read/write speed, small size, low operating voltage, and, most importantly, multibit storage per device. However, these features have not been realized due to variations in dot size and lack of uniform insulator cladding layers on the dots [167].

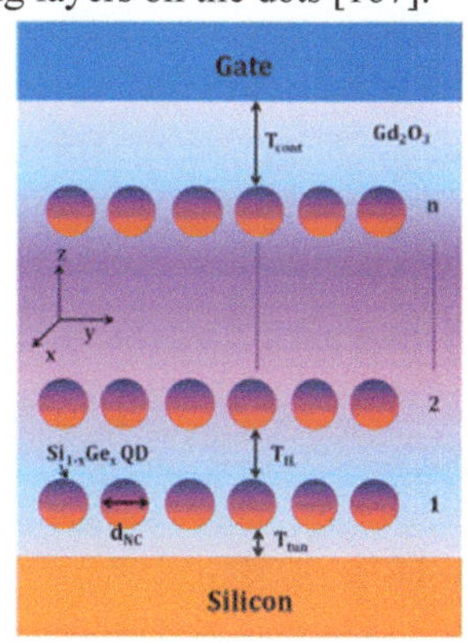

Figure 26 Structure of quantum dot memory. Adopted from ref. [168].

Incorporating QDs into the floating gate results in a reduction in charge leakage and power dissipation with enhanced programming speed. Researchers in India and

Germany have now unveiled the memory characteristics of silicon and silicon-germanium QDs embedded in epitaxial rare-earth oxide gadolinium oxide (Gd_2O_3) grown on Si (111) substrates as shown in the DQM structure in Figure 26. Multilayer Si as well as single-layer $Si_{1-x}Ge_x$ (where $x=0.6$) QDs show excellent memory characteristics, making them attractive for next-generation Flash-floating-gate memory devices [168, 169].

3-D cross-point memory

Memory producers are also trying to develop alternative technologies that may be scalable beyond 20-nm lithography. For true scalability beyond 20-nm technology nodes, it is necessary to design a cross-point memory array which does not require diodes for access elements [170]. The cross-point memory architecture could be designed such that it can be easily fabricated in multiple layers to form a stacked 3-D memory [171]. The 3-D technology has brought to high volume an NVM where arrays of memory cells are stacked above control logic circuitry in the third dimension, and stacking 3-D memory directly over CMOS allows for high array efficiency and very small die size [172]. The 3-D technology uses no new materials, processes, or fabrication equipment, which control logic circuitry composed of typical CMOS. The memory construction uses typical back-end processing tools, and each memory layer is a repeat of the layers below it. The basic design of the 3-D cell consists of a vertical diode in series with a memory element as shown in Figure 27. Building integrated circuits vertically allows for a reduced chip footprint when compared to a traditional 2-D design, by an approximate factor of the number of layers used. This offers significant advantages in terms of reduced interconnect delay when routing to blocks that otherwise would have been placed laterally. The process for the 2-D cross-point array can be built into a multilayer 3-D architecture.

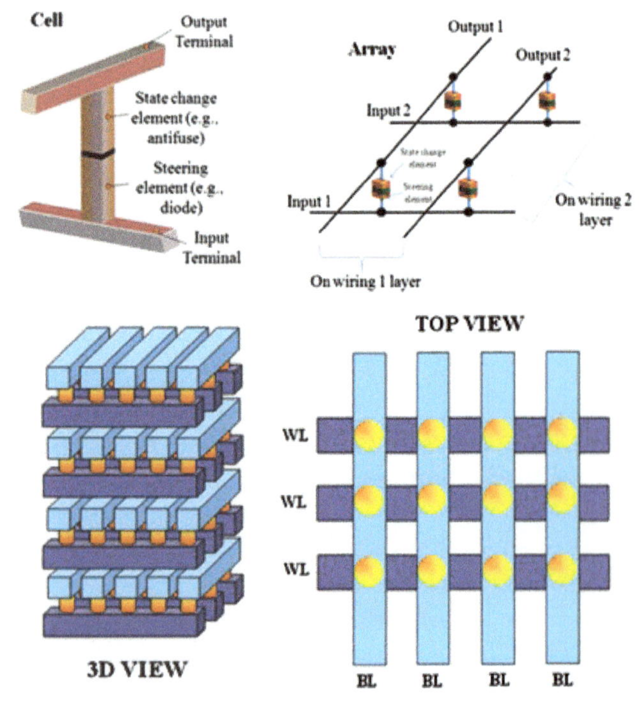

Figure 27 The basic design of a 3-D cell that consists of a vertical diode in series. (top) Side view, (bottom right) top view, and (bottom left) 3-D view.

Traditionally, a 3-D integrated circuit (3-D-IC) has used more than one active device layer. While resistance-change memory cells are not active devices, they function as rectifying devices in design. Further characterization of the resistance-change material is also necessary in order to guarantee that the 3-D cross-point memory will be practical for data storage. Also, the scalability of metal-oxide resistance-change materials beyond 20-nm technology nodes still needs to be studied. Moreover, the programming operation is expected to be competitive with both NAND and NOR Flash in terms of speed because of the relatively low voltage requirements of resistance-change materials. If the peripheral circuitry for accommodating the write operation can be made sufficiently compact, then the 3-D cross-point memory will indeed be a viable replacement for NAND and NOR Flash in future process generations.

TFM

Transparent and flexible electronics (TFE) is, today, one of the most advanced topics for a wide range of device applications, where the key component is transparent conducting oxides (TCOs), which are unique materials that oxides of different origin play an important role, not only as a passive component but also as an active component [173]. TFE is an emerging technology that employs materials (including oxides, nitrides, and carbides) and a device for the realization of invisible circuits for implementing next-generation transparent conducting oxides in an invisible memory generation [174]. In general, the TF-RRAM device is based on a capacitor-like structure (e.g., ITO/transparent resistive material/ITO/transparent and flexible substrate), which provides transmittance in the visible region [175]. For such new class of memory technology, data retention is expected to be about 10 years. The basic structural design of the new memory chips is configured, namely with two terminals per bit of information on a transparent and flexible substrate rather than the standard three terminals per bit on a rigid and opaque substrate (Figure 28). They are much better suited for the next revolution in electronic 3-D memory than Flash memory. These new memory chips that are transparent, are flexible enough to be folded like a sheet of paper, shrug off 1,000°F temperatures twice as hot as the max in a kitchen oven, and survive other hostile conditions could usher in the development of next-generation Flash-competitive memory for tomorrow's keychain drives, cell phones, and computers, a scientist reported today. Speaking at the 243rd National Meeting and Exposition of the American Chemical Society, the world's largest scientific society, he said that devices with these chips could retain data despite an accidental trip through the drier or even a voyage to Mars. And with a unique 3-D internal architecture, the new chips could pack extra gigabytes of data while taking up less space [176]. Despite the recent progress in TF-RRAM, it needs lots of work to satisfy the dual requirements of resistance to repeated bending stress and transparent properties. Thus, it is supposed that an achievement of such TF-RRAM device will be the next step towards the realization of transparent and flexible electronic systems. We hope that FT-RRAM devices will mark a milestone in the current progress of such unique and invisible electronic systems in the near future.

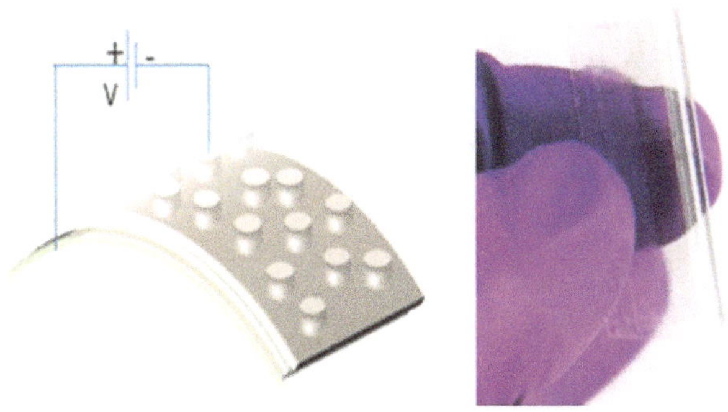

Figure 28 A schematic design of FT-RRAM and a flexible, transparent memory chip image created by researchers at Rice University. Reproduced from Tour Lab/Rice University.

1T1R-RRAM

One-transistor one-resistor (1T1R)-RRAM is also one class of emerging memory technology with impressive characteristics. It meets the demands for next-generation memory systems. It is expected that 1T1R-RRAM could be able to meet the demand of high-speed (e.g., performance) memory technology. The 1T1R structure is chosen because the transistor isolates current to cells, which are selected from cells which do not. The basic cell structure of 1T1R is depicted in Figure 29. 1T1R-RRAM consists of an access transistor and a resistor as a storage element. Zangeneh and Joshi have also mentioned that the 1T1R cell structure is similar to that of a DRAM cell in that the data is stored as the resistance of the resistor and the transistor serves as an access switch for reading and writing data [177, 178]. In reference to this, they revealed the 1T1R cell as the basic building block of a NVRRAM array as it avoids sneak path problem to ensure reliable operation. Moreover, the 1T1R structure is more compact and may enable vertically stacking memory layers, ideally suited for mass storage devices. But, in the absence of any transistor, the isolation must be provided by a 'selector' device, such as a diode, in series with the memory element, or by the memory element itself. Such kinds of isolation capabilities have been inferior to the use of transistors, limiting the ability to operate very large RRAM arrays in 1T1R architecture. 1T1R memory polarity can be either binary or unary. Bipolar effects cause polarity to reverse reset operation to set operation. Unipolar switching leaves polarity unaffected but uses different voltages.

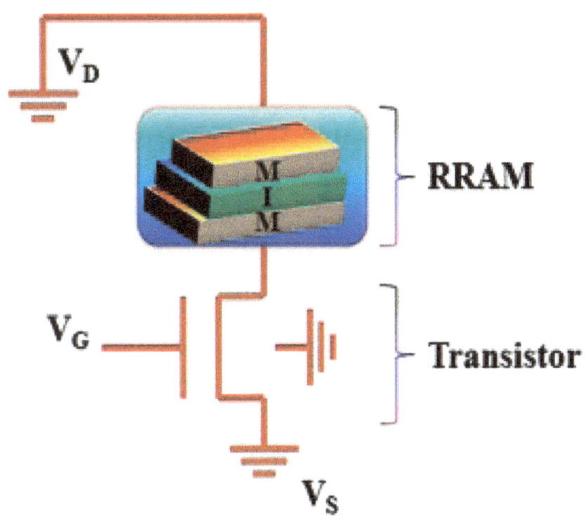

Figure 29 The basic cell structure of 1T1R-RRAM.

MTM, PFRAM, SPBMM, and CMORRAM - future alternate NVMs

Other potential emerging classes of memory technologies, we are describing in short, are molecular tunnel memory (MTM), polymeric ferroelectric RAM (PFRAM), spin-polarized beam magnetic memory (SPBMM), light memory, and complex metal-oxide RRAM (CMORRAM). We can say that these are sister memory technologies of molecular memory, ferroelectric/polymer memory, magnetic memory, and metal-oxide RRAM, respectively. Although these new technologies will almost certainly result in more complex memory hierarchies than their family memories, they are likely to allow the construction of memory chips that are nonvolatile, have low energy, and have density and development close to or better than those of DRAM chips, with improved performance and allowing memory systems to continue to scale up.

CONCLUSIONS

This article reviewed the historical development to the recent advancement on memory architecture and scaling trend of several conventional types of Flash within the MOS family and projected their future trends. With great progress being made in the emerging memory technologies, current trends and limitations were discussed before leading to some insight into the next generation of memory products. For the past three and a half decades in existence, the family of semiconductor memories has expanded greatly and achieved higher densities, higher speeds, lower power, more functionality, and lower costs. In the past 40 to 50 years, NVSM has grown from the FG concept to FAMOS, SAMOS, Flash memory, multilevel cells, RRAM, 3-D structures, and TF-RRAM. Since 1990, NVSM is an inspired technology, which has ushered in the digital age, enabled the development of all modern electronic systems, and brought unprecedented benefit to humankind. At the same time, some of the limitations within each type of memory are also becoming more realized. As the device dimension is reduced to the deca-nanometer regime, NVSM faces many serious scaling challenges such as the interface of neighboring cells, reduction of stored charges, and random telegraph noise. As such, we hope and are confident that there are several emerging technologies aiming to go beyond those limitations and potentially replace all or most of the existing semiconductor memory technologies to become a USM. Despite these limitations, the field of conventional semiconductor memories would continue to flourish and memory device scientists will find the way to meet these challenges and may even develop a 'unified memory' with low cost, high performance, and high reliability for future electronic systems. Progress towards a viable new resistive memory technology relies on fully understanding the mechanisms responsible for switching and charge transport, the failure mechanisms, and the factors associated with materials reliability. Moreover, the development of current redox-based resistive switching will help to improve our old technologies, and further research will produce more impressive results that will benefit industries and society to improve the quality of life for billions of people around the world.

AUTHORS' INFORMATION

JSM received his bachelor's degree (Physics Honors) from the Department of Physics, Aligarh Muslim University, Aligarh,

India, and Master's degree in solid-state technology from the Department of Physics and Meteorology, Indian Institute of Technology (IIT), Kharagpur, India, in 2007. He received his Ph.D. degree in Nanotechnology from National Chiao Tung University (NCTU), Taiwan, in February 2012. From March 2012 to December 2013, he has been a Postdoctoral Research Associate in the Department of Photonics and Display Institute, NCTU, Taiwan. He is currently a Postdoctoral Research Associate in the Department of Electronics and the Institute of Electronics, NCTU, Taiwan. His current research interests include designing, fabrication, and testing of a transparent and flexible random-access memory (RRAM) for application in invisible and rollable nonvolatile memory devices. He has published various research papers in reputed journals and presented his research in international conferences over flexible substrate-based thin-film transistors and capacitor devices for their applications in display and RF identification tags.

SMS received his B.S. degree from National Taiwan University, M.S. degree from the University of Washington, and Ph.D. degree from Stanford University, all in Electrical Engineering. He was with Bell Telephone Laboratories from 1963 to 1989 as a member of the Technical Staff. He joined National Chiao Tung University (NCTU) from 1990 to 2006 as a Distinguished Professor. At present, he is a National Endowed Chair Professor at NCTU. He has served as a Visiting Professor or Consulting Professor to many academic institutions including the University of Cambridge, Delft University, Beijing Jiaotong University, Tokyo Institute of Technology, Swiss Federal Institute of Technology, and Stanford University. He has made fundamental and pioneering contributions to semiconductor devices, especially metal-semiconductor contacts, microwave devices, and submicron MOSFET technology. Of particular importance is his coinvention of the *nonvolatile semiconductor memory* (NVSM) which has subsequently given rise to a large family of memory devices including Flash memory and EEPROM. The NVSM has enabled the development of *all modern electronic systems* such as the digital cellular phone, ultrabook computer, personal digital assistant, digital camera, digital television, smart IC card, electronic book, portable DVD, MP3 music player, antilock braking system (ABS), and Global Positioning System (GPS). He has authored or coauthored over 200 technical papers. He has written and edited 16 books. His book *Physics of Semiconductor Devices* (Wiley, 1969; 2nd ed, 1981; 3rd ed, 2007) is one of the *most* cited works in contemporary engineering and applied science publications (over 24,000 citations according to ISI Press). He has received the IEEE J.J. Ebers Award, the Sun Yat-sen Award, the National Endowed Chair Professor Award, and the National Science and Technology Prize. He is a Life Fellow of IEEE, an Academician of the Academia Sinica, a foreign member of the Chinese Academy of Engineering, and a member of the US National Academy of Engineering.

UC received his MS degree in Solid State Electronics from the Indian Institute of Technology, Roorkee, India, in 2010. He is currently a Ph.D. candidate of the Institute of Electronics, National Chiao Tung University, Taiwan. He is currently working on the fabrication and characterization of resistive switching memory devices with a focus on memory structure innovations and reliability. TYT is now a Lifetime Chair Professor in the Department of Electronics Engineering, National Chiao Tung University. He was the Dean of the College of Engineering and Vice Chancellor of the National Taipei University of Technology. He received numerous awards, such as the Distinguished Research Award from the National Science Council, Academic Award of the Ministry of Education, National Endowed Chair Professor, and IEEE CPMT Outstanding Sustained Technical Contribution Award.

ACKNOWLEDGEMENTS

First of all, the authors would like to thank and gratefully acknowledge all corresponding publishers for the kind permission to reproduce their figures and related description, used in this review article. This work was supported by the National Science Council, Taiwan, under Project No. NSC 102-2221-E-009-134-MY3.

Competing interests

The authors declare that they have no competing interests.

Authors' contributions

JSM designed the structure and modified the manuscript. SMS and TYT participated in the sequence alignment and editing the manuscript. UC participated in its design and coordination and helped to draft the manuscript. All authors read and approved the final manuscript.

REFERENCES

1. Kahng D, Sze SM: A floating gate and its application to memory devices. Bell Syst Tech J 1967, 46(6):1288–1295.
2. Ayling JK, Moore RD, Tu GK: A high-performance monolithic store. In Dig Tech Papers IEEE Int Solid-State Circuits Conf: February 19–21 1969; Philadelphia. Piscataway: IEEE; 1969:36–37.
3. Bez R, Camerlenghi E, Modelli A, Visconti A: Introduction to flash memory. Proc IEEE 2003, 91(4):489–502.
4. Aritome S: Advanced flash memory technology and trends for files storage application. In IEEE IEDM Tech Dig: December 10–13 2000; San Francisco. Piscataway: IEEE; 2002:763–766.
5. SNIA: Solid State Storage 101: An Introduction to Solid State Storage. San Francisco: SNIA; 2009.
6. Kim K, Koh G: The prospect on semiconductor memory in nano era. IEEE 7th Inter Conf Solid-State Integer Circuits Technol 2004, 1: 662–667.
7. Lam CH: The quest for the universal semiconductor memory. In IEEE Conf Electron Devices and Solid-State Circuits: December 19–21 2005; Hong Kong. Piscataway: IEEE; 2005:327–333.
8. Tseng TY, Sze SM: Nonvolatile Memories: Materials, Devices and Applications. Lewis Way: American Scientific Publishers; 2012:25650.
9. Lin CW, Wang DY, Tai Y, Jiang YT, Chen MC, Chen CC, Yang YJ, Chen YF: Type-II heterojunction organic/inorganic hybrid non-volatile memory based on FeS2 nanocrystals embedded in poly(3-hexylthiophene). J

Phys D Appl Phys 2011, 44(29):292002.

10. Burr GW, Kurdi BN, Scott JC, Lam CH, Gopalakrishnan K, Shenoy RS: Overview of candidate device technologies for storage-class memory. IBM J Res & Dev 2008, 524: 449–464.

11. Intel is hot on plastic memory [http://www.geek.com/chips/intel-is-hot-on-plastic-memory-548187] []

12. Lin HT, Pei Z, Chen JR, Hwang GW, Fan JF, Chan YJC: A new nonvolatile bistable polymer-nanoparticle memory device. IEEE Electron Device Lett 2007, 28(11):951–953.

13. Cho SH, Lee DI, Jung JH, Kim TW: Electrical bistabilities and memory stabilities of nonvolatile bistable devices fabricated utilizing C60 molecules embedded in a polymethyl methacrylate layer. Nanotechnology 2009, 20(34):345204.

14. Tang W, Shi HZ, Xu G, Ong BS, Popoyic ZD, Deng JC, Zhao J, Rao GH: Memory effect and negative differential resistance by electrode-induced two-dimensional single-electron tunneling in molecular and organic electronic devices. Adv Mater 2005, 17(19):2307–2311.

15. Intel brings plastic memory one step closer to reality [http://www.geek.com/chips/intel-brings-plastic-memory-one-step-closer-to-reality-553620] []

16. Wang X: Electrical characterization of novel charge-trapping non-volatile semiconductor memory devices. In Theses and Dissertations. Volume 932. Lehigh University; 2006.

17. Lankhorst MHR, Ketelaars BWSMM, Wolters RAM: Low-cost and nanoscale non-volatile memory concept for future silicon chips. Nat Mater 2005, 4: 347–352.

18. Viking Technology: The Expanding Role of Non-volatile Memory in High-Performance Embedded Architecture. Foothill Ranch: Viking Technology; 2012:643–7255.

19. Coburn J, Caulfield AM, Akel A, Grupp LM, Gupta RK, Jhala R, Swanson S: NV-heaps: making persistent objects fast and safe with next-generation, non-volatile memories. In Proceedings of the Sixteenth International Conference on Architectural Support for Programming Languages and Operating Systems (ASPLOS'11): March 5–11 2011; Newport Beach. California: ACM; 2011:1–13.

20. Brewer JE: Lest we forget: NVSM from origins to the "beyond CMOS" era. In IEEE Non-volatile Memory Technology Symposium (NVMTS): November 15–17 2004; Orlando. Florida: IEEE; 2004.

21. Global next generation memory markets (2012–2017), by Technology (DRAM, SRAM, Flash Memory, Memristor, Magneto Resistive RAM (MRAM), Phase Change RAM (PCRAM), Ferroelectric RAM) by Technology Application & Geography; 2013 [http://www.marketsandmarkets.com/Market-Reports/Memristor-Memory-Market-632.html] []

22. Lu CY: International Symposium on Non-volatile Memory: the Technology Driver of the Electronics Industry. Volume 2. 1st edition. Singapore: World Scientific; 2012.

23. Proctor A: Non-volatile memory & its use in enterprise applications. In Viking Technology Understanding Non-volatile Memory Technology Whitepaper. Irvine: Viking Technology; 2012:1–8.

24. Ouyang J, Chu CW, Szmanda CR, Ma L, Yang Y: Programmable polymer thin film and non-volatile memory device. Nat Mater 2004, 3: 918–922.

25. Bozano LD, Kean BW, Deline VR, Salem JR, Scott JC: Mechanism for bistability in organic memory elements. Appl Phys Lett 2004, 84(4):607–609.

26. Tung VC, Chen LM, Allen MJ, Wassei JK, Nelson K, Kaner RB, Yang Y: Low-temperature solution processing of graphene-carbon nanotube hybrid materials for high-performance transparent conductors. Nano Lett 2009, 9(5):1949–1955.

27. Zirkl M, Haase A, Fian A, Schön H, Sommer C, Jakopic G, Leising G, Stadlober B, Graz I, Gaar N, Schwödiauer Gogonea RSB, Bauer S: Low-voltage organic thin-film transistors with high-k nanocomposite gate dielectrics for flexible electronics and optothermal sensors. Adv Mater 2007, 19(17):2241–2245.

28. Koeppe R, Bartu P, Bauer S, Sariciftci NS: Light- and touch-point localization using flexible large area organic photodiodes and elastomer waveguides. Adv Mater 2009, 21(34):3510–3514.

29. Lin HT, Pei Z, Chen JR, Kung CP, Lin YC, Tseng CM, Chan YJ: A 16-byte nonvolatile bistable polymer memory array on plastic substrates. IEEE Int Eletron Devices Meet Tech Dig 2007, 1/2: 233–236.

30. Kinoshita K, Okutani T, Tanaka H, Hinoki T, Agura H, Yazawa K, Ohmi K, Kishida S: Flexible and transparent ReRAM with GZO memory layer and GZO-electrodes on large PEN sheet. Solid-State Electronics 2011, 58(1):48–53.

31. Hubert A, Nowak E, Tachi K, Maffini-Alvaro V, Vizioz C, Arvet C, Colonna JP, Hartmann JM, Loup V, Baud L, Pauliac S, Delaye V, Carabasse C, Molas G, Ghibaudo G, De Salvo B, Faynot O, Ernst T: A stacked SONOS technology, up to 4 levels and 6 nm crystalline nanowires, with Gate-All-Around or independent gates (φ-Flash), suitable for full 3D integration. In IEDM Tech Dig: December 7–9 2009; Baltimore. Piscataway: IEEE; 2009:637–640.

32. Yun JG, Lee JD, Park BG: 3D NAND flash memory with laterally-recessed channel (LRC) and connection gate architecture. Solid-State Electron 2011, 55(1):37–43.

33. Mearian L: Non-volatile memory's future is in software: new memory technology to serve dual roles of mass storage and system memory. Computerworld: Framingham; 2012.

34. Eleftheriou E, Haas R, Jelitto J, Lantz MA, Pozidis H: Trends in storage technologies. IEEE Data Eng Bull 2010, 33(4):4–13.

35. Ji Y, Cho B, Song , Kim TW, Choe M, Kahng YH, Lee T: Stable switching characteristics of organic nonvolatile memory on a bent flexible substrate. Adv Mater 2010, 22(28):3071–3075.

36. Ballhaus W, Pagella A, Vogel C: A Change of Pace for the Semiconductor Industry?. Düsseldorf: Price Waterhouse

Coopers; 2009.

37. Choi J, Seol KS: 3D approaches for non-volatile memory. In IEEE Symp VLSI Tech Digest: June 14–16 2011, Honolulu. Piscataway: IEEE; 2011:178–179.

38. Das R, Harrop P: Printed, Organic & Flexible Electronics: Forecasts, Players & Opportunities 2013–2023. Cambridge: IDTechEx; 2013.

39. Doe P: Options for adding memory and logic to printed or flexible electronics. [http://electroiq.com/blog/2013/07/options-for-adding-memory-and-logic-to-printed-or-flexible-electronics/] []

40. Park KT, Kim D, Hwang S, Kang M, Cho H, Jeong Y, Seo Y, Jang J, Kim HS, Jung SM, Lee YT, Kim C, Lee WS: A 45nm 4Gb 3-dimensional double-stacked multi-level NAND flash memory with shared bitline structure. In Digest of Technical Papers. IEEE International Solid-State Circuits Conference: February 3–7 2008; San Francisco. California: IEEE; 2008:510–632.

41. Elliott J, Jung ES: Ushering in the 3D memory era with V-NAND. In Flash Memory Summit: August 12–15 2013. Santa Clara: Santa Clara Convention Center; 2013.

42. Innovative Research and Products Inc: Advanced Solid-State Memory Systems and Products: Emerging Non-Volatile Memory Technologies, Industry Trends and Market Analysis. Stamford: Innovative Research and Products Inc; 2010:146.

43. Yole Développement: Emerging Non-volatile Memory. Lyon: Yole Développement; 2013:275.

44. Global Industry Analysts, Inc: Advanced Solid-State Memory Systems - Global Strategic Business Report. San Jose: Global Industry Analysts, Inc; 2012:731.

45. Groeseneken G, Maes HE, Van Houdt J, Witters JS: Basics of nonvolatile semiconductor memory devices. In Nonvolatile Semiconductor Memory Technology. Edited by: Brown WD, Brewer JE. New York: IEEE; 1998:1–88.

46. Nonvolatile memories for 90 nm SoC and beyond. Electronic Engineering (EE) Times 2003 [http://www.eetimes.com/document.asp?doc_id=1201899] []

47. Arreghini A, Driussi F, Vianello E, Esseni D, van Duuren MJ, Golubovi›c DS, Akil N, van Schaijk R: Experimental characterization of the vertical position of the trapped charge in Si nitride-based nonvolatile memory cells. IEEE Trans Electron Devices 2008, 55(5):1211–1219.

48. Salvo BD, Gerardi C, Schaijk RV, Lombardo SA, Corso D, Plantamura C, Serafino T, Ammendola G, Duuren MV, Goarin P, Mei WY, Jeugd KVD, Baron H, Gély M, Mur P, Deleonibus S: Performance and reliability features of advanced nonvolatile memories based on discrete traps. IEEE Trans Device Mater Rel 2004, 4(3):77–389.

49. Toshiba NAND vs. NOR flash memory technology overview. Toshiba America Electronic Components Inc; 1–4 [http://umcs.maine.edu/~cmeadow/courses/cos335/Toshiba%20NAND_vs_NOR_Flash_Memory_Technology_Overviewt.pdf] []

50. Micron Technology Inc: Technical Note: NAND Flash 101: An Introduction to NAND Flash and How to Design It in to Your Next Product. Boise: Micron Technology Inc; 2006:1–27.

51. Yan Y, Chen W, Fan R, Guo X, Guo H, Zhang F, Ding L, Zhang K, Lin D, Wang Y: Analysis of the TID induced failure modes in NOR and NAND flash memories. IEEE Trans Nucl Sci 2013, 60(1):224–229.

52. Cooke J: Flash Memory 101: An Introduction to NAND Flash. Micron Technology Inc: Boise; 2006.

53. Lai SK: Brief history of ETOX NOR flash memory. J Nanosci Nanotech 2012, 12(10):7597–7603.

54. Jose ST, Pradeep C: Design of a multichannel NAND Flash memory controller for efficient utilization of bandwidth in SSDs. In Proc Intern Multi-Conf Autom Computing Commun Control and Comp Sensing (IMAC4S): March 22–23 2013; Kottayam. Piscataway: IEEE; 2013:235–923.

55. Zheng M, Tucek J, Qin F, Lillibridge M: Understanding the robustness of SSDs under power fault. In 11th USENIX Conference on File and Storage Technologies (FAST›13): February 12–15 2013: San Jose. Berkeley: USENIX Association; 2013:271–284.

56. Crippa L, Micheloni R, Motta I, Sangalli M: Nonvolatile memories: NOR vs. NAND Architectures: Memories in Wireless Systems. Signal Communication Technology. Berlin, Heidelberg: Springer; 2008:29.

57. Tal A: Two Flash Technologies Compared: NOR vs NAND. 91-SR-012–04–8 L REV. 1.0. M-Systems Flash Disk Pioneers: Newark; 2002.

58. Hu CW, Chang TC, Tu CH, Chiang CN, Lin CC, Sze SM, Tseng TY: NiSiGe nanocrystals for nonvolatile memory devices. Appl Phys Lett 2009, 94(6):062102.

59. Fazio A: Flash memory scaling. MRS Bull 2004, 29: 814–817.

60. Tu CH: Electrical Switching and Memory Behaviors in Organic-based Devices. Ann Arbor: ProQuest LLC; 2008.

61. Prall K: Scaling non-volatile memory below 30 nm. In 22nd IEEE Non-volatile Semiconductor Memory Workshop: August 26–30 2007; Monterey. Piscataway: IEEE; 2007:5–10.

62. Nicollian PE, Rodder M, Grider DT, Chen P, Wallace RM, Hattangady SV: Low voltage stress-induced-leakage-current in ultrathin gate oxides. In 37th IEEE Annual Intern Reliab Phys Symp: March 23–25 1999: San Diego. Piscataway: IEE; 1999:400.

63. King Y: Thin dielectric technology and memory devices. In PhD dissertation. University of California, Berkeley; 1999:60–05. 2265 2265

64. Brewer JE, Gill M: Nonvolatile Memory Technologies with Emphasis on Flash: A Comprehensive Guide to Understanding and Using Flash Memory Devices. Hoboken: Wiley; 2008:750.

65. Zhao W, Dong G, Sun H, Zheng N, Zhang T: Reducing latency overhead caused by using LDPC codes in NAND flash memory. EURASIP J Adv Signal Proc 2012, 203: 1–9.

66. Ielmimi D: Reliability issues and modeling of flash and post-flash memory. Microelectron Enger 2009, 86: 1870–1875.

67. Cappelletti P, Golla C, Olivo P, Zanoni E: Flash Memories. New York: Springer; 1999.
68. Ramkumar K, Prabhakar V, Kouznetsov I, Geha S: Cypress SONOS Technology. San Jose: Cypress Semiconductor; 2013:1–13.
69. White MH, Adams DA, Bu J: On the go with SONOS. IEEE Circuit Devices Magazine 2000, 16(4):22–31.
70. You HC, Hsu TH, Ko FH, Huang JW, Yang WL, Lei TF: SONOS-type flash memory using an HfO2 as a charge trapping layer deposited by the sol–gel spin-coating method. IEEE Electron Device Lett 2006, 27(8):653–655.
71. Sematech International: Test equipment. The International Technology Roadmap for Semiconductors (ITRS). Austin: Sematech International; 2001:27–28.
72. Bu J, White MH: Design considerations in scaled SONOS nonvolatile memory devices. Solid State Electron 2001, 45: 113–120.
73. French ML, Chen CY, Sathianathan H, White MH: Design and scaling of SONOS multidielectric device for NVSM application. IEEE Trans Comp Pack Manu Tech Part A 1994, 17(3):390–397.
74. Steimle RF, Muralidhar R, Rao R, Sadd M, Swift CT, Yater J, Hradsky B, Straub S, Gasquet H, Vishnubhotla L, Prinz EJ, Merchant T, Acred B, Chang K Jr, White BE Jr: Silicon nanocrystal non-volatile memory for embedded memory scaling. Microelectron Reliab 2007, 47(4):585–592.
75. Hsu H, Lee JY: Electrical characterization of metal-oxide-high-k dielectric-oxide-semiconductor (MOHOS) structures for memory applications. Microelectron Reliab 2007, 47(4):606–609.
76. Watanabe H, Aritome S, Hemink GJ, Maruyama T, Shirota R: Scaling of tunnel oxide thickness for flash EEPROMs realizing stress-induced leakage current reduction. In Symp VLSI Tech Dig Tech Papers: June 7–9 1994; Honolulu. Piscataway: IEEE; 1994:47–48.
77. Chowdhury NA, Misra D: Charge trapping at deep states in Hf–silicate based high-k gate dielectrics. J Electrochem Soc 2007, 154(2):G30-G37.
78. Yang Y, White MH: Charge retention of scaled SONOS nonvolatile memory devices at elevated temperatures. Solid-State Electron 2000, 44(6):949–958.
79. Xuan P, She M, Harteneck B, Liddle A, Bokor J, King TJ: FinFET SONOS flash memory for embedded applications. In IEDM Tech Dig: December 8–10 2003; Washington, DC. Piscataway: IEEE; 2003:609–613.
80. Sugizaki T, Kobayashi M, Ishidao M, Minakata H, Yamaguchi M, Tamura Y, Sugiyama Y, Nakanishi T, Tanaka H: Novel multi-bit SONOS type flash memory using a high-k charge trapping layer. In VLSI Symp Tech Dig: June 10–12 2003; Kyoto. Piscataway: IEEE; 2003:27–28.
81. Ostraat ML, De Blauwe JW, Green ML, Bell LD, Brongersma ML, Casperson J, Flagan RC, Atwater HA: Synthesis and characterization of aerosol silicon nanocrystal nonvolatile floating-gate memory devices. Appl Phys Lett 2001, 79(3):433–435.
82. Wong HSP: Emerging memory devices. In Intern Semicond Dev Res Symp (ISDRS): December 7–9 2011; College Park. Piscataway: IEEE; 2011:1.
83. Kryder MH, Kim CS: After hard drives—what comes next? IEEE Trans Magn 2009, 45(10):3406–3413.
84. Wang B: Emerging Technology Analysis: The Future and Opportunities for Next-Generation Memory. Gartner, Inc: Stamford; 2011.
85. Gallagher WJ, Parkin SSP: Development of the magnetic tunnel junction MRAM at IBM: from first junctions to a 16-Mb MRAM demonstrator chip. IBM J Res & Dev 2006, 50(1):5–23.
86. Julliere M: Tunneling between ferromagnetic films. Physics Letters A 1975, 54(3):225–226.
87. Wang D, Tondra M, Pohm AV, Nordman C, Anderson J, Daughton JM, Black WC: Spin dependent tunneling devices fabricated for magnetic random access memory applications using latching mode. J Appl Phys 2000, 87(9):6385–6387.
88. Daughton JM: Magnetoresistive memory technology. Thin Solid Films 1992, 216(1):162–168.
89. Naji PK, Durlam M, Tehrani S, Calder J, Deherrera MF: A 256 kb 3.0 V 1T1MTJ nonvolatile magnetoresistive RAM. In IEEE Intern Solid-State Circ Conf Dig Tech Papers: February 7 2001; San Francisco. Volume 438. Piscataway: IEEE; 2001:122–123.
90. Tehrani S, Engel B, Slaughter JM, Chen E, Deherrera M, Durlam M, Naji P, Whig R, Jenesky J, Calder J: Recent developments in magnetic tunnel junction MRAM. IEEE Trans Magn 2000, 36(5):2752–2757.
91. Everspin Technologies-MR2A16AYS35-MRAM 4MB. 2009 [http://www.everspin.com/PDF/EST_MR2A16A_prod.pdf] []
92. Chen E, Apalkov D, Diao Z, Driskill-Smith A, Druist D, Lottis D, Nikitin V, Tang X, Watts S, Wolf SA, Ghosh AW, Lu JW, Poon SJ, Stan M, Butler WH, Gupta S, Mewes CKA, Mewes T, Visscher PB: Advances and future prospects of spin-transfer torque random access memory. IEEE Trans Magn 2010, 46(6):1873–1878.
93. Khvalkovskiy AV, Apalkov D, Watts S, Chepulskii R, Beach RS, Ong A, Tang X, Driskill-Smith A, Butler WH, Visscher PB, Lottis D, Chen E, Nikitin V, Krounbi M: Basic principles of STT-MRAM cell operation in memory arrays. J Phys D Appl Phys 2013, 46(7):074001.
94. Kawahara T, Ito K, Takemura R, Ohno H: Spin-transfer torque RAM technology: review and prospect. Microelectron Reliab 2012, 52(4):613–627.
95. Hosomi M, Yamagishi H, Yamamoto T, Bessho K, Higo Y, Yamane K, Yamada H, Shoji M, Hachino H, Fukumoto C, Nagao HH, Kano H: A novel nonvolatile memory with spin torque transfer magnetization switching: spin-ram. In Int Elec Dev Meet (IEDM) Tech Dig: December 5 2005; Washington, DC. Piscataway: IEEE; 2005:459–462.
96. Apalkov D, Khvalkovskiy A, Watts S, Nikitin V, Tang X, Lottis D, Moon K, Luo X, Chen E, Ong A, Driskill-Smith A, Krounbi M: Spin-transfer torque magnetic random access memory (STT-MRAM). ACM J Emerg

Tech Comp Sys (JETC) 2013, 9(2):13:2–13:35.

97. Auciello O, Scott JF, Ramesh R: The physics of ferroelectric memories. Phys Today 1998, 51: 22–27.

98. Ramtron: Ramtron Announces 8-Megabit Parallel Nonvolatile F-RAM Memory. Colorado Springs: Ramtron International Corporation; 2009.

99. Ramtron: F-RAM Technology Brief. Colorado Springs: Ramtron International Corporation; 2007.

100. Moise TS, Summerfelt SR, McAdams H, Aggarwal S, Udayakumar KR, Celii FG, Martin JS, Xing G, Hall L, Taylor KJ, Hurd T, Rodriguez J, Remack K, Khan MD, Boku K, Stacey G, Yao M, Albrecht MG, Zielinski E, Thakre M, Kuchimanchi S, Thomas A, McKee B, Rickes J, Wang A, Grace J, Fong J, Lee D, Pietrzyk C, Lanham R, et al.: Demonstration of a 4 Mb, high density ferroelectric memory embedded within a 130 nm, 5 LM Cu/FSG logic process. In Int Elec Dev Meet (IEDM) Tech Dig: December 8–11 2002; San Francisco. Piscataway: IEEE; 2002:535–538.

101. Toshiba develops world›s highest bandwidth, highest density non-volatile RAM. 2009 [https://www.toshiba.co.jp/about/press/2009_02/pr0902.htm] []

102. Hudgens S, Johnson B: Overview of phase-change chalcogenide nonvolatile memory technology. MRS Bull 2004, 29(11):829–832.

103. Pirovano A, Lacaita AL, Benvenuti A, Pellizzer F, Hudgens S, Bez R: Scaling analysis of phase-change memory technology. In Int Elec Dev Meet (IEDM) Tech Dig: December 8–10 2003; Washington, DC. Piscataway: IEEE; 2003:29.6.1–29.6.4.

104. Ahn SJ, Song YJ, Jeong CW, Shin JM, Fai Y, Hwang YN, Lee SH, Ryoo KC, Lee SY, Park JH, Horii H, Ha YH, Yi JH, Kuh BJ, Koh GH, Jeon GT, Jeong HS, Kim K, Ryu BI: Highly manufacturable high density phase change memory of 64 Mb and beyond. In Int Elec Dev Meet (IEDM) Tech Dig: December 13–15 2004; San Francisco. Piscataway: IEEE; 2004:907–910.

105. Raoux S, Burr GW, Breitwisch MJ, Rettner CT, Chen YC, Shelby RM, Salinga M, Krebs D, Chen SH, Lung HL, Lam CH: Phase-change random access memory: a scalable technology. IBM J Res & Dev 2008, 52(4/5):465–479.

106. Ryu SW, Oh JH, Choi BJ, Hwang SY, Hong SK, Hwang CS, Kim HJ: SiO incorporation effect in Ge2Sb2Te5 films prepared by magnetron sputtering for phase change random access memory devices. Electrochem Solid-State Lett 2006, 9(8):G259-G261.

107. Raoux S, Salinga M, Jordan-Sweet J, Kellock A: Effect of Al and Cu doping on the crystallization properties of the phase change materials SbTe and GeSb. J Appl Phys 2007, 101(4):44909–44915.

108. Atwood G, Bez R: Current status of chalcogenide phase change memory. In 63rd Device Research Conf Dig (DRC ‹05): June 22 1005; Santa Barbara. Volume 1. Piscataway: IEEE; 2005:29–33.

109. Intel N: Intel News Release: STMicroelectronics Deliver Industry›s First Phase Change Memory Prototypes. Intel: Santa Clara; 2008.

110. Wire B: Samsung Electronics and Numonyx Join Forces on Phase Change Memory. San Francisco: Business Wire; 2009.

111. Chua L: Memristor—the missing circuit element. IEEE Trans Circuit Theory 1971, 18(5):507–519.

112. Kim S: Resistive RAM (ReRAM) Technology for High Density Memory Applications. In 4th Workshop Innovative Memory Technol MINATEC 2012; June 21–24 2012. Grenoble: MINATEC; 2012:4.

113. Strukov DB, Snider GS, Stewart DR, Williams RS: The missing memristor found. Nature 2008, 453: 80–83.

114. Xu C, Dong X, Jouppi NP, Xie Y: Design implications of memristor-based RRAM cross-point structures. In Desig Auto Test Eur Conf Exhib (DATE): March 14–18 2011; Grenoble. Piscataway: IEEE; 2011:1–6.

115. Seo S, Lee MJ, Seo DH, Jeoung EJ, Suh DS, Joung YS, Yoo IK, Hwang IR, Kim SH, Byun IS, Kim JS, Choi JS, Park BH: Reproducible resistance switching in polycrystalline NiO films. Appl Phys Lett 2004, 85(23):5655–5657.

116. Wu MC, Lin YW, Lin CH, Tseng TY: Low-power and highly reliable multilevel operation in ZrO2 1T1R RRAM. IEEE Electron Dev Lett 2011, 32(8):1026–1028.

117. Huang CY, Huang CY, Tsai TL, Lin CA, Tseng TY: Switching mechanism of double forming process phenomenon in ZrOx/HfOy bilayer resistive switching structure with large endurance. Appl Phys Lett 2014, 104(6):062901(4).

118. Lin MH, Wu MC, Lin CH, Tseng TY: Effects of vanadium doping on resistive switching characteristics and mechanisms of SrZrO3-based memory films. IEEE Trans Electron Devices 2010, 57(8):1801–1808.

119. Yang JJ, Pickett MD, Li X, Ohlberg DAA, Stewart DR, Williams RS: Memristive switching mechanism for metal/oxide/metal nanodevices. Nat Nanotech 2008, 3: 429–433.

120. Kamiya K, Yang MY, Park SG, Magyari-Köpe B, Nishi Y, Niwa M, Shiraishi K: ON-OFF switching mechanism of resistive–random–access–memories based on the formation and disruption of oxygen vacancy conducting channels. Appl Phys Lett 2012, 100: 073502.

121. Lee JS: Progress in non-volatile memory devices based on nanostructured materials and nanofabrication. J Mater Chem 2011, 21: 14097–14112.

122. Wang SY, Tseng TY: Interface engineering in resistive switching memory. J Adv Dielectrics 2011, 1(2):141–162.

123. Lin CY, Wu CY, Wu CY, Hu C, Tseng TY: Modified resistive switching behavior of ZrO2 memory films based on the interface layer formed by using Ti top electrode. J Appl Phys 2007, 102(9):094101–094105.

124. Goux L, Fantini A, Kar G, Chen YY, Jossart N, Degraeve R, Clima S, Govoreanu B, Lorenzo G, Pourtois G, Wouters DJ, Kittl JA, Altimime L, Jurczak M: Ultralow sub-500nA operating current high-performance TiNAl2O3HfO2HfTiN bipolar RRAM achieved through understanding-based stack-engineering. In VLSI Tech Dig: June 12–14 2002; Honolulu. Piscataway: IEEE;

2012:159–160.

125. Sawa A: Resistive switching in transition metal oxides. Mater Today 2008, 11(6):28–36.

126. Seo YJ, An HM, Kim HD, Kim TG: Improved performance in charge-trap-type flash memories with an Al2O3 dielectric by using bandgap engineering of charge-trapping layers. J Korean Phys Soc 2009, 55(6):2679–2692.

127. Prakash A, Ouyang J, Lin JL, Yang Y: Polymer memory device based on conjugated polymer and gold nanoparticles. J Appl Phys 2006, 100(5):054309–054314.

128. Ling QD, Lim SL, Song Y, Zhu CX, Chan DSH, Kang ET, Noeh KG: Nonvolatile polymer memory device based on bistable electrical switching in a thin film of poly(N-vinylcarbazole) with covalently bonded C60. Langmuir 2007, 23(1):312–319.

129. Moller S, Perlov C, Jackson W, Taussing C, Forrest SR: A polymer/semiconductor write-once read-many-times memory. Nature 2003, 426: 166–169.

130. Stikeman A: Polymer memory: computer memory could soon earn the ultimate commercial validation: the cheap plastic knock-off. MIT Techno Rev 2002.

131. Hubert A, Schäfer R: Magnetic Domains: The Analysis of Magnetic Microstructures. Heidelberg, Berlin: Springer; 1998.

132. Parkin SSP, Jiang X, Kaiser C, Panchula A, Roche K, Samant M: Magnetically engineered spintronic sensors and memory. Proc IEEE 2003, 91(5):661–680.

133. Parkin SSP, Hayashi M, Thomas L: Magnetic domain-wall racetrack memory. Science 2008, 320(5873):190–194.

134. Parkin SSP: (2004–2007) U.S. Patents 6,834,005, 6,898,132, 6,920,062, 7,031,178, and 7,236,386.

135. Shipway AN, Katz E, Willner I: Molecular memory and processing devices in solution and on surfaces. Structure and Bonding 2001, 99: 237–281.

136. Nonvolatile molecular memory. Ames Research Center, National Aeronautics and Space Administration; 2008 [http://www.nasa.gov/centers/ames/research/technology-onepagers/nonvolatile_memory.html] []

137. Islam MS, Li Z, Chang SC, Ohlberg DAA, Stewart DR, Wang SY, Williams RS, Stanley R: Dramatically improved yields in molecular scale electronic devices using ultra-smooth platinum electrodes prepared by chemical mechanical polishing. In Proc 5th IEEE Conf Nanotech: July 11–15 2005; Nagoya. Volume 1. Piscataway: IEEE; 2005:80–83.

138. Oliveira ON Jr: Langmuir-Blodgett films - properties and possible applications. Braz J Phys 1992, 22(2):60–69.

139. Plafke J: Individual molecules: storage devices of the future. [http://www.extremetech.com/computing/146690-individual-molecules-storage-devices-of-the-future] []

140. Lee SH, Jung Y, Agarwal R: Highly scalable non-volatile and ultra-low-power phase-change nanowire memory. Nat Nanotech 2007, 2: 626–630.

141. Lieber CM: Nanoscale science and technology: building a big future from small things. MRS Bull 2003, 28(7):486–491.

142. Agarwal R, Lieber CM: Semiconductor nanowires: optics and optoelectronics. Appl Phys A Mater Sci Proc 2006, 85(3):209–215.

143. Reed M, Chen J, Rawlett AW, Price DW, Tour JM: Molecular random access memory cell. Appl Phys Lett 2001, 78(23):3735–3737.

144. Li C, Zhang D, Liu X, Han S, Tang T, Zhou C, Fan W, Koehne J, Han J, Meyyappan M, Rawlett AW, Price DW, Tour JM: Fabrication approach for molecular memory arrays. Appl Phys Lett 2003, 82(4):645–647.

145. Li C, Lei B, Fan W, Zhang D, Meyyappan M, Zhou C: Molecular memory based on nanowire–molecular wire heterostructures. J Nanosci Nanotechnol 2007, 7(1):138–150.

146. Atwood A: Future directions and challenges for ETox flash memory scaling. IEEE Trans Device Mater Reliab 2004, 4(3):301–305.

147. Govoreanu B, Brunco DP, Van Houdt J: Scaling down the interpoly dielectric for next generation Flash memory: challenges and opportunities. Solid State Electron 2005, 49(11):1841–1848.

148. Patel P: Computing News. Bendable memory made from nanowire transistors. MIT Technol Rev Rep 2010.

149. Li Q, Zhu X, Xiong HD, Koo SM, Ioannou DE, Kopanski JJ, Suehle JS, Richter CA: Silicon nanowire on oxide/nitride/oxide for memory application. Nanotechnology 2007, 18: 235204.

150. Sohn JI, Choi SS, Morris SM, Bendall JS, Coles HJ, Hong WK, Jo G, Lee T, Welland ME: Novel nonvolatile memory with multibit storage based on a ZnO nanowire transistor. Nano Lett 2010, 10(11):4316–4320.

151. Kang JW, Lee JH, Lee JH, Hwang HJ: A study on carbon nanotube bridge as a electromechanical memory device. Physica E 2005, 27(3):332–340.

152. Nantero, imec collaborate on development of carbon-nanotube-based memory. Nantero: Nantero Inc; 2012 [http://www.azom.com/news.aspx?newsID=34644#] []

153. Rueckes T, Kim K, Joselevich E, Tseng GY, Cheung CL, Lieber CM: Carbon nanotube-based nonvolatile random access memory for molecular computing. Science 2000, 289(5476):94–97.

154. Cui JB, Sordan R, Burghard M, Kern K: Carbon nanotube memory devices of high charge storage stability. Appl Phys Lett 2002, 81(17):3260–3262.

155. Vettiger P, Cross G, Despont M, Drechsler U, Dürig U, Gotsmann B, Häberle W, Lantz MA, Rothuizen HE, Stutz R, Binnig GK: The "millipede"—nanotechnology entering data storage. IEEE Trans Nanotechnol 2002, 1(1):39–55.

156. Walsh C: IBM›s Millipede. [http://www.ewp.rpi.edu/hartford/~walshc2/FWM/Research%20Project/Final/IBM%27s%20Millipede.pdf] []

157. Gotsmann B, Knoll AW, Pratt R, Frommer J, Hedrick JL, Duerig U: Designing polymers to enable nanoscale thermomechanical data storage. Adv Funct Mater 2010,

20(8):1276–1284.

158. Jones MR, Osberg KD, Macfarlane RJ, Langille MR, Mirkin CA: Templated techniques for the synthesis and assembly of plasmonic nanostructures. Chem Rev 2011, 111(6):3736–3827.

159. Hagen JA, Li W, Steckl AJ, Grote JG: Enhanced emission efficiency in organic light-emitting diodes using deoxyribonucleic acid complex as an electron blocking layer. Appl Phys Lett 2006, 88: 171109.

160. Yu Z, Li W, Hagen JA, Zhou Y, Klotzkin D, Grote JG, Steckl AJ: Photoluminescence and lasing from deoxyribonucleic acid (DNA) thin films doped with sulforhodamine. Appl Opt 2007, 46(9):1507–1513.

161. Hung YC, Hsu WT, Lin TY, Fruk L: Photoinduced write-once read-many-times memory device based on DNA biopolymer nanocomposite. Appl Phys Lett 2011, 99: 253301.

162. Monson CF, Woolley AT: DNA-templated construction of copper nanowires. Nano Lett 2003, 3(3):359–363.

163. Pashley RD, Lai SK: Flash memories: the best of two worlds. IEEE Spectrum 1989, 26: 30–33.

164. Pavan P, Bez R, Olivo P, Zanoni E: Flash memory cells-an overview. Proc IEEE 1997, 85(8):1248–1271.

165. Geller M, Marent A, Nowozin T, Bimberg D, Akcay N, Öncan N: A write time of 6 ns for quantum dot-based memory structures. Appl Phys Lett 2008, 92(9):092108.

166. Cui K, Ma W, Zhang Y, Huang J, Wei Y, Cao Y, Guo X, Li Q: 540-meV hole activation energy for GaSb/GaAs quantum dot memory structure using AlGaAs barrier. IEEE Elect Dev Lett 2013, 34(6):759–761.

167. Tiwari S, Rana F, Hanafi H, Hartstein A, Crabbe EF, Chan K: A silicon nanocrystals based memory. Appl Phys Lett 1996, 68(10):1377–1379.

168. Manna S, Aluguri R, Katiyar A, Das S, Laha A, Osten HJ, Ray SK: MBE-grown Si and Si1-xGex quantum dots embedded within epitaxial Gd2O3 on substrate for floating gate memory device. Nanotechnology 2013, 24: 505709.

169. Geller MP: Investigation of carrier dynamics in self-organized quantum dots for memory devices. PhD thesis. Technical University of Berlin; 2007.

170. Jung SM, Jang J, Cho W, Cho H, Jeong J, Chang Y, Kim J, Rah Y, Son Y, Park J, Song MS, Kim KH, Lim JS, Kim K: Three dimensionally stacked NAND Flash memory technology using stacking single crystal Si layers on ILD and TANOS structure for beyond 30 nm node. In Proc Intern Elect Devices Meet (IEDM): December 11–13 2006; San Francisco. Piscataway: IEEE; 2006.

171. Wong S, El-Gamal A, Griffin P, Nishi Y, Pease F, Plummer J: Monolithic 3D integrated circuits. In IEEE Inter Symp VLSI Technol Systems and Applications (VLSI-TSA): April 23–25 2007; Hsinchu. Piscataway: IEEE; 2007:1–4.

172. Vyyoda MA: 3-Dimensional Monolithic Nonvolatile Memories and the Future of Solid-State Data Storage. Milpitas: SanDisk Corporation; 2008.

173. Wager JF: Transparent electronics. Science 2003, 300(5623):1245–1246.

174. Kiriakidis G, Nathan A, Keszler DA, Fortunato E, Nomura K: E-MRS 2011 Fall Symposium I: Advances in Transparent Electronics, from Materials to Devices III. IOP Conf Ser Mater Sci Eng 2012, 34: 011001.

175. Seo JW, Park JW, Lim KS, Kang SJ, Hong YH, Yang JH, Fang L, Sang GY, Kim HK: Transparent flexible resistive random access memory fabricated at room temperature. Appl Phys Lett 2009, 95: 133508.

176. Transparent, flexible 3-D memory chips may be the next big thing in small memory devices. Featured Research. American Chemical Society 2012 [http://www.kurzweilai.net/transparent-flexible-3d-memory-chips-may-be-the-next-big-thing-in-small-memory-devices]

177. Zangeneh M, Joshi A: Performance and energy models for memristor-based 1T1R RRAM cell. In Proc. Great Lakes Symposium on Very Large Scale Integration (GLSVLSI): May 3–4 2012; Salt Lake City. Utah: ACM; 2012:9–14.

178. Zangeneh M, Joshi A: Design and optimization of nonvolatile multibit 1T1R resistive RAM. IEEE Trans VLSI System 2014, 22(8):1815–1822.

CHAPTER 6

ROOM-TEMPERATURE ELECTRICALLY PUMPED NEAR-INFRARED RANDOM LASING FROM HIGH-QUALITY M-PLANE ZNO-BASED METAL-INSULATOR-SEMICONDUCTOR DEVICES

Chao Chen, Ti Wang, Hao Wu, He Zheng , Jianbo Wang , Yang Xu and Chang Liu

Keywords: m-plane ZnO, MIS, Random lasing

BACKGROUND

Due to its wide direct bandgap and large exciton binding energy (60 meV), ZnO is an attractive material for optoelectronics, such as ultraviolet (UV) light-emitting diodes [1-3], photodetectors [4], and lasers [5]. Because of the lowest surface free energy of the wurtzite structure {0001} plane, ZnO thin films tend to grow along the <0001>direction in most cases. Note that the wurtzite structure is polarized along the c-axis caused by spontaneous and piezoelectric polarizations, resulting in the quantum-confined Stark effect and decrease of the internal quantum efficiency of light-emitting devices. Therefore, semi-polar and nonpolar structures are proposed to diminish or even eliminate the polarization effects, and the nonpolar m-plane is one of the common structures [6]. Unfortunately, due to the coexistence of minor domains with <0001>, <11-22>, or <10-13>growth directions in m-plane ZnO thin films [7-9], it is difficult to deposit high-quality and smooth m-plane ZnO thin films by several normal growth techniques, such as molecular beam epitaxy [7,8], metalorganic chemical vapor deposition [9], and pulsed laser deposition [10]. In addition, with different growth conditions, various striped features are usually presented on the surface of m-plane ZnO thin films deposited on foreign substrates [11,12]. These disadvantages limit the applicability of m-plane ZnO thin films. Furthermore, epitaxial growth of single-crystal ZnO thin films always needs a very high growth temperature. The higher growth temperature may cause defects such as oxygen vacancies in epitaxial films. Featured by layer-by-layer growth, atomic layer deposition (ALD) is one of the most promising techniques that possess several practical advantages including accurate thickness control, good reproducibility, excellent conformity, capability to produce sharp interfaces, and low growth temperature. Especially, among these advantages, low growth temperature is the most attractive one, which makes this technology possible on soft substrates. Although several works have presented epitaxial m-plane ZnO thin films on m-plane sapphire substrates by ALD, the optoelectronic applications of this material still remain an open topic [13,14].

Due to the potential applications in photonics, random lasing has received considerable attention, and ZnO is ideally suited for random lasing because of its high refractive index. Many researchers have already reported random lasing action in ZnO powder and polycrystalline thin films [15-18]. However, almost all of the devices exhibit UV random lasing, and it is seldom reported that a material possesses near-infrared (NIR) random lasing. With these investigations, it is undoubted that electrically pumped ZnO random lasing has promising applications.

Citation: Chen, C., Wang, T., Wu, H. et al. Room-temperature electrically pumped near-infrared random lasing from high-quality m-plane ZnO-based metal-insulator-semiconductor devices, doi:10.1186/s11671-015-0816-4.

Here, m-plane ZnO thin films have been deposited on m-plane sapphire substrates by ALD at 200°C. X-ray diffraction (XRD) and transmission electron microscopy (TEM) results show that high-crystalline-quality m-plane ZnO thin films can be fabricated at such a low growth temperature. In addition, a 50-nm Al_2O_3 thin film is deposited on the top of the m-plane ZnO thin films by ALD to fabricate metal-insulator-semiconductor (MIS) devices, and NIR random lasing is observed from the MIS devices.

METHODS

ZnO thin films were grown on m-plane sapphire substrates at 200°C by ALD (Beneq TSF-200, Beneq Oy, Espoo, Finland) with diethyl zinc (DEZn) and deionized water (H_2O) as precursors. The detailed experimental processes can be found in our previous work [19,20]. The thickness of the ZnO films was about 600 nm. Afterwards, a 50-nm Al_2O_3 thin film was deposited on the ZnO thin films at 200°C. Trimethyl aluminum and H_2O were used as the aluminum and oxygen sources, respectively [21]. A 50-nm-thick Au thin film was deposited on the Al_2O_3 thin film as the electrode, and In was used as the electrode for the ZnO thin film.

XRD (Bede D1, Bede Scientific Instruments, Durham, England, UK), TEM (JEOL JEM2010, JEOL Ltd., Akishima-shi, Tokyo, Japan), and high-resolution transmission electron microscopy (HRTEM; JEOL JEM2010FEF, JEOL Ltd., Akishima-shi, Tokyo, Japan) were used to evaluate the crystallization, microstructure, and interface properties. The surface morphology and roughness of the ZnO thin films were characterized by atomic force microscopy (AFM; SHIMADZU SPM-9500 J3, SHIMADZU Corporation, Nakagyo-ku, Kyoto, Japan). The chemical compositions and the defect states of the ZnO thin films are investigated by X-ray photoelectron spectroscopy (XPS; Thermo Scientific ESCALAB 250Xi, Thermo Fisher Scientific, Waltham, MA, USA). Photoluminescence (PL; HORIBA LabRAM HR800, HORIBA, Ltd., Minami-ku, Kyoto, Japan) measurements were conducted at room temperature to analyze the optical properties of the ZnO thin films in a wavelength range of 350 to 700 nm. Electroluminescence (EL) measurements were carried out at room temperature to assess the performance of the MIS devices.

RESULTS AND DISCUSSION

Figure 1a shows the XRD spectra of the ZnO thin films deposited on the m-plane sapphire substrates, in which the ZnO (10–10) peak is observed besides the sapphire (30–30) peak, clearly indicating that the ZnO thin films are predominantly grown along the [10] direction. The φ scans of the sapphire (10–12) peak and ZnO (10–13) peak are presented in Figure 1b. Both of them show a binary symmetry, revealing that the ZnO thin films are epitaxial on the sapphire substrates. An angle of 90° between ZnO (10–13) peaks and sapphire (10–12) peaks in φ scan spectra is noted. Therefore, it can be deduced that there exists a 90° rotation between the m-plane ZnO thin films and m-plane sapphire substrates. The in-plane orientations are ZnO [0001]//sapphire [1–210] and ZnO [−12-10]//sapphire [0001].

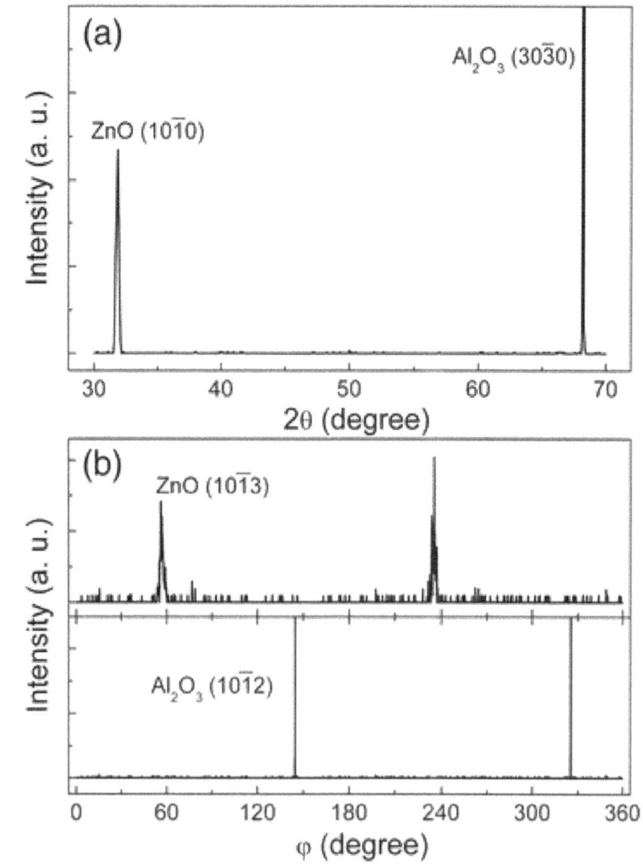

Figure 1 XRD scans of the ZnO thin films grown on m-plane sapphire. (a) XRD spectra of the ZnO thin films grown on m-plane sapphire substrates. (b) XRD φ scans of sapphire (10–12) and ZnO (10–13).

To further analyze the crystalline quality and the in-plane orientation, TEM studies were performed. As is shown in Figure 2, the ZnO thin film deposited by ALD has a very sharp interface. Figure 2b,c and the inset of Figure 2a show the selected area electron diffraction (SAED) patterns of the sapphire substrate, the ZnO thin film, and the interface region, respectively. The growth direction of the ZnO thin film is along [10] as presented in Figure 2b, in full agreement with the XRD result. The orientation relationships between the ZnO thin film and the sapphire substrate can be manifested as follows: ZnO [0001]//sapphire [1–210] and ZnO (−12-10)//sapphire (0006), which are demonstrated in the inset of Figure 2a. Furthermore, it is found that the ZnO thin film epitaxially grows on the m-plane sapphire substrate and has a high crystalline quality as demonstrated by the HRTEM image shown in Figure 3a. The HRTEM image is taken along ZnO [0001], and the epitaxial relationship between ZnO and sapphire is ZnO [10]//sapphire [10] and ZnO [−12-10]//sapphire [0001], consistent with the SAED results. The interface is observed to be atomically sharp. The mismatch-induced strains are usually relaxed through the generation of misfit dislocations (MDs) at the interface in heteroepitaxial structures. Figure 3b shows the magnified image of the white boxed area in Figure 2a. Figure 3b presents the inverse fast Fourier transform (IFFT)-filtered image of the red dashed area in Figure 3a by using ZnO (−12-10) and (1–210) reflections and sapphire (0006) and (000–6) reflections. The MDs are clearly visible in Figure 3b, and extra half-planes within the sapphire substrate are marked by the arrows. In each dashed boxed area, an extra half-plane is showed and the lattice fringes of ZnO and sapphire present 4:3. Considering the interplanar distances of ZnO (−12-10) (0.1625 nm) and sapphire (0006) (0.2165 nm), the lattice misfit ( ) along the ZnO [−12-10] direction is calculated to be 24.94%. However, with four ZnO (−12-10) planes and three sapphire (0006) planes, it can be matched nearly perfectly with a mismatch only about 0.08%. In this way, the residual strain along the ZnO [−12-10] direction is released. Moreover, the equilibrium spacing (*D*) is calculated to be 0.6516 nm via the following formula:$D=|b|/$, where $|b|$ is the magnitude of the Burgers vector component parallel to the interface, i.e., 0.1625 nm (=d-12-10ZnO) (=dZnO-12-10). The calculated equilibrium spacing of the MDs is close to the observed average spacing between the MDs marked by arrows in Figure 3b (i.e., 0.65 nm).

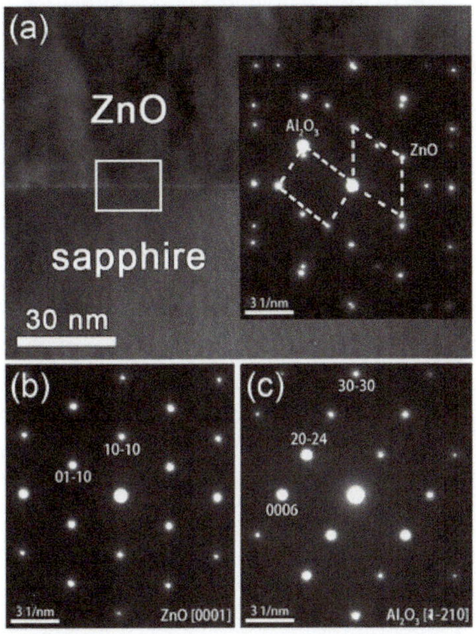

Figure 2 Bright-field images and SAED patterns. (a) Bright-field images of the m-plane ZnO on m-plane sapphire. The inset of (a) shows the SAED pattern at the interface. (b) SAED pattern obtained from the ZnO thin film. (c) SAED pattern obtained from the m-plane sapphire substrate.

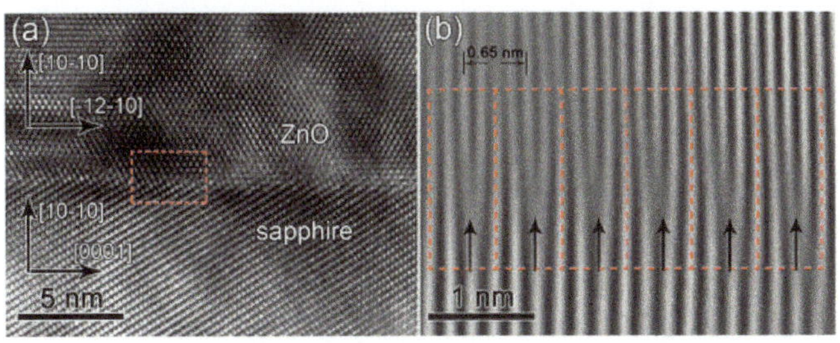

Figure 3 Cross-sectional HRTEM and IFFT-filtered images. (a) Cross-sectional HRTEM image of the interface (the magnified image of the boxed area in Figure 2a). **(b)** IFFT-filtered image of the red dashed box of **(a)** using ZnO (−12-10) and (1–210) reflections and sapphire (0006) and (000–6) reflections.

In order to analyze the chemical compositions and the defect states of the ZnO thin films, XPS measurement was performed. From the statistical results of XPS, the atom ratio of Zn:O is about 1.06:1, in which it can be observed that the ZnO thin films are a little Zn-rich. Figure 4a shows the O 1*s* XPS spectra of the ZnO films. The binding energies of the XPS spectra have been calibrated by taking the carbon C 1*s* peak (284.5 eV) as a reference. The O 1*s* peak is deconvoluted into three peaks at 529.8, 530.9, and 531.9 eV by using Gaussian fitting. The peaks centered at 529.8 and 530.9 eV are assigned to oxygen atoms in the oxide lattice without and with oxygen vacancies, respectively. The peak at 531.9 eV is usually attributed to chemisorbed oxygen on the surface of ZnO thin films, such as carbonyl and hydroxyl groups. The relative integrated intensity percentages of the different oxygen species calculated from fitting peaks are used to analyze the quantities of the oxygen lattice, oxygen vacancy, and hydroxyl group [22,23]. The percentages of peaks centered at 529.8, 530.9, and 531.9 eV are 69.53%, 16.68%, and 13.8%, respectively. Compared with the XPS results from other works [23,24], our ZnO films have a high percentage of the oxygen lattice-related peak and a small ratio of the oxygen vacancy-related peak. Therefore, from the XPS results, it can be concluded that there are only a few oxygen vacancies in the ZnO thin films deposited at 200°C by ALD.

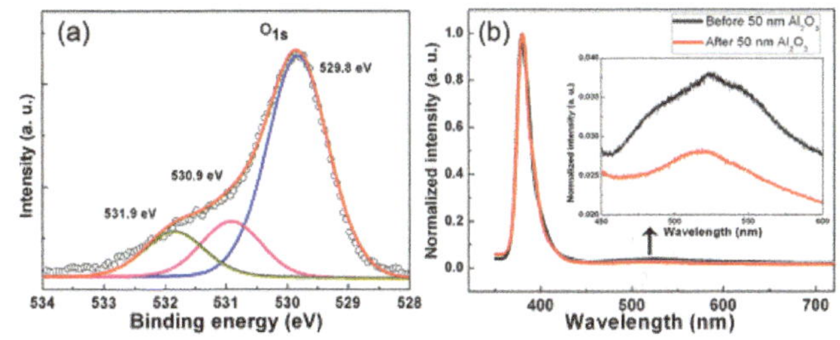

Figure 4 O 1 *s*XPS spectra and PL spectra of the ZnO thin films. (a) O 1*s* XPS spectra of the ZnO thin films. Dots are the experimental data and lines are the fitting results. **(b)**The PL spectra of the ZnO thin films before and after depositing a 50-nm Al_2O_3 cap layer. The inset of **(b)** is the magnified area of the PL spectra from 450 to 600 nm.

To fabricate MIS devices, a 50-nm Al_2O_3 thin film was deposited on the m-plane ZnO thin film by ALD. Figure 4b shows the PL spectra of the ZnO thin films with and without the Al_2O_3 thin film. The intensities of the peaks have been normalized. Both spectra display a dominant sharp near-band-edge (NBE) emission and a weak deep-level emission around 520 nm. The deep-level emission has been reported to be caused by oxygen vacancies. Hence, it indicates that few oxygen vacancies exist in the ZnO thin films, which is consistent with the XPS results. Note that they show similar UV emissions though the deep-level emission has decreased after depositing the Al_2O_3 thin film as shown in the inset of Figure 4b. The only difference between the two samples is with or without the 50-nm Al_2O_3 thin film. Therefore, the decrease of the deep-level emission intensity can be attributed to the decrease of oxygen vacancies at the interface between ZnO and Al_2O_3.

The inset of Figure 5a shows the schematic diagram of the m-plane ZnO-based MIS devices. Au and In were used as electrodes on the Al_2O_3 thin films and the ZnO thin films, respectively. Figure 5a presents the EL spectra of the MIS devices under various forward biases. NIR emission peaks are observed, whereas no UV emission peaks are found in the EL spectra. As is known, the MIS structures can increase carrier density in the radiative recombination region by confining carriers in the semiconductor-insulator interface. Consequently, the main recombination occurred at the interface in MIS devices. Wang et al. have observed NIR emissions from the PL spectra of ZnO microspheres and attributed the NIR emissions to oxygen interstitials as deep acceptors [25]. With the 50-nm Al_2O_3 thin film, the amount of oxygen vacancies reduces at the interface demonstrated by PL measurements. Therefore, due to the 50-nm Al_2O_3 cap layer, the surface of the m-plane ZnO thin film has been changed and oxygen interstitials emerged at the interface. As a result, NIR emissions are generated due to the radiative recombinations between shallowly trapped electrons and the deeply trapped holes at oxygen interstitials [25]. In addition, it is interesting that NIR random lasing has been observed above the forward bias of 12 V, which appears as randomly distributed and intensive sharp peaks in the EL spectra. When the bias increases, more electrons accumulate at the interface between ZnO and Al_2O_3, which leads to more transitions directly from the conduction band edge to the oxygen interstitial level. As a consequence, photons with higher energy are emitted from the luminescent region, and the EL emission center is blueshifting with the increase of the bias. The generating mechanism of NIR random lasing needs to be further systematically studied. However, a similar random lasing has been observed in some works, and they deduced that the random lasing phenomenon was caused by light scattering in the luminescent region. The origin of the scattering can be attributed to total internal reflection at grain boundaries [5], fluctuations of refractive indexes at the interface of heterostructures [26], refractions among the grains [15], etc. To confirm the possibility of this mechanism in our system, the surface morphology of the ZnO thin films was characterized by AFM measurement.

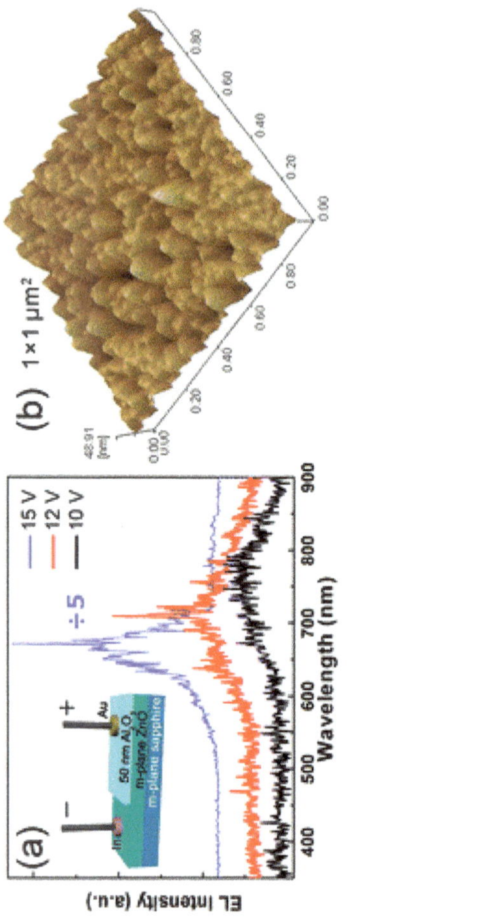

Figure 5 EL spectra of the MIS devices and AFM image of the ZnO thin film. **(a)** EL spectra of the MIS devices. The inset of **(a)** is the schematic diagram of the MIS devices. **(b)** AFM image of the ZnO thin film.

As is shown in Figure 5b, the surface is smooth and uniform, with a root-mean-square roughness of only 5.7 nm. But it presents that there are many grains at the surface and the size of the grains is generally below 100 nm. These grains and grain boundaries can be the reasons for light scattering. As a consequence, through multiple scattering, light may return to the scatterer from which it has been scattered before and results in the self-formation of a closed loop path. The loop path with coherent feedback of the scattering light can achieve optical gain, and thus, it serves as a resonator for that light frequency. Distinguished from the other light-emitting devices, radiative transitions in the MIS devices are confined at the interface between the semiconductor and insulator. Such concentrated luminescence can greatly increase the chance to form closed loop paths. Due to the randomness of the formation of closed loop paths, their respective resonance frequencies result in discrete sharp random peaks in the emission spectrum.

CONCLUSIONS

In summary, high-crystalline-quality m-plane ZnO thin films have been deposited on m-plane sapphire substrates with a low growth temperature. XRD results denote that the in-plane orientations are ZnO [0001]//sapphire [1–210] and ZnO [−12-10]//sapphire [0001], which can also be demonstrated by SAED images. It can be speculated from the HRTEM results that the residual strain along the ZnO [−12-10] direction was nearly zero. Moreover, to fabricate MIS devices, a 50-nm Al_2O_3 thin film was deposited on the m-plane ZnO thin films by ALD, and it was interesting to observe NIR random lasing from the MIS devices.

ACKNOWLEDGEMENTS

This work is supported by the NSFC under Grant Nos. 10904116, 11074192, 11175135, and J1210061, the Hubei Provincial Program of Science and Technology (2012BAA19009), the foundation from CETC No. 46 Research Institute, and the Fundamental Research Funds for the Central Universities (201220202021 5 and 201220202021 0).

Competing interests

The authors declare that they have no competing interests.

Authors' contributions

CC and TW carried out the experiments and drafted the manuscript. HW conceived the study and participated in its design. HZ and JBW participated in the design of the study and performed the analysis. YX participated in the measurements. CL supervised the overall study. All authors read and approved the final manuscript.

REFERENCES

1. Yang Q, Wang WH, Xu S, Wang ZL. Enhancing light emission of ZnO microwire-based diodes by piezo-phototronic effect. Nano Lett. 2011;11:4012–7.
2. Lupan O, Pauporte T, Viana B. Low-voltage UV-electroluminescence from ZnO-nanowire array/p-GaN light-emitting diodes. Adv Mater. 2010;22:3298–302.
3. Wang T, Wu H, Wang Z, Chen C, Liu C. Ultralow emission

threshold light-emitting diode of nanocrystalline ZnO/p-GaN heterojunction. Electron Device Lett, IEEE. 2012;33:1030–2.

4. Law JBK, Thong JTL. Simple fabrication of a ZnO nanowire photodetector with a fast photoresponse time. Appl Phys Lett. 2006;88:133114.
5. Zhu H, Shan CX, Yao B, Li BH, Zhang JY, Zhang ZZ. Ultralow-threshold laser realized in zinc oxide. Adv Mater. 2009;21:1613–7.
6. Waltereit P, Brandt O, Trampert A, Grahn HT, Menniger J, Ramsteiner M, et al. Nitride semiconductors free of electrostatic fields for efficient white light-emitting diodes. Nature. 2000;406:865–8.
7. Lee JW, Kim JH, Han SK, Hong SK, Lee JY, Hong SI, et al. Interface and defect structures in ZnO films on m-plane sapphire substrates. J Cryst Growth. 2010;312:238–44.
8. Kim JH, Han SK, Hong SI, Hong SK, Lee JW, Lee JY, et al. Growth and structural properties of ZnO films on (10–10) m-plane sapphire substrates by plasma-assisted molecular beam epitaxy. J Vac Sci Technol, B: Microelectron Nanometer Struct. 2009;27:1625–30.
9. Moriyama T, Fujita S. Growth behavior of nonpolar ZnO on M-plane and R-plane sapphire by metalorganic vapor phase epitaxy. Jpn J Appl Phys. 2005;44:7919–21.
10. Ho YT, Wang WL, Peng CY, Chen WC, Liang MH, Tian JS, et al. Substrate engineering of $LaAlO_3$ for non-polar ZnO growth. Thin Solid Films. 2010;518:2988–91.
11. Chou MMC, Chang L, Chung HY, Huang TH, Wu JJ, Chen CW. Growth and characterization of nonpolar ZnO (10–10) epitaxial film on γ-$LiAlO_2$ substrate by chemical vapor deposition. J Cryst Growth. 2007;308:412–6.
12. Matsui H, Tabata H. Correlation of self-organized surface nanostructures and anisotropic electron transport in nonpolar ZnO (10–10) homoepitaxy. J Appl Phys. 2006;99:124307..
13. Ku CS, Lee HY, Huang JM, Lin CM. Epitaxial growth of m-plane ZnO thin films on (10–10) sapphire substrate by atomic layer deposition with interrupted flow. Cryst Growth Des. 2010;10:1460–3.
14. Yang S, Lin BH, Kuo CC, Hsu HC, Liu WR, Eriksson MO, et al. Improvement of crystalline and photoluminescence of atomic layer deposited m-plane ZnO epitaxial films by annealing treatment. Cryst Growth Des. 2012;12:4745–51.
15. Ma XY, Chen PL, Li DS, Zhang YY, Yang DR. Electrically pumped ZnO film ultraviolet random lasers on silicon substrate. Appl Phys Lett. 2007;91:251109.
16. Liang HK, Yu SF, Yang HY. Directional and controllable edge-emitting ZnO ultraviolet random laser diodes. Appl Phys Lett. 2010;96:101116.
17. Cao H, Zhao YG, Ho ST, Seelig EW, Wang QH, Chang RPH. Random laser action in semiconductor powder. Phys Rev Lett. 1999;82:2278–81.
18. Yu SF, Yuen C, Lau SP, Park WI, Yi GC. Random laser action in ZnO nanorod arrays embedded in ZnO epilayers. Appl Phys Lett. 2004;84:3241.
19. Wang T, Wu H, Chen C, Liu C. Growth, optical, and electrical properties of nonpolar m-plane ZnO on p-Si substrates with Al2O3 buffer layers. Appl Phys Lett. 2012;100:011901.
20. Wang T, Wu H, Zheng H, Wang J, Wang Z, Chen C, et al. Nonpolar light emitting diodes of m-plane ZnO on c-plane GaN with the Al2O3 interlayer. Appl Phys Lett. 2013;102:141912.
21. Zhang G, Wu H, Chen C, Wang T, Wang P, Mai L, et al. Transparent capacitors based on nanolaminate Al2O3/TiO2/Al2O3 with H2O and O3 as oxidizers. Appl Phys Lett. 2014;104:163503.
22. Kim Y-H, Heo J-S, Kim T-H, Park S, Yoon M-H, Kim J, et al. Flexible metal-oxide devices made by room-temperature photochemical activation of sol–gel films. Nature. 2012;489:128–32.
23. Zhu Q, Xie C, Li H, Yang C, Zhang S, Zeng D. Selectively enhanced UV and NIR photoluminescence from a degenerate ZnO nanorod array film. J Mater Chem C. 2014;2:4566–80.
24. Kim YJ, Yang BS, Oh S, Han SJ, Lee HW, Heo J, et al. Photobias instability of high performance solution processed amorphous zinc tin oxide transistors. ACS Appl Mat Interfaces. 2013;5:3255–61.
25. Wang MS, Zhou YJ, Zhang YP, Kim EJ, Hahn SH, Seong SG. Near-infrared photoluminescence from ZnO. Appl Phys Lett. 2012;100:101906.
26. Tian Y, Ma XY, Jin L, Yang D. Electrically pumped ultraviolet random lasing from ZnO films: compensation between optical gain and light scattering. Appl Phys Lett. 2010;97:251115.

CHAPTER 7

AN IONIC BOTTLE FOR HIGH-SPEED, LONG-RETENTION MEMRISTIVE DEVICES

Dmitri B. Strukov, R. Stanley Williams

A promising device for 'universal' memory and storage is the bipolar resistance switch, of which there are now many known examples [3, 8]. Recently, we demonstrated [5] that such switches were physical examples of the memristor [1] or memristive device [2], which were introduced as a new fundamental passive two-terminal circuit element. An essential issue regarding these devices is the "pinched hysteresis loop", i.e. the observation that the current/voltage (I–V) characteristic always passes through the origin for any periodic applied excitation. This property means that the device does not store charge or energy, but it does store information; the value of the resistance of the device at time t is a function of the total integrated current (equivalently, the total charge) that has passed through the device. Thus, an ideal memristor should be nonvolatile; it should be able to hold its state indefinitely. However, since the mechanism for real bipolar switches often involves the drift of charged species under the influence of an applied electric field [5, 9], the actual state lifetime of a device may be limited by the diffusion of the mobile species back to their original configuration. For example, the complete and partial OFF states for a thin-film memristive device simulated in Ref. [7] are not thermodynamically stable. After the device is switched to the OFF state and the external bias is removed the mobile ions will slowly diffuse toward the middle of the film, turning the device back to the ON state. The rate of such relaxation can be determined by plotting the zero-bias conductance as a function of time (black line in Fig. 1). The absolute value of the relaxation rate, which depends on device geometry, mobility μI and diffusivity DI of the ions in the film, does not conveniently reflect the retention characteristics of the device. Instead, it is more useful to consider the ratio of the relaxation rate, equivalent to the retention time τ store, with respect to the time required to turn the device from the OFF state to the ON state by applying a constant external bias v (of the polarity opposite to that used to turn the device to the OFF state), i.e. the time τ write to write the device. For the model considered in Ref. [6], a bias v applied to an OFF state device first turns the device to the ON state and then to a second OFF state (i.e., characterized by the reversed ionic concentration profile), and τ write is determined by the peak position of zero-bias conductance (Fig. 1). When only linear ionic drift is considered, the ratio of τ store/τ write is not large (generally less than 1000) and clearly proportional to v/v_0 [7] (here $v_0 = e/k_B T$ is the characteristic thermal voltage, where e, k_B, T are the electron charge, the Boltzmann constant and the temperature, respectively). The retention characteristics could be dramatically improved by choosing an appropriate mobile species and host material (e.g. positively charged oxygen vacancies that act as dopants in TiO_2). Because of internal Joule heating during switching and the exponential dependence of the drift velocity at very high electric fields [7], it is possible to obtain a large dynamic ratio between switching time and state retention time (as many as 15 orders of magnitude). However, this requires the existence of particular and perhaps rare materials properties.

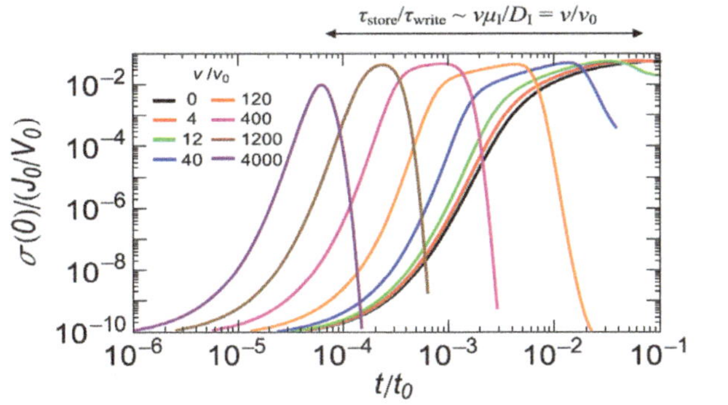

Figure. 1 Zero-bias conductance evolution as a function of time for several external biases calculated using the model considered in Ref. [6]. Here all axes are dimensionless with $t0, J_0$, and v_0 constants defined similarly to Ref. [6]

Citation: Dmitri B. Strukov, R. Stanley Williams, An ionic bottle for high-speed, long-retention memristive devices, DOI 10.1007/s00339-011-6321-4.

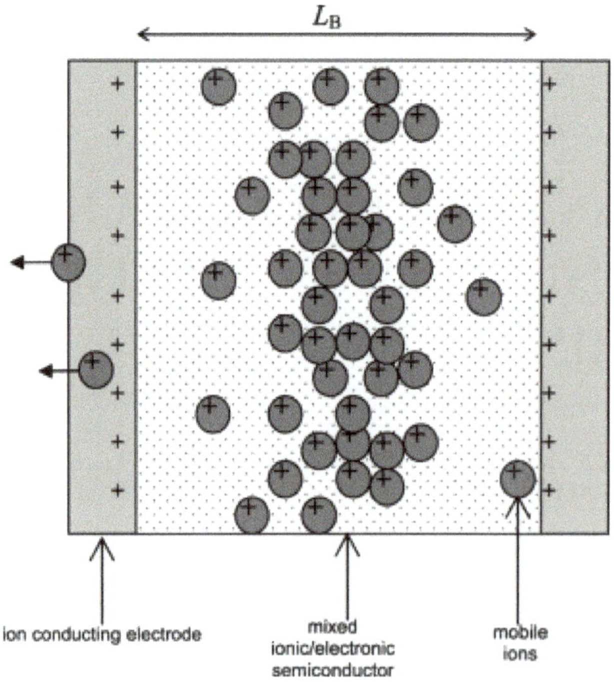

Fig. 2 "Ionic bottle" effect

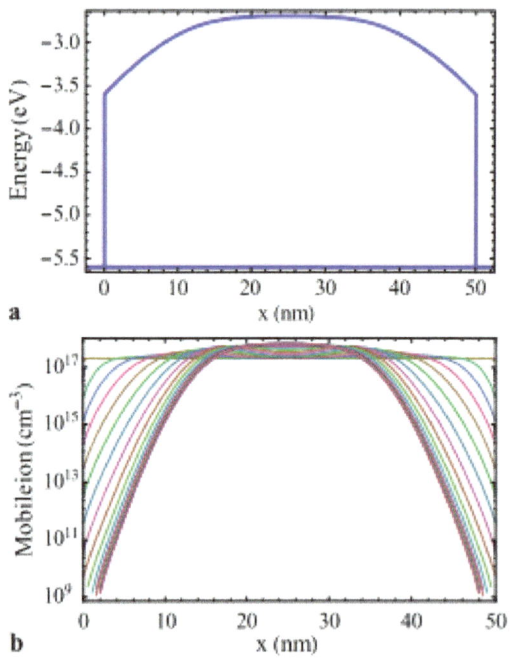

Figure. 3 Equilibrium mobile-ion concentration profile based on the self-consistent simulation model described in Ref. [6]. Panel (a) shows the bending of the bottom of the conduction band with energy levels corresponding to that of rutile TiO_2, while (b) shows the relaxation of the mobile ions starting from a uniform concentration $n(x) = 2\times10^{17}$. The background acceptor doping is 3×10^{18} cm^{-3}, while the energy levels of the donors and acceptors are assumed to be equal to the bottom of the conduction band and the top of the valence band, respectively

In this paper, we describe a more general structure, the ion bottle, which has the potential to realize bipolar memristance with low power dissipation, high switching speed and long state retention time that depends only weakly on materials properties. Consider a semiconductor film of thickness L_B sandwiched between two conductive electrodes and containing a significant concentration of charged mobile species, e.g. positive O vacancies in TiO_2 (Fig. 2). In general, we are interested in keeping the ions inside the semiconductor film even if the electrodes are permeable (e.g. metallic oxides such as RuO_2 or IrO_2) and the concentration of mobile dopants outside the device is negligible. To achieve this goal, assume that the bulk Fermi level of the semiconductor layer is lower than that of the electrodes by the value U. This could be due to additional doping of the semiconductor with immobile acceptors (e.g. C or N in TiO_2) or to the intrinsic alignment between the electrochemical potentials of the semiconductor (e.g. WO_3) and the metallic electrodes. In this ionic bottle, the electrodes are positively charged and repel the positively charged mobile ions toward the middle of the device. The mobile ionic concentration profile at equilibrium resembles an upside down U. (Alternatively, for negatively charged acceptor-like mobile dopants, the electrodes could be negatively charged if the Fermi energy of the semiconductor is above that of the electrode metal.) Figure 3 shows self-consistent numerical simulation results for the steady-state solution of the Boltzmann–Poisson equations assuming Fermi–Dirac statistics for the electron distribution inside semiconductor layer [6]. Neglecting the space charge effects of the mobile dopants (which is valid if the background immobile acceptor doping is much larger), the dependence of the mobile-ion concentration n(x) on the potential ϕ near the electrode is given by the Poisson–Boltzmann equation:

$$n(x) \approx N_0 \exp\left[e\varphi(x)/(k_B T)\right]. \qquad (1)$$

where N_0 is a constant that is determined by the normalization condition $\int_0^L n(x)\,dx = N_{\text{total}}$. When the voltage across the device is zero, the concentrations of mobile dopants near the interfaces are close to N_0. Assuming that the concentration of mobile ions outside the metallic electrodes is negligibly small, the rate of escape of ions from the volume is proportional to the ion diffusion current

$$J_{\text{diff}} \approx eD_M N_0/a, \qquad (2)$$

where a is the lattice constant and D_M is the diffusion constant of the ions in the electrode material. When an external positive voltage V is applied to the right electrode of the device, the mobile-ion concentration near the left electrode/semiconductor interface increases exponentially. The ratio of ion escape rates from the device when the applied voltage changes from zero to V is

$$R = J_{\text{diff}}(v=0)/J_{\text{diff}}(v=V) = \exp\left[eV/(2k_B T)\right]. \qquad (3)$$

For T = 300 K, R can be as high as 1017 at V = 2 V and 109 at 1 V, and is independent of the kinetic properties of the device. This exponential modulation of the mobile-ion concentration near the electrodes by an external voltage bias, and the resulting exponential change of the rate of inward or outward ionic currents is the main result of this paper. We now describe an example of a memristive device based on this ionic bottle effect (Fig. 4). The device consists of two semiconductor films connected by a permeable metal electrode and confined by two impermeable outer metallic electrodes (e.g. P_t). The right semiconductor film of the device is assumed to be heavily doped with immobile acceptors with a uniform concentration $N_A^{(B)}$, which is much larger than that of the mobile ions so that the electrical resistance of this film is independent of small changes in the mobile dopant concentration. The left film of the device is also doped with immobile acceptors, but the smaller concentration $N_A^{(D)}$ is such that the presence of mobile dopants significantly changes the resistance of the device. Here, we assume that diffusive transport of electrons determines the resistance of the semiconductor films [6]. When no bias is applied to the device, the mobile ions are effectively confined in both layers and only redistribute between the two reservoirs with a rate determined by the concentration difference across the middle electrode, i.e. described by (2). Application of a voltage bias forces exponential changes in the concentrations near the two interfaces of the permeable electrode, and hence a dramatically faster

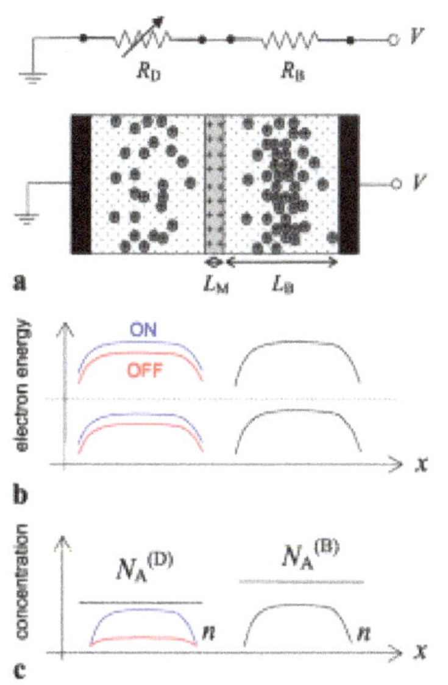

Figure. 4 An example of the memristive device: (a) cartoon of the device and equivalent circuit, (b) band diagram, and (c) mobile-ion concentration profile (schematically)

redistribution of ions with the direction depending on the bias polarity. Therefore, (3) effectively describes the ratio of state retention time to the switching time of the device. We can estimate practical values of the parameters required to operate the device, and thus determine the plausible energy and speed performance. First, to suppress the concentration of mobile ions at the interface, a large band bending (built-in voltage) of the order of 1 eV is required. The depletion width near the interface should be smaller than or comparable to the semiconductor film thickness. For the system shown in Fig. 3, the film thickness and doping should satisfy the condition $L_B \approx (\varepsilon\varepsilon_0 U k_B T/N_A^{(B)})^{1/2}/q$. Alternatively, $N_A^{(B)}$ could be replaced with $N_D \times \exp[\Delta U_{\text{gap}}/(k_B T)]$, where N_D is the effective density of states at the appropriate edge of the band and ΔU_{gap} is the band gap of the semiconductor if no background doping is required to achieve sufficient band bending. For example, for L = 10 nm, doping of the order $N_A^{(B)} \approx 5 \times 10^{18}$ cm^{-3} is needed (for a dielectric constant of the film ε = 10). For the two-terminal configuration described above, the metallic membrane is galvanically connected via the resistance of the right semiconductor film R_B to the right electrode. When a potential is applied across the full device, a portion of the voltage drops across the right hand side. Since R_B is much smaller than R_D (Fig. 4), this does not affect the on/off ratio of the device significantly. This could be achieved with nonlinear I –V characteristics, i.e. if at low bias $R_B \approx R_D$ while at high biases R_B R_D. Our previous work shows that in fact such nonlinearity might be achieved in the considered device structure due to the induced potential barrier for the electrons and/or holes caused by the dopant ion distribution [6]. On the other hand, there is no such limitation for a three-terminal configuration, for which an additional terminal is connected to the middle electrode, similar to the structure of a bipolar transistor. The threeterminal devices would be practical for applications that do not have a strict area density requirement and could accommodate more wires, such as an electrochromic device for a display application. The three processes involved in mobile-ion redistribution when a bias voltage is applied are (1) establishing a new mobile dopant ion concentration at the permeable electrode interface, (2) diffusion of these ions through the electrode (the potential difference across the electrode is zero since it is metallic), and (3) redistribution of the ions in the volume of the target film. Since (2) should be the rate-limiting step, we can neglect (1) and (3) in the switching time estimate and thus $t_{write} \approx L_M^2/D_M$. Obviously, the switching time will decrease by reducing the thickness of the permeable electrode, which should be thicker than the Thomas– Fermi screening length. For example, for few-nanometer membranes and $D_M = 10^{-5}\ cm^2 V^{-1}$, the switching time could be less than 100 ns. Long state retention time and fast switching speed may be combined with small switching energy and currents. For $R_D = 1\ G\Omega, V = 5\ V$ and t_{write} = 100 ns, the energy required to switch a bit should be less than 1 f_J, which is comparable with the best end of the roadmap semiconductor memories [4]. In summary, we proposed a device structure that can potentially combine high-speed switching 1015. Moreover, the device operation is based on a linear mode for the ionic drift, which at the same time allows for high endurance because of the low energy (and thus Joule heating) required for device operation. In principle, the device can function with nanocrystalline or amorphous materials and therefore be compatible with backend processes enabling cost efficient multilevel integration.

ACKNOWLEDGEMENT

Work at UCSB is funded by NSF grant CCF- 1028336.

REFERENCES

1. L.O. Chua, Memristor – the missing circuit element. IEEE Trans. Circuit Theory 18, 507–519 (1971)
2. L.O. Chua, S.M. Kang, Memristive devices and systems. Proc. IEEE, 64(2), 209–223 (1976)
3. G. Dearnaley, A.M. Stoneham, D.V. Morgan, Electrical phenomena in amorphous oxide films. Rep. Prog. Phys. 33, 1129–1192 (1970)
4. International Technology Roadmap for Semiconductors 2009 Edition, available online at http://ww.itrs.net
5. D.B. Strukov, G. Snider, D. Stewart, R.S. Williams, The missing memristor found. Nature 453, 80–83 (2008)
6. D.B. Strukov, J.L. Borghetti, R.S. Williams, Coupled ionic and electronic transport model of thin-film semiconductor memristive behavior. Small 5(9), 1058–1063 (2009)
7. D.B. Strukov, R.S. Williams, Exponential ionic drift: Fast switching and low volatility of thin film memristors. Appl. Phys. A, Mater. Sci. Process. 94(3), 515–519 (2009)
8. R. Waser, M. Aono, Nanoionics-based resistive switching memories. Nat. Mater. 6, 833–840 (2007)
9. J.J. Yang, M.D. Pickett, X. Li, D.A.A. Ohlberg, D.R. Stewart, R.S. Williams, Memristive switching mechanism for metal/oxide/metal nanodevices. Nat. Nanotechnol. 3, 429–433 (2009)

CHAPTER 8

AN ANALYTICAL STUDY ON BISTABILITY OF FABRY PEROT SEMICONDUCTOR OPTICAL AMPLIFIERS

Gang WANG, Shuqiang CHEN, and Huajun YANG

Keywords: Bistability, Fabry-Perot resonator, semiconductor optical amplifiers

INTRODUCTION

Optical bistability (OB) has been found to be a beneficial technology for designing optical sensors [1–2]. It is well known that Fabry Perot semiconductor optical amplifiers (FPSOAs) can operate in a bistable regime with proper parameters. Therefore, further studies on bistability of FPSOAs are worthwhile from a practical point of view. In [3], the authors presented a commonly-used model for investigating the bistability in SOAs. By use of this model, the static and dynamic bistable characteristics in different types of SOAs have been investigated substantially [3–8]. However, these studies have been performed either numerically or experimentally. So far, few analytical results on the bistability in SOAs have been presented. In this paper, based on Adams's model [3], the necessary condition for bistability in SOAs is deduced. Moreover, analytical expressions of the switch-up power, switch-down power of the incident beam, and the width of hysteresis loop in SOAs are derived. Numerical calculations are carried out to verify the analytical results. The remainder of this paper is organized as follows. Section 2 gives detailed derivations of the necessary condition for bistability and the expression of hysteresis loop width in an FPSOA. Section 3 affords the numerical experiments to verify the correctness of the analytical results. Discussion and conclusions are provided in Section 4.

Citation: Gang WANG, Shuqiang CHEN, and Huajun YANG, "An Analytical Study on Bistability of Fabry Perot Semiconductor Optical Amplifiers," Photonic Sensors, DOI: 10.1007/s13320-016-0332-x.

THEORETICAL ANALYSIS

Considering an SOA, it is assumed that in the cavity the optical intensity is uniform and the spontaneous emission is neglected. In the following, the analysis is carried out according to the relationships derived by Adams [3], in which the input intensity I_{in}, the output intensity I_{out}, and the z-independent intracavity intensity I_{av} could be expressed as follows.

$$\frac{I_{out}}{I_s} = \frac{I_{in}}{I_s} \cdot \frac{(1-R_1)(1-R_2)e^{gL}}{(1-\sqrt{R_1R_2}e^{gL})^2 + 4\sqrt{R_1R_2}e^{gL}\sin^2\phi} \tag{1}$$

$$\frac{I_{av}}{I_s} = \frac{I_{out}}{I_s} \cdot \frac{(1-e^{-gL})(1+R_2e^{gL})}{(1-R_2)gL} \tag{2}$$

where R_1 and R_2 are the reflectivities of the two cavity mirrors, g is the net gain per unit length, ϕ is the single-pass phase change, and L is the cavity length. In the steady state, g and ϕ can be expressed as

$$g = \frac{\Gamma g_0}{1 + I_{av}/I_s} - \alpha \tag{3}$$

$$\phi = \phi_0 + \frac{g_0 L b}{2}\left(\frac{I_{av}/I_s}{1 + I_{av}/I_s}\right) \tag{4}$$

$$g_0 = a(n - n_0) \tag{5}$$

where Γ is the optical confinement factor, a is the effective loss coefficient, g0 is the unsaturated material gain, ϕ0 is the initial phase detuning, I_s is the saturation optical intensity, b is the linewidth enhancement factor, n is the carrier density, a is the gain coefficient, and n0 is the carrier density at transparency. From (1) to (5), the input-output (IO) hysteresis loop can be observed by numerical simulation under certain parameter values [3]. In this paper, we analytically calculate the bistability in SOAs. From the numerical results and the definition of the bistability, it is obvious that in the bistable operation region, for a given input intensity I_{in}, there are three corresponding solutions of the output intensity Iout to (1)–(5). Namely, one is the unstable-state solution, and the other two are the stable-state solutions. Equation (2) shows that I_{out} is a single valued function of I_{av}. Thereby, it can be inferred that in the region where the hysteresis loop appears, there exist three values of I_{av} corresponding to one value of I_{in}. To ensure this feature, the function $I_{in}(I_{av})$ should have two extreme values in this region. Hence, from the mathematical point of view, the necessary condition for bistability of SOA is that the first-order derivative of the function $I_{in}(I_{av})$ should have two zero points. From (1) and (2), one can obtain the relationship between I_{in} and I_{aV}

$$\frac{I_{in}}{I_s} = \frac{(1-\sqrt{R_1R_2}e^{gL})^2 + 4\sqrt{R_1R_2}e^{gL}\sin^2\phi}{(1-R_1)(e^{gL}-1)(1+R_2e^{gL})}\cdot gL\cdot\frac{I_{av}}{I_s}. \qquad (6)$$

For convenience, let x = I_{av}/Is, y = I_{in}/Is. Then (6) can be rewritten as

$$y = \frac{\ln(G)\cdot[(1-\sqrt{R_1R_2}G)^2 + 4\sqrt{R_1R_2}G\sin^2\phi]}{(1-R_1)(G-1)(1+R_2G)}\cdot x \qquad (7)$$

where G=exp(gL) represents the single pass gain in the cavity. Clearly, (7) owns a complicated form, which makes it difficult to obtain the zero point of the first-order derivative of y. In order to theoretically find the necessary condition for the occurrence of hysteresis loop of SOAs and further achieve analytical results on the bistability, some approximations are demanded to simplify the form of (7). Naturally, the right hand of (7) can be expanded into Taylor series in terms of x. Since we try to examine whether the first-order derivative of y has two zero points, it is reasonable that three-order approximation of the Taylor series of y is adopted. Thus y can be rewritten as follows:

$$y \approx F_0 + F_1\cdot(x-x_0) + \frac{1}{2!}F_2\cdot(x-x_0)^2 + \frac{1}{3!}F_3\cdot(x-x_0)^3 \qquad (8)$$

where $F_0=y(x_0)$ and $F_i = \frac{d^iy}{dx^i}\big|_{x=x_0}\ (1\le i\le 3)$ is the ith order derivative of y at the reference point x=x0. In the following, (8) is a start point for further deduction and analysis. After straight and complex computation, the coefficients in (8) can be derived in the following form:

$$\frac{dy}{dx} = \frac{dT_1}{dx}x + T_1 + \frac{dT_2}{dx}\cdot\sin^2\phi\cdot x + T_2\left(\sin^2\phi + \sin 2\phi\cdot x\cdot\frac{d\phi}{dx}\right) \qquad (9)$$

$$\frac{d^2y}{dx^2} = \frac{d^2T_1}{dx^2}\cdot x + 2\frac{dT_1}{dx} + \frac{d^2T_2}{dx^2}\cdot\sin^2\phi\cdot x + \frac{dT_2}{dx}(2\sin^2\phi + \sin 2\phi\cdot x) + T_2\left[\frac{d^2\phi}{dx^2}\cdot\sin 2\phi\cdot x + 2\left(\frac{d\phi}{dx}\right)^2\cdot\cos 2\phi\cdot x + 2\frac{d\phi}{dx}\cdot\sin 2\phi\right] \qquad (10)$$

$$\frac{d^3y}{dx^3} = \frac{d^3T_1}{dx^3}x + 3\frac{d^2T_1}{dx^2} + \frac{d^3T_2}{dx^3}\cdot\sin^2\phi\cdot x + \frac{d^2T_2}{dx^2}(3\sin^2\phi + \frac{d\phi}{dx}\cdot\sin 2\phi\cdot x + \sin 2\phi\cdot x) + \frac{dT_2}{dx}\left[\frac{d^2\phi}{dx^2}\sin 2\phi\cdot x + 2\left(\frac{d\phi}{dx}\right)^2\cdot\cos 2\phi\cdot x + 2\frac{d\phi}{dx}(2\sin 2\phi + \cos 2\phi\cdot x) + \sin 2\phi\right]\times T_2\left[\frac{d^3\phi}{dx^3}\sin 2\phi\cdot x + 3\frac{d^2\phi}{dx^2}\sin 2\phi + 6\frac{d^2\phi}{dx^2}\frac{d\phi}{dx}\cos 2\phi\cdot x - 4\left(\frac{d\phi}{dx}\right)^3\sin 2\phi\cdot x + 6\left(\frac{d\phi}{dx}\right)^2\cos 2\phi\right] \qquad (11)$$

Where

$$T_1 = \frac{\ln(G)\cdot(1-\sqrt{R_1R_2}G)^2}{(1-R_1)(G-1)(1+R_2G)}$$

$$T_2 = \frac{4\ln(G)\cdot\sqrt{R_1R_2}G}{(1-R_1)(G-1)(1+R_2G)}$$

$$\frac{dT_1}{dx} = \frac{dT_1}{dG}\cdot\frac{dG}{dx},\quad \frac{dT_2}{dx} = \frac{dT_2}{dG}\cdot\frac{dG}{dx}$$

$$\frac{dG}{dx} = -\frac{\Gamma\cdot g_0\cdot L}{\exp\left[\left(\alpha - \frac{\Gamma\cdot g_0}{x+1}\right)L\right](x+1)^2}$$

$$\frac{d\phi}{dx} = \frac{g_0Lb}{2(x+1)^2}.$$

The selection of the reference point x0 for the Taylor series expansion is of importance as well. Here more attention should be paid to the fact that the bistability in SOA occurs at the long wavelength side of the cavity resonance wavelength, as pointed out by aforementioned literatures [3, 6]. Therefore, it can be inferred that in the region that bistability occurs the phase detuning factor ϕ is negative. This characterizes the interval of x in which we want to examine the feature of the first order derivative of y. Of course, the reference point should be chosen such that the phase detuning factor is less than zero. In this paper, x_0 is taken as half of the point at which ϕ is equal to 0. Using (4), it can be derived as

$$x_0 = \frac{-\phi_0}{\Gamma g_0Lb + 2\phi_0}. \qquad (12)$$

Substitute (9)–(12) into (8), y is re-expressed as a cubic function of x. Differentiating both sides of (8), we can get

$$\frac{dy}{dx}=\frac{F_3}{2}x^2+(F_2-F_3\cdot x_0)x+\left(\frac{F_3}{2}\cdot x_0^2-F_2\cdot x_0+F_1\right). \quad (13)$$

Let (13) equal to 0, then a quadratic equation is achieved in the form

$$\frac{F_3}{2}x^2+(F_2-F_3\cdot x_0)x+\left(\frac{F_3}{2}\cdot x_0^2-F_2\cdot x_0+F_1\right)=0. \quad (14)$$

which owns the discriminant Δ as

$$\Delta=F_2^2-F_1\cdot F_3. \quad (15)$$

Equation (14) would own two real roots when Δ is greater than 0. Thus the necessary condition for bistability of SOAs can be described as

$$\Delta=F_2^2-F_1\cdot F_3>0. \quad (16)$$

When this condition is satisfied, an IO hysteresis loop can be acquired. Next, based on the above results, we would derive the analytical expression of the hysteresis loop width. When Δ>0, the two real roots of (14) are

$$x_1=\frac{F_3\cdot x_0-F_2+\sqrt{\Delta}}{F_3} \quad (17)$$

$$x_2=\frac{F_3\cdot x_0-F_2-\sqrt{\Delta}}{F_3}. \quad (18)$$

Substitute (17) and (18) into (7), the switch-up power Pup and switch-down power P_{down} of the input beam for the hysteresis loop in the SOA can be easily obtained:

$$P_{up}=\frac{\ln(G_2)\cdot[(1-\sqrt{R_1R_2}G_2)^2+4\sqrt{R_1R_2}G_2\sin^2\phi_2]}{(1-R_1)(G_2-1)(1+R_2G_2)}\cdot x_2\cdot I_s\cdot A_{in} \quad (19)$$

$$P_{down}=\frac{\ln(G_1)\cdot[(1-\sqrt{R_1R_2}G_1)^2+4\sqrt{R_1R_2}G_1\sin^2\phi_1]}{(1-R_1)(G_1-1)(1+R_2G_1)}\cdot x_1\cdot I_s\cdot A_{in} \quad (20)$$

where G_1 and ϕ1 denote the values of G and ϕ at $x=x_1$, G_2 and ϕ_2 denote the values of G and ϕ at $x=x_2$, and Ain represents the area of the input beam. Combining (19) and (20), we can derive the bistable loop width H as

$$H=P_{up}-P_{down}. \quad (21)$$

Numerical simulations

In the numerical simulation, the parameter values are taken as follows, $A_{in}=0.8\mu m^2$, $L=500\mu m$, $R_1=R_2=0.3$, $n0=1.5\times10^{24}$ m−3, $a=2.7\times10^{-20}$ m², α=1000m−1, b=4.8, and Γ=0.1. Figs.1 and 2 plot the dependence of Δ on the initial phase factor and the normalized carrier concentration, respectively. According to the above derived criterion for the occurrence of hysteresis, when Δ is greater than 0, the bistable loop could be found. In Fig.1, it can be seen that Δ is positive when ϕ0 is in the interval [−0.78π, −0.18π] as the normalized carrier concentration is 0.95. Further numerical simulations are carried out under the same parameters, which show that in this interval the SOA would appear the bistable behaviors. Similarly, Fig. 2 reveals that if the normalized carrier density is greater than 0.88, Δ is positive when the initial phase detuning factor ϕ_0 is −0.22π. Numerical calculations also exhibit that just in this region the bistable loop can be observed. All these confirm the effectiveness of the derived criterion (16) for the hysteresis in the SOAs. Figure 3 displays the numerical and analytical results of the switch-up power as a function of the normalized carrier concentration for various initial phase detuning factors. Obviously, the discrepancy between the simulations and the derived formulae is negligible. Moreover, it can be seen that Pup dwindles as the carrier density increases, and with the enhancement of the initial phase detuning, P_{up} rises up drastically. In Fig. 4, the tendency of the switch-down power P_{down} is plotted, as the normalized carrier concentration grows up for various initial phase detuning factors. The analytical results show a good agreement with the numerical ones. Furthermore, P_{down} possesses a similar tendency to Pup, when the carrier concentration or the initial phase detuning factor increases. In Figs.5 and 6, the bistable loop width is shown as a function of the normalized carrier density and the initial phase detuning factors, respectively. It is observed that analytical calculations are slightly smaller than numerical simulations. Meanwhile, the hysteresis loop becomes wider with the increment of either the carrier density or the initial phase detuning factor.

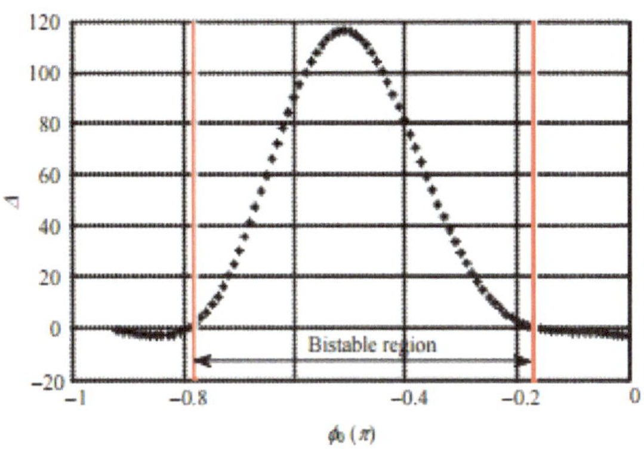

Figure.1 Dependence of Δ on the initial phase detuning factor ϕ_0, $n/n_{th}=0.95$.

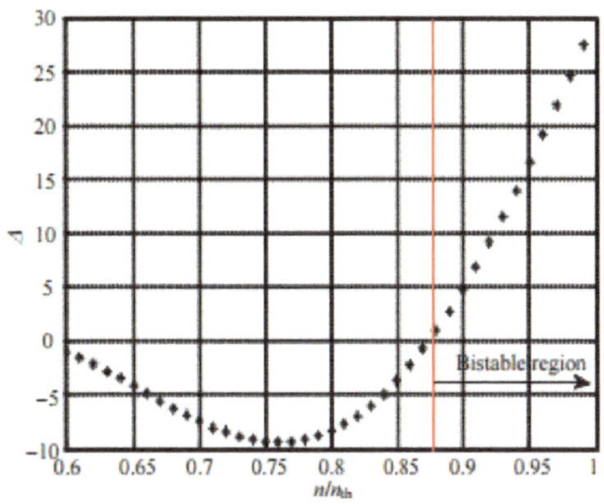

Figure. 2 Dependence of Δ on the normalized carrier concentration n/n_{th}, $\phi_0=-0.22\pi$

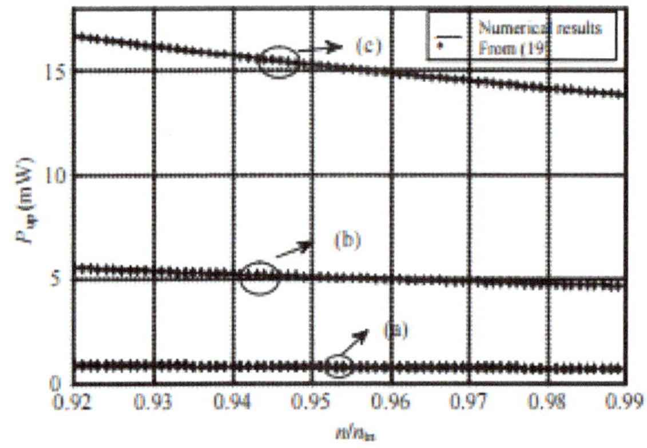

Figure.3 Jump-up power Pup versus normalized carrier density n/nth for three cases: (a) $\phi_0 = -0.22\pi$, (b) $\phi_0 = -0.42\pi$, and (c) $\phi_0 = -0.62\pi$. The solid lines and the symbols represent the numerical simulations and the analytical results, respectively.

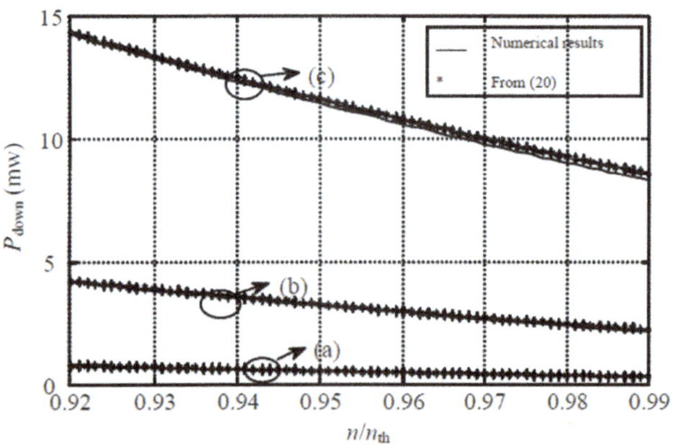

Fig.4 Comparison between analytical and numerical results of the jump-down input power P_{down} for three cases: (a) $\phi_0 = -0.22\pi$, (b) $\phi_0 = -0.42\pi$, and (c) $\phi_0 = -0.62\pi$. Other parameters are the same as Fig.3.

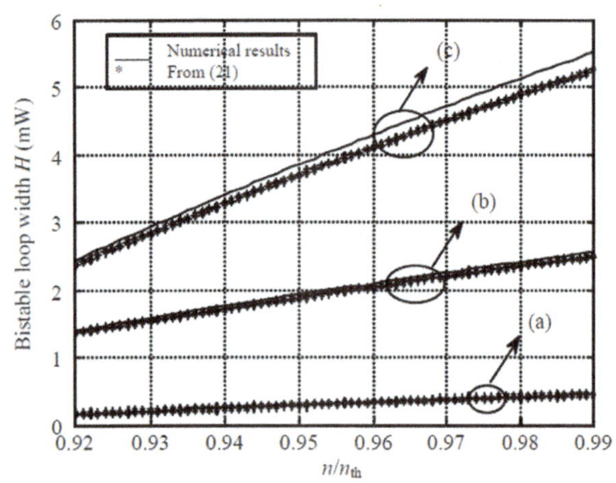

Fig. 5 Bistable loop width as a function of the normalized carrier density n/nth for three cases: (a) $\phi_0 = -0.22\pi$, (b) $\phi_0 = -0.42\pi$, and (c) $\phi_0 = -0.62\pi$. The solid lines and the symbols represent the numerical simulations and the analytical results, respectively.

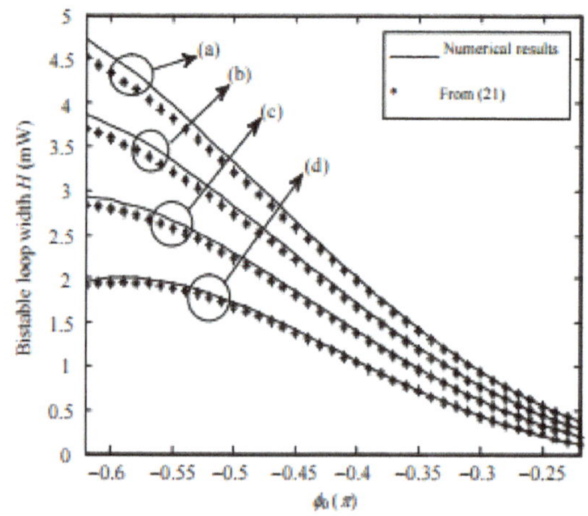

Figure.6 Analytical and numerical results of bistable loop width as a function of the initial phase detuning factor ϕ_0 for four cases: (a) $n/n_{th}=0.97$, (b) n/nth=0.95, (c) $n/n_{th}=0.93$, and (d) $n/n_{th}=0.91$.

CONCLUSIONS

In summary, based on Adams's relationship between the z-independent intracavity intesnsity Iav and the incident intensity Iin in an SOA, we deduce the criterion for the hysteresis in SOAs by expanding Iin as a Taylor series in terms of Iav. As a result, the analytical expressions of the switch-up power, the switch-down power of the incident beam, and the width of the hysteresis loop in SOAs are derived for the first time. Numerical simulations are performed to test the effectiveness of the criterion and the accuracy of the expressions. The comparisons reveal that our analytical formulae are in good agreement with the numerical results. These conclusions would afford a valuable way for future investigations into the bistability in SOAs and other devices with a similar structure.

REFERENCES

1. P. M. Mejias and R. M. Herrero, "Bistable fiber-optics interferonetric sensor," Applied Optics, 1988, 27(5): 811–813.
2. S. H. Shehadeh, D. Steup, and J. Weinzierl, "Cascaded linear and nonlinear optical resonators: Towards a smart deflection sensor," in Microsystems and Nanoelectronics Research Conference, Canada, Oct. 13–14, 2009.
3. P. Pakdeevanich and M. J. Adams, "Measurements and modeling of reflective bistability in 1.55-µm laser diode amplifiers," IEEE Journal of Quantum Electronics, 1999, 35(12): 1894–1903.
4. N. F. Mitchell, J. O'Gorman, and J. Hegarty, "Optical bistability in asymmetric Fabry-Perot laser-diode amplifiers," Optics Letters, 1994, 19(4): 269–271.
5. P. Wen, Michael S, M. Gross, and S. Esener, "Observation of bistability in vertical-cavity semiconductor optical amplifier (VCSOA)," Optics Express, 2002, 10(22): 1273–1278.
6. Hurtado, A. Gonzalez-Marcos, and J. A. Martin-Pereda, "Modeling reflective bistability in vertical-cavity semiconductor optical amplifiers," IEEE Journal of Quantum Electronics, 2005, 41(3): 376–383.
7. F. Marki, D. R. Jorgesen, H. J. Zhang, P. Wen, and S.

C. Esener, “Observation of counterclockwise, clockwise and butterfly bistability in 1550 nm VCSOAs,” Optics Express, 2007, 15(8): 4953–4959.

8. W. L. Zhang and S. F. Yu, “Bistabilities of birefringent vertical-cavity semiconductor optical amplifiers with antiresonant reflecting optical waveguide,” IEEE Journal of Quantum Electronics, 2010, 46(1): 11–18.

CHAPTER 9

APPLICATION OF MEH-PPV/SnO2 BILAYER AS HYBRID SOLAR CELL

Jose Antonio Ayllon, Monica Lira-Cantu

INTRODUCTION

The application of conducting organic polymers (COPs) in organic electronics is based on their ability to transport, emit or absorb light under specific conditions [1]. Thus, we can find COPs as part of the active layer of electronic devices such as transistors, LEDs or solar cells. In organic photovoltaic (OPVs) devices, COPs have been part of extensive research work, specially the combination of poly(3-hexylthiophene) (P3HT) and the C60 derivative ([6, 6]-phenyl-C61-butyric acid methyl ester) PCBM which has shown excellent PV properties with efficiencies beyond 5% [1–4]. The potential of OPVs resides in their low cost, not because of the low price of the materials applied, but due to the printing techniques applied for their fabrication [4]. Among the challenges to overcome for this technology are the improvement of solar cell efficiency and lifetime. Lifetime limitation resides in the degradation of the organic semiconductors due to the presence of oxygen or moisture from the atmosphere [5–7], thus encapsulation of these devices is a requirement [7]. Several attempts have been made in order to find new electron transport materials with better transport properties and higher stability towards oxygen. In this respect, semiconductor oxides emerged as an attracting alternative to replace PCBM in OPVs due to their appealing properties. Besides lower cost these oxides are wide bandgap materials able to accept electrons effectively. Furthermore, recent studies have shown the possibility of semiconductor oxides to work under ambient atmospheres when they are used in direct contact with conducting organic polymers in a electronic devices like solar cells [8–12] or more recently as air-stable OLEDs [13, 14]. We have recently analyzed different semiconductor oxides in HSC [8–12] and have documented a fine interaction between the semiconductor oxide and oxygen from the atmosphere. In an effort to broaden our knowledge on the interaction of these systems with air, we present here our results on the application of thin films of SnO_2 as electron acceptor in HSC. Analyses of the interaction of SnO_2 with conjugated polymers have been limited to a fundamental level [15–17], and other studies reported that the electron transfer from MEH-PPV to SnO_2 is energetically favorable and can be photoinduced [15]. Although, these processes are known to be slower than for TiO_2 (800 fs vs. less than 100 fs) the advantage for SnO2 utilization is based on a more favorable position of its conduction band [15]. It is also known that the electron transfer rate from the polymer MEH-PPV and the SnO2 is on the order of microseconds, yielding a possible candidate to construct HSCs [15]. Finally, the application of nanocrystalline thin films rather than isolated nanoparticles could be advantageous, as it increases the ability of the semiconductor to transport the charge to the electrode through the connected inorganic network [16]. We present in this work HSCs applying a device configuration of ITO/SnO_2 thin film/MEH-PPV/Ag. We analyzed the effect of polymer thickness and solar cell performance by I–V-curves, photophysical analyses, and especially lifetime studies. A brief discussion on the application of different semiconductor oxides already studied in our laboratories and their comparison with SnO_2 are also presented.

EXPERIMENTAL

Materials

MEH-PPV was synthesized as described in [8–12]. Indium tin oxide (ITO) substrates were purchased from Delta Technologies, Lt. (U.S.A.). The substrates dimensions were 25 × 50 × 1.1 mm (coated on one surface, Rs = 10 ohms ± 5 ohms), and substrates were etched and prepared following a method described earlier [8–12]. Once etched, the substrates were washed with acetone and ethanol in an ultrasonic bath for 10 min, respectively. $SnCl_2$ $2H_2O$ and absolute ethanol were obtained from Panreac and used as received.

Citation: Jose Antonio Ayllon, Monica Lira-Cantu, Application of MEH-PPV/SnO2 bilayer as hybrid solar cell, DOI 10.1007/s00339-008-5023-z.

Preparation of SnO_2 thin films Tin oxide sol precursor was prepared by dissolving 13.5 g of $SnCl_2$ $2H_2O$ (5.98×10^{-2} mol) in 100 ml of absolute ethanol and then refluxing during 2.5 h [18]. This solution was allowed to cool to room temperature, filtered and spin coated on top of ITO substrates. The as-prepared films were annealed at 450◦C for two hours to completely oxidize the precursor and induce the crystallization of SnO_2.

Polymer photovoltaic preparation and characterization

The ITO/thin film semiconductor oxide substrate was first subjected to 100◦C for 10 min to eliminate any possible absorbed water on the surface. The substrates were then cleaned with chloroform (via spin coating) and then a microfiltered MEH-PPV polymer solution (0.45 μm pore size) from chlorobenzene was spincoated at 1000 rpm or 1500 rpm on top of the ITO/semiconductor oxide substrates. The absorbance of the spincoated polymer films was controlled to be about 1.0 absorbance units unless otherwise specified. The active area of the devices was 3.2 ± 0.1 cm^2. Silver electrodes were thermally evaporated onto the polymer film at a pressure of $<10^{-5}$ mBar though a mask. The device was mounted using conductive silver epoxy glue for the electrical contacts. Final device configuration was ITO/$SnO_{2thin\ film}$/MEH-PPV/Ag. The positive terminal of the Keithley Sourcemeter was attached to the glass/ITO/oxide, and the negative terminal was attached to the metal electrode. This deliberate choice was made in order to distinguish the function of the transparent conducting oxide (TCO) and metal electrode as compared to traditional polymer photovoltaics where the TCO is the hole collector and the metal is the electron collector. Samples were then analyzed in the ambient atmosphere. As a measure of the decay we recorded the short circuit current (J_{sc}) as a function of time during illumination since the efficiency of the device and Jsc are linked as described earlier. Values of V_{oc} and J_{sc} were measured for the devices under ambient conditions and finally I–V-curves, action spectra and solar decay analyses were determined in that order for the devices. The wavelength dependence of the photovoltaic response was characterized using a set-up described earlier [8–12], and the photovoltaic response under AM1.5 simulated sunlight was performed using a SolarConstant 575 from Steuernagel Lichttechnik GmBH, Germany. The spectral distribution and quality of the solar simulator were monitored following the ASTM E927 standard recommendations using a spectrometer from AvaSpec 2048 from Avantes, and a Precision Pyranometer from Eppley Laboratories was used to monitor the total power that was set to 1000 W m−2 in all experiments. Polymer film thicknesses were analyzed by AFM.

RESULTS AND DISCUSSION

General characterization Figure 1 shows the grazing incidence X-ray diffraction analysis of the SnO_2 thin film applied as electron acceptor for the hybrid solar cells reported in this work. For comparison purposes the diffractogram of the ITO, [$91(In_2O_3)$: $9(SnO_2)$], substrate is also shown. These results confirm the correct formation of SnO_2 on the ITO substrate. HSCs applying semiconductor oxides are stable in ambient atmospheres, and have been shown to stabilize and improve PV properties after storage in the dark under ambient atmosphere conditions [8–12]. Usually a few hours up to a couple of days is required for these solar cells to reach maximum performance. Storage in the dark in air can be made for several months, even years (up to 1.5 years in our laboratory) without significant degradation of photovoltaic properties. Thus, the HSCs analyzed in this work were stabilized in the dark for 1 h and 24 h under ambient atmosphere conditions before analyses. Figure 2 shows the I–V-curves carried out to HSCs applying two polymer thicknesses of 90 nm and 110 nm. In order to obtain different thicknesses, the same MEH-PPV polymer solution was coated on top of the SnO_2 layer at different spin coating

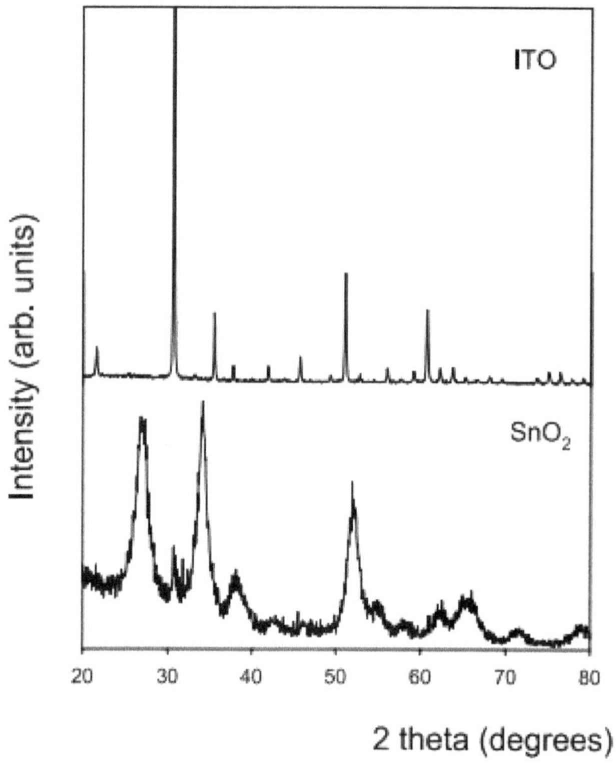

Fig. 1 Grazing incidence X-ray diffraction analysis of the SnO_2 thin film semiconductor oxide synthesized on ITO from sol–gel solution (diffraction angle 0.3 degrees). For comparison proposes the diffractogram of the ITO, [$91(In2O_3)$: $9(SnO_2)$], substrate is also shown

speeds. The use of 1500 rpm resulted in a 90 nm polymer thicknesses, while the spin coating velocity fixed at 1000 rpm gave polymer films of 110 nm. Figures 2(a) and 2(b) show the I–V-curves obtained for devices with polymer thicknesses of 90 nm and 110 nm, respectively. We observed a small difference in shape between both graphs but, in general, both HSC showed almost the same V_{oc} and J_{sc} values around 0.3 V and 0.15–0.2 mA/cm^2; FFs were also in the same 26–28% range, as expected due to the similarity of the polymer thickness. These similar results show that a difference of 20 nm in polymer thickness between layers is not enough to produce meaningful effects on photovoltaic performance. Thus, a polymer thickness of 90 nm was chosen to study the HSC photovoltaic properties after the device was stabilized in the dark for 24 h. The HSC (90 nm polymer thickness), shown in Fig. 2(c), revealed that a 24 h period of stabilization in the dark improved V_{oc} values from 0.3 to 0.5 V, while J_{sc} was observed to be on the same range as before stabilization. The V_{oc} and J_{sc} values observed after stabilization were 0.5 V and 0.18 mA/cm2, respectively, with a FF of 33%. As already reported for other HSC applying semiconductor oxides, we have attributed this behavior to the oxygen requirement of the semiconductor oxide [8–12]. It has been reported that a small oxygen loss from the semiconductor oxide structure can reduce and modify its photoconductivity properties [19], a condition that is present in our HSC during electrode evaporation which takes place under high vacuum conditions (1 × 10−6 mbar). In the case of SnO_2, however, this effect could also be influenced by its high sensitivity to different atmospheres [20]. Complete lifetime studies showed only 8 h of lifetime before device degradation.

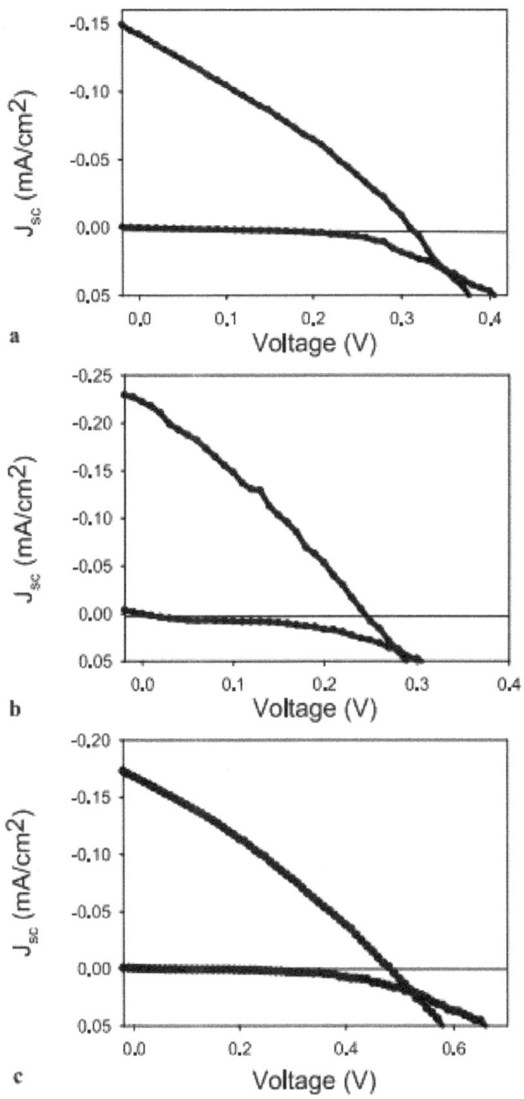

Fig. 2 I–V-curves for a hybrid solar cell (HSC) of the type ITO/SnO_2 thin film/MEH-PPV/Ag with polymer thickness of (a) 110 nm, (b) 90 nm. Both cells were stabilized in the dark under ambient atmospheres for 1 h. (c) HSC stabilized for 24 h in the dark undercambient atmosphere, polymer thickness of 90 nm. HSCs measured in ambient atmosphere with AM1.5 spectral distribution and 1000 W/m^2. Active area is 3.2 ± 0.1 cm^2

Photophysical analyses

Photophysical analyses performed to the HSCs applying two polymer thicknesses are shown in Fig. 3(a) (labeled as I and II). For comparison purposes the UV–VIS spectra of the MEH-PPV polymer are also shown (Fig. 3(b)). The shape of the polymer spectra and that of the photoaction spectra of the HSC are similar, which suggests that the absorption of the polymer MEH-PPV is mainly responsible for the photocurrent generation in the cell. Moreover, the absorption band below 350 nm corresponds to the contribution of SnO_2. A comparison of the absorption spectra of the MEHPPV polymer (Fig. 3(b)) with the maxima of the photoaction spectra reveals a shift in the position of the wavelength band which corresponds to the MEH-PPV polymer. This shift has been observed in our HSCs independently of the thin film semiconductor oxide applied [10]. The presence of a blue shift in the action spectra has been observed before as an indication of the formation of a charge transfer complex (CTC) between the polymer and oxygen from the atmosphere as already reported [10]. Nevertheless, the photoaction spectra shown in Fig. 3(a) indicate a shift towards the UV–VIS (lower wavelengths) invalidating the possible formation of a CTC. The observed shift has been attributed to thin film interferences as also observed for other HSC applying other semiconductor oxides [8–12].

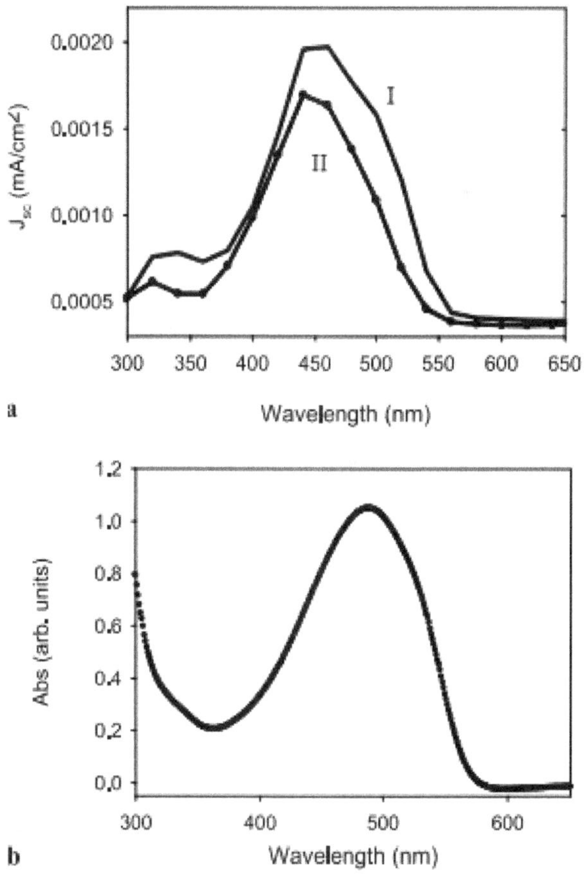

Fig. 3 Photophysical analyses of a hybrid solar cell of the type ITO/SnO_2 thin film/MEH-PPV/Ag with two different polymer thicknesses (aI) 110 nm and (aII) 90 nm. For comparison purposes we show the UV–VIS absorbance spectra of the MEH-PPV (b). Active area is 3.2 ± 0.1 cm^2

Effect of UV-light

A well-known property of semiconductor oxides, like SnO2, is their ability to absorb in the UV-wavelength range. The major advantage observed for SnO_2 is its electron affinity when compared with one of the best oxides, TiO_2 (CB edge TiO_2: −4.2 eV vs. vacuum, and SnO_2: −4.77 eV vs. vacuum) [21, 22]. Thus, we seek to understand if the degradation of the polymer observed when semiconductor oxides are applied in HSC as reported earlier [18, 22] can be reduced by the application of SnO2 thin films, also to analyze its effect on overall solar cell performance. With this in mind, we carried out I–V-curve analyses of the HSC applying SnO_2, under conditions of light and under ambient atmosphere. By blocking the UV-light (applying a UV-filter) we were able to analyze the effect of the SnO_2 on the performance of the HSC. A decrease in current density is expected during the application of the filter since a part of the UV-wavelength range at which the semiconductor is excited accounts for overall solar cell photocurrent. Nevertheless, unimportant

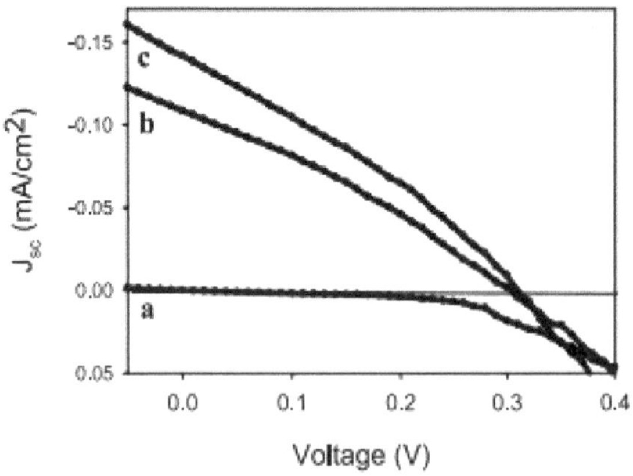

Figure. 4 I–V-curves performed to an HSC of the type ITO/SnO_2/MEH-PPV/Ag under ambient atmosphere: (a) dark, (b) with a UV filter (

changes were observed on the I–V-curves as observed in Figs. 4(b) and 4(c). Almost the same voltage, about −0.32 V, and current densities between 0.14 mA/cm2 and 0.17 mA/cm2 were observed when the samples were analyzed with and without filter, respectively. The latter could indicate that the application of SnO_2 could be a good choice as electron transport material which neglects or decreases the degradation of the MEH-PPV polymer during irradiation under ambient atmospheres. In order to prove the latter, we carried out analyses of the HSC under long-term irradiation as described in the following section.

Stability under Long-Term Irradiation

Stability analyses under constant irradiation conditions (1000 mW/cm^2) were performed to HSC of the type ITO/SnO2 thin film/MEH-PPV/Ag after the solar cell was stabilized in the dark. As shown in Fig. 5, we observed an initial increase in current density from 0.13 mA/cm2 to about 0.18 mA/cm2 in less than 1 hour. This improvement is characteristic of HSC applying thin film semiconductor oxides and has been linked to the re-adsorption of oxygen from the atmosphere into the semiconductor oxide during irradiation [8–12]. After reaching a maximum, Jsc stabilized and decreased steadily with time. This behavior is observed in all semiconductor oxides, and, in the case of HSC applying SnO_2, it was observed that the degradation was worse (faster degradation) than with Nb_2O_5 or TiO_2, CeO_2 or CeO_2–TiO_2 as can be seen from our previous work [8]. Only ZnO has been observed to degrade the polymer faster than SnO_2.

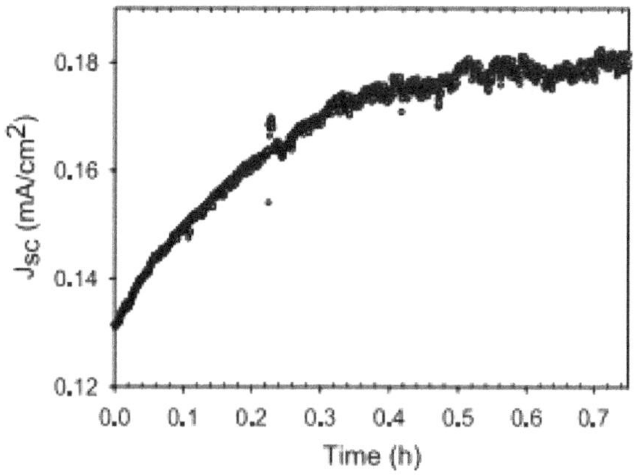

Fig. 5 An increase in J_{sc} observed during the first hour of solar decay analyses carried out to an HSC of the type ITO/SnO_2 thin film/MEH-PPV/Ag. HSCs measured in ambient atmosphere with AM1.5 spectral distribution and 1000 W/m^2. Active area is 3.2 ± 0.1 cm^2. Polymer thickness is 110 nm.

Comparison of SnO_2 with TiO_2, Nb_2O_5 and ZnO thin films

Finally, we would like to compare the response of HSC applying different thin film semiconductor oxides as analyzed before in our laboratory, TiO2, Nb2O5, ZnO, and CeO-base oxides [8–12], and the one presented in this work applying SnO2. In general, we have observed that the voltage of the HSC is dependent on the type of semiconductor oxide applied. The highest V_{oc} and J_{sc} have been observed for HSC applying TiO_2 and $Nb2O_5$ (about −0.7 V) followed by SnO2 with −0.5 V, ZnO with V_{oc} of about −0.3 V, and on the positive side CeO_2 with a value of +0.12 V and CeO_2– TiO2 with a value of +0.42 V as shown in Table 1. Similarities between HSCs are observed also on all I– V-curves which are characterized by low fill factors (never higher than 44%). This could be an indication of high internal resistances and leakage currents present in these systems, usually observed in devices with high recombination problems. Current leakages could come from the inhomogeneity of the oxide thin film or due to their thickness that are not more than a couple of tenths of nanometers (between 15 and 30 nm thick). In all cases, the degradation of the polymer with time and the decrease of photovoltaic properties have been inevitable when the HSC had the configuration applied in this work. The time it takes to degrade the polymer depends on the semiconductor oxide applied as follows (from the oxide showing the fastest degradation to the slowest): ZnO > SnO_2 > Nb_2O_5 > TiO_2 > CeO_2 > CeO_2– TiO_2.

CONCLUSIONS

Our results show the possibility to apply SnO_2 thin films as electron acceptors in HSC. Current densities of about

Semiconductor oxide	Bandgap	Stabilization time (h)	J_{sc} (mA/cm^2)[c]	V_{oc} (V)[d]	FF (%)
TiO_2	3.0–3.2 [23, 24]	2	0.42 ± 0.02	-0.74 ± 0.06	0.30 ± 0.03
Nb_2O_5[a]	2.3–3.4 [25]	48	0.28 ± 0.11	-0.64 ± 0.12	0.40 ± 0.04
Nb_2O_5[b]	2.3–3.4 [25]	48	0.34 ± 0.05	-0.40 ± 0.02	0.36 ± 0.03
SnO_2	3.1–3.0 [23]	1	0.20 ± 0.06	-0.50 ± 0.01	0.33 ± 0.03
ZnO	3.2–3.3 [28, 29]	2	0.30 ± 0.02	-0.38 ± 0.04	0.35 ± 0.05
CeO_2	3.0–3.2 [30–32]	120	0.004 ± 0.05	$+0.12 \pm 0.06$	0.25 ± 0.04
CeO_2–TiO_2	3.2–3.6 [26, 27]	50	0.007 ± 0.03	$+0.42 \pm 0.07$	0.32 ± 0.05

[a]ITO substrates on aluminosilicate glass

[b]ITO substrates on sodalime glass

[c]At maximum J_{sc}

[d]V_{oc} at the beginning, before stabilization. Active area is 3.3 ± 0.2 cm^2

~0.17 mA/cm2, V_{oc} of ~0.45 V, and 33% fill factors were achieved. Our HSC showed that an improvement in Voc and fill factors can be obtained after the device is subjected to a stabilization period in the dark under ambient atmospheres. We have compared our results with other HSC applying different semiconductor oxides: TiO_2, Nb_2O_5 and ZnO, and found similar trends in photovoltaic behavior. We have suggested that this type of HSCs requires modification of device configuration to reach better performance and polymer stability. Nevertheless, the application of thin film semiconductor oxides in related electronic applications, such as hybrid transistors or hybrid OLEDs, is possible.

ACKNOWLEDGEMENTS

We thank Dr. Frederik C. Krebs for a permission to analyze our samples in his laboratory. Also thanks to the Agència de Gestió d'Ajuts Universitaris i de Reserca (AGAUR, Catalunya Goverment), to the Canon Foundation in Europe, to the projects OTT2007-0423, OTT2007-0422 and OTT2007-1014 from the Spanish Ministry of Science and Innovation (MSI) and CDTI, and to the Consoider NANOSELECT project CSD2007-00041.

REFERENCES

1. F. So, J. Kido, P. Borrows, Mater. Res. Soc. Bull. 33, 663 (2008)
2. R.A.J. Janssen, J.C. Hummelen, N.S. Sariciftci, Mater. Res. Soc. Bull. 30, 33 (2005)
3. C.J. Brabec, J.A. Hauch, P. Schilinsky, C. Waldauf, Mater. Res. Soc. Bull. 30, 50 (2005)
4. C.J. Brabec, J.R. Durrant, Mater. Res. Soc. Bull. 33, 670 (2008)
5. F.C. Krebs, Sol. Energy Mater. Sol. Cells 92(7), 685 (2008)
6. F.C. Krebs, Sol. Energy Mater. Sol. Cells 92(7), 715 (2008)
7. F.C. Krebs, K. Norrman, Prog. Photovolt. Res. Appl. 15(8), 697 (2007)
8. M. Lira-Cantu, F.C. Krebs, Sol. Energy Mater. Sol. Cells 90, 2076 (2006)
9. M. Lira-Cantu, F.C. Krebs, in Rec. Res. Dev. Appl. Phys., ed. By S.G. Pandalai (Transworld Res. Network, 2005). ISBN:81-7895- 187-8
10. M. Lira-Cantu, K. Norrman, J.W. Andreasen, F.C. Krebs, Chem. Mater. 18, 5684–5690 (2006)
11. M. Lira-Cantu, K. Norrman, J.W. Andreasen, N. Casan-Pastor, F.C. Krebs, Electrochem. Trans. 3(22), 1 (2007)
12. M. Lira-Cantu, P. Gómez-Romero, in Hybrid Nanocomposites for Nanotechnology: Electronic, Optical, Magnetic and Bio/Medical Applications, ed. by L. Merhari (Springer, Berlin, 2008). ISBN: 978-0-387-72398-3
13. H.J. Bolink, E. Coronado, D. Repetto, M. Sessolo, E.M. Barea, J. Bisquert, G. Garcia-Belmonte, J. Prochazka, L. Kavan, Adv. Funct. Mater. 18(1), 145 (2008)
14. K. Morii, T. Kawase, S. Inoue, Appl. Phys. Lett. 92(21), 213304 (2008)
15. N.A. Anderson, E. Hao, X. Ai, G. Hastings, T. Lian, Physica E Low-Dimens. Syst. Nanostruct. 14, 215–218 (2002)
16. N.A. Anderson, E. Hao, X. Ai, G. Hastings, T. Lian, Chem. Phys. Lett. 347304-310 (2001)
17. H. Miyaoka, G. Mizutani, H. Sano, M. Omote, K. Nakatsuji, F. Komori, Solid State Commun. 123, 399–404 (2002)
18. L.L. Diaz-Flores, R. Ramirez-Bon, A. Mendoza-Galvan, E. Prokhorov, J. Gonzalez-Hernandez, J. Phys. Chem. Solids 64, 1037–1042 (2003)
19. H. Neugebauer, C.J. Brabec, J.C. Hummelen, R.A.J. Janssen, N.S. Sariciftci, Synth. Met. 121, 35 (2001)
20. H. Meixner, U. Lampe, Sens. Actuators B 33, 198–202 (1996)
21. J.E. Kroeze, T.J. Savenije, Thin Solid Films 451–452, 54–59 (2004)
22. G. Benko, P. Myllyperkio, J. Pan, A.P. Yartsev, V. Sundstrom, J. Am. Chem. Soc. 125(5), 1118–1119 (2003)
23. K. Kalyanasundaram, M. Grätzel, Coord. Chem. Rev. 177, 347– 414 (1998)
24. S.J. Pearton, C.R. Abernathy, M.E. Overberg, G.T. Thaler, D.P. Norton, J. Appl. Phys. 93, 1–13 (2003)
25. N. Özer, D.-G. Chen, C.M. Lampert, Thin Solid Films 277, 162– 168 (1996)
26. F.E. Ghodsi, F.Z. Tepehan, G.G. Tepehan, Sol. Energy Mater. Sol. Cells. 68, 355–364 (2001)

CHAPTER 10

LOW-FREQUENCY FLICKER NOISE IN A MSM DEVICE MADE WITH SINGLE SI NANOWIRE (DIAMETER ≈ 50 NM)

Sudeshna Samanta, Kaustuv Das and Arup Kumar Raychaudhuri

Keywords: Flicker noise, MSM device, Single Si NW

BACKGROUND

Exploring the fundamental properties of an individual silicon nanowire (Si NW) is important as it forms the backbone of the fabrication of single-nanowire nanoelectronic devices. There are reports on the development of Si NW-based nanoscale devices such as field-effect transistors (FETs) [1, 2] with wrap-around gates, surface-gated sensitive chemical and biomolecular sensors [3, 4], as well as nanoscale opto-electronic devices [5]. In the context of such nanowire-based device, one important physical parameter is the low-frequency flicker noise, which has a direct impact on the device performance. In recent publications, it has been argued that flicker noise in qubits can lead to decoherence and can be the limiting factor in increasing the coherence time [6]. While flicker noise in a sub-micron metal oxide semiconductor field-effect transistor (MOSFET) with varying channel width has been investigated for some time [7], there are no reports of measurements of the low-frequency flicker noise in Si NWs and nanowire-based devices particularly with diameters much less than 100 nm.

In this paper, we report the measurement of flicker noise in a metal-semiconductor-metal (MSM) device made from a single strand of a Si NW. In such a device, the flicker noise can come from the junction region where the metals make contacts with the semiconductor (MS junction) as well as from the single Si NW. The noise arising from the junction region can be large and can even mask the noise from the Si NW by a few orders. This is because the flicker noise is likely to arise from charge carrier density fluctuations due to trapping-detrapping in the junction region. By an innovative application of direct current (dc) bias (used for biasing the device) mixed with an alternating current (ac) bias (used for the noise measurements), we could suppress the noise from the junction region and observe the noise which likely arises from the single Si NW. The enabling physics that leads to suppression of the noise in the junction region on application of the dc bias is the collapse of the depletion region at the junction region by the applied dc bias.

The low-frequency flicker noise in most materials has a power spectral density (PSD) with 1/f frequency dependence and can serve as a diagnostic of the presence of structural defects arising from mobility fluctuations. In semiconductors, the 1/f noise can also arise from recombination-generation process [8]. For the Si NW devices, proper estimation of the generic noise arising from nanowire itself is an essential device parameter for the better performance of low-noise electronics. The fluctuations in this cases arise from resistance fluctuations in a current biased system which shows up voltage fluctuations with PSD $S_V(f)$.

Measurement of the 1/f noise on a single Si NW has a number of challenges that involve fabrication and manipulation of a single NW, and aligning and connecting NW with low-resistance contacts. We measured the noise in the configuration where two metal electrodes have been fabricated by nanolithography on a single Si NW. A schematic diagram of the Si NW-based device and the corresponding MSM structure are depicted in Figure 1a,b, respectively. For most of the devices, including opto-electronic devices, fabricated on a single Si NW, the basic configuration is the MSM configuration. In such cases, the contact resistance at the Schottky junction plays an important role in carrier transport through the NW. This can also lead to a substantial flicker noise at the junction regions due to the existence of traps in the depletion region. In this report, we show the noise measurement carried on with an ac excitation (V_{ac}) with a superimposed independent dc bias ((V_{dc}), more than the Schottky barrier height (□) formed at the metal-semiconductor (MS) junction region) which can

Citation: Sudeshna Samanta, Kaustuv Das and Arup Kumar Raychaudhuri, Low-frequency flicker noise in a MSM device made with single Si nanowire (diameter ≈ 50 nm), DOI: 10.1186/1556-276X-8-165.

lead to severe suppression of the noise arising at the junction region by few orders of magnitude. This suppression of the junction noise enables us to estimate of the noise arising from the single Si NW. In the case of a single Si NW MSM device, such experiments do not exist, and the report here may provide an independent tool to reduce the junction noise by applying an external dc bias.

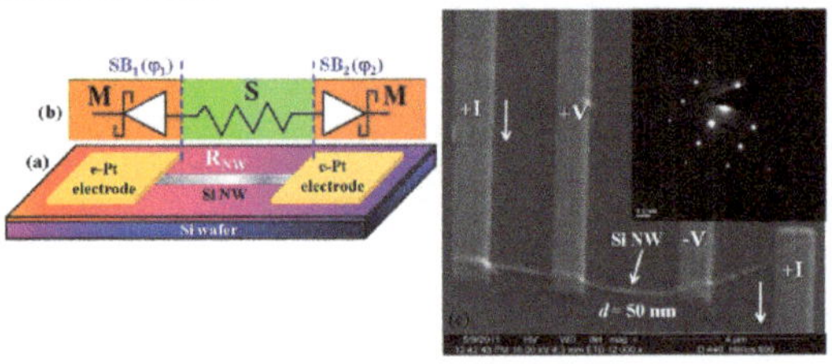

Figure 1 Schematic diagram, MSM structure and SEM image. (**a**) Schematic diagram of a single Si NW with e-beam-deposited Pt contact electrodes. (**b**) A representative MSM structure of the NW device, consisting of two Schottky diodes connected back to back with a series resistance R_{NW}. (**c**) SEM image of the single Si NW device with four electrical leads, and the inset shows a HRTEM image of the wire itself.

METHODS

Synthesis and device fabrication

The Si NWs used in this experiment were fabricated by metal-assisted chemical etching [9] technique. The method leads to a dense array of single crystalline Si NWs with a diameter ranging from approximately 20 to 100 nm and lengths of more than 10 μm. A high-resolution transmission electron microscope (HRTEM) image shows the probable existence of an oxide layer with a thickness ≤ 2 nm at the surface. The Pt contacts (in the configuration of the MSM device) for the noise measurement were made by using e-beam-assisted local deposition of methylcyclopentadienyl platinum trimethyl precursor at a bias of 15 kV in a dual beam system FEI-HELIOS 600 (FEI Co., Hillsboro, OR, USA). The scanning electron microscopy (SEM) image of a single NW connected with four electrical contacts is shown in Figure 1c. The four electrical contacts allow us four-probe measurements of the resistance of the individual NW and hence its resistivity (ρ). The inner two electrodes were used for current-voltage (I−V) measurements in the MSM device configuration. P_t in a dual beam machine can also be deposited using G_a ions. However, to avoid damage as well as contamination from implanted G_a ions, we used e-beam-assisted deposition. We note that the P_t deposited from the decomposition of the high carbon-containing precursor is not pure Pt. Instead, it is a composite of carbon and Pt, which has been analysed before by our group for its physical characteristics and compositional details [10].

Electrical measurements

The metallic contacts at the ends lead to the Schottky barrier (SB) formation in the junction region (see Figure 1b). The resulting MSM device can be modelled as two back-to-back Schottky diodes (SB_1 and SB_2) at the ends with a Si NW with resistance R_{NW} connecting them. The current passing through such a device is mainly controlled by the barrier heights φ_1 and φ_2 at the two contacts SB_1 and SB_2, respectively. This device configuration also enabled us to do two-probe as well as four-probe measurements on the same Si NW, which then allows us to find the contact resistance R_C, an important device parameter. The area of contact, A_C, can be obtained from the SEM image of a given device from which a reliable estimate of specific contact resistivity $\rho_C = A_C R_C$ can be obtained.

Figure 2a shows the non-linear and asymmetrical I−V characteristics of a typical device made from a single Si NW with diameter of approximately 50 nm. At the highest device current of 10 μA, the current density is $\approx 2.5 \times 10^4 \text{A/cm}^2$, which is much less than the electromigration damage threshold. The nanowire used has a resistivity at room temperature $\rho_{300K} = 290$ m Ω.cm. Comparison of the ρ with the resistivity of bulk Si gives us an estimate of carrier density $n \approx \times 10^{17}/\text{cm}^3$. The non-linearity at low bias is a signature of the Schottky-type contacts. The asymmetric nature of the I−V curves arises because of $\varphi_1 \neq \varphi_2$. This inequality arises from the likely differences in the surface conditions at the two contacts (M-S) that will determine the actual value of the barriers. The bias-dependent currentI has been fitted with the equation for back-to-back Schottky diodes connected by a resistor [11]

$$I(V) = I_0 \exp\left(\frac{qV'}{\eta kT} - 1\right) \times \frac{\exp\left(\frac{-q(\varphi_1+\varphi_2)}{kT}\right)}{\exp\left(\frac{-q\varphi_2}{kT}\right) + \exp\left(\frac{-q\varphi_1}{kT}\right)\exp\left(\frac{qV'}{\eta kT}\right)} \qquad (1)$$

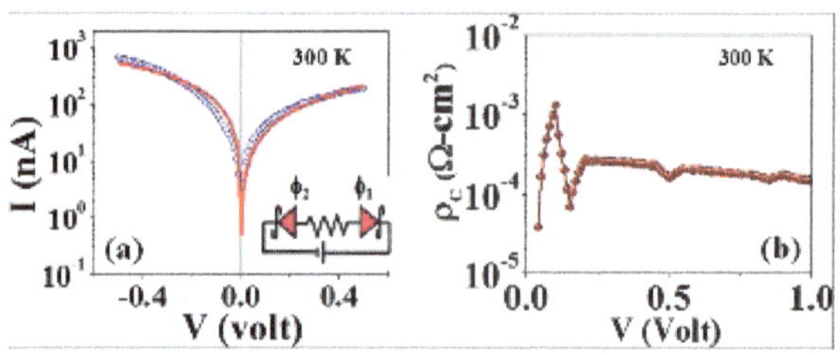

Figure 2 *I−V* characteristics and specific contact resistance. (**a**) The I−V characteristics at 300 K where the solid line shows a fitted curve using Equation 1 (see text). (**b**) The variation of specific contact resistivity with bias voltage.

where $V' = V - I\,R_{NW}$, R_{NW}. (In the equation above, φ_1 is related to the terminal with V+ve.) I_0 arises from thermoionic emission. The I−V data at low bias (<0.5 V) as well as the fit to the data are shown in Figure 2a (solid line). Equation 1 fits the I−V data well, and we could obtain the barrier heights. For the data shown in Figure 2a,$\varphi_1 \approx 0.1$ eV and $\varphi_2 \approx 0.04$ eV. From the contact resistance R_C measured as a function of bias, as depicted before, we obtained the bias-dependent specific contact resistance ρ_C in Figure 2b. With increase of bias, ρ_C is substantially reduced (by nearly a factor of 2). We limited the analysis to bias up to 1 V, because the variation of ρ_C saturates for bias that is much higher than the barrier heights φ_1 and φ_2.

Low-frequency noise measurements on MSM device

Measurement of low-frequency noise (resistance fluctuation) at room temperature (300 K) was done using the ac detection scheme [12] shown in Figure 3a. The ac bias V_{ac} is used to measure the fluctuation, while the dc bias $V_{dc}$was applied independently for tuning the device at a given point on the I−V curve [13, 14, 15]. The applied V_{dc} lowers the contact resistance as well as the noise from the junction region. The separate control of the V_{ac} and V_{dc} is important because it decouples the biasing needed for sending current through the MSM device from the noise measurement. Our measurement allows us, even at a relatively high level of V_{dc}, to maintain V_{ac} at a low level such that $S_V(f) \propto V_{ac}^2$. This makes the noise measurement process ohmic, and one can obtain the correct value of the relative fluctuations. The noise spectra were taken in the window $f_{min}=0.01$ Hz to $f_{max}=10$ Hz. The normalized variance of resistance noise (mean square fluctuation) can be obtained as $\langle(\Delta R)^2\rangle/R^2 \equiv \left(1/V_{ac}^2\right)\int_{f_{min}}^{f_{max}} S_V(f)\,df$, where $f_{min} \rightarrow f_{max}$ is the bandwidth of measurements. For $f>f_{max}$, background noise (mostly Nyquist noise) dominates, and for $f<f_{min}$, long-term drifts interfere with the measurement because of long data acquisition time [15]. The magnitude as well as the PSD shows a large dependence on the dc bias. Figure 3b shows the typical time series of resistance fluctuations for two representative dc bias voltages but with the same V_{ac}.

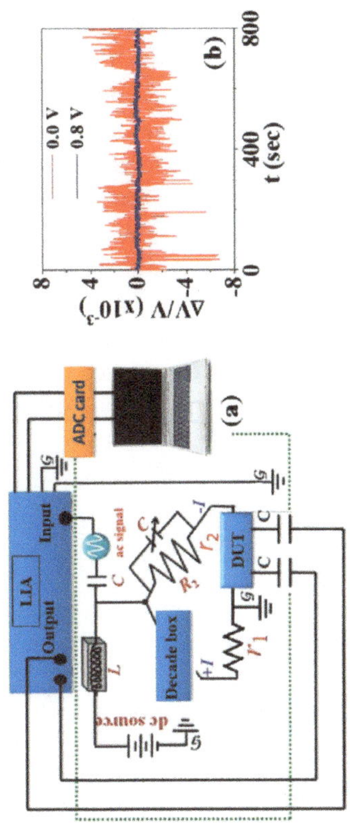

Figure 3 Noise detection scheme and time series of resistance fluctuations. (**a**) The schematic diagram of the ac noise detection scheme with the application of dc bias. (**b**) The typical time series of resistance fluctuations for two representative dc bias voltages but with the same V_{ac}.

The noise data reported here were taken with the contact with larger barrier height (φ_1) forward biased. The dominant contribution to the contact noise as well as the contact resistance arises from this contact. On applying forward bias to this junction, the noise (as well as the contact resistance) is severely reduced. The other contact with much smaller barrier (φ_2) has much less contribution to the contact noise. Thus, even if it is reversed biased (and the depletion width increases due to the reverse bias), its contribution still remains low.

RESULTS AND DISCUSSION

The normalised PSD $S_V(f)/V_{ac}^2$ is shown in Figure 4 which is $\propto 1/f^{\alpha}$. The data has been taken with varying dc bias. The superimposed dc bias reduces the magnitude of $S_V(f)/V_{ac}^2$, and the change is approximately five orders of magnitude. The dc bias also changes the nature of frequency dependence. For $V_{dc}=0$, $\alpha \approx 2$. However, α becomes approximately 1 for $V_{dc} \geq 0.2$ V, which is larger than the barrier heights. Thus, the applied V_{dc} reduces not only the specific contact resistance ρ_C, but also the magnitude of PSD significantly along with a change in the frequency dependence. Our measurement also allows independent measurement of the frequency-independent background noise S_{bg}. The inset of Figure 4 shows the S_{bg} with different applied V_{dc}. We find that S_{bg} is also reduced with increased V_{dc}, although it is much less than the suppression of the flicker noise. The S_{bg} was found to be the same as the Nyquist noise $S_{nyq}=4k_B T R$, where R is the total resistance $=R_C+R_{NW}$. The reduction of the Nyquist noise occurs mainly due to reduction of R_C by the dc bias. This analysis separates out the noise due to the contact resistance which appears in the frequency-independent Nyquist noise. The observed flicker noise ($S_V(f)$) occurring on top of the Nyquist noise has two components: one arising from the junction region at the M-S interface and the other likely from the bulk of the Si NW. This can be intrinsic for the NW and can arise either from the defect-mediated mobility fluctuation or the carrier density fluctuation which arises from recombination-generation process [16]. The superimposed bias V_{dc} dependence of the flicker noise cleanly separates out the above two contributions.

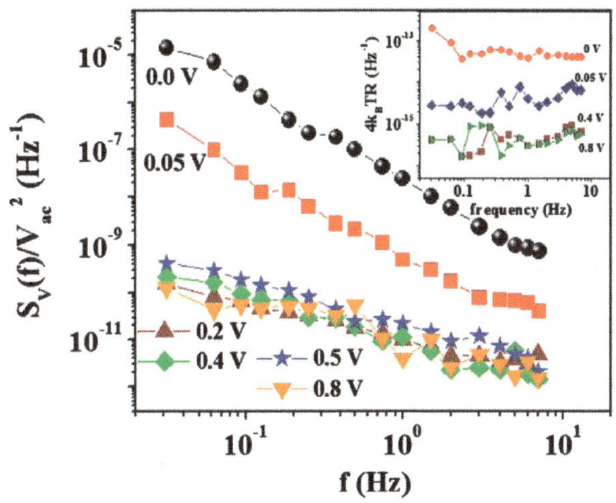

Figure 4 The power spectral density $S_V(f)/V_{ac}^2$ a function of frequency *f* at few representative superimposed V_{dc}. The inset shows the Nyquist noise for different V_{dc}.

To elucidate further, we have plotted the normalized mean square fluctuation $\langle(\Delta R)^2\rangle/R^2$ as a function of V_{dc} in Figure 5a. There is a steep decrease of $\langle(\Delta R)^2\rangle/R^2$ by more than four orders, when $V_{dc}>0.2$ V. At low V_{dc} (< barrier height), the noise is predominantly dominated by the junction noise. For higher V_{dc}, the junction noise is suppressed substantially, and residual observed noise gets dominant contribution likely from the intrinsic noise due to the Si NW. The changing spectral character of PSD is quantified by α plotted against V_{dc} in Figure 5b. We found that α is nearly 2 for low V_{dc} and can arise from the depletion region at the M-S contact. For $V_{dc}>0.2$

V, αdecreases and reaches a bias-independent value of 0.8±0.1. α≈1 is an indication of conventional 1/f noise spectrum which arises from the Si NW.

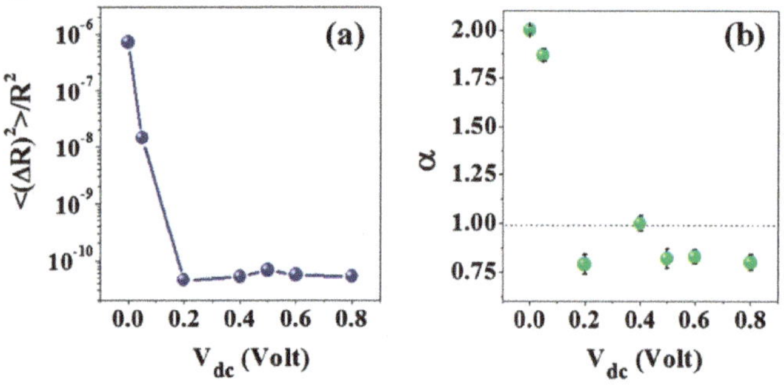

Figure 5 The variation of (a) $\langle(\Delta R)^2\rangle/R^2$ and (b) α as a function of V_{dc} at 300 K.

Evaluation of the noise in a single Si NW needs to be put in perspective and compared with bulk systems. In noise spectroscopy, one often uses a quantitative parameter for noise comparison is the Hooge parameter [17]. The spectral power of 1/f noise in many conductors often follows an empirical formula [17] $S_V(f) = \frac{\gamma_H V_{ac}^2}{N.f^{\alpha}}$ where γ_H is the Hooge's parameter, and N is the number of carriers in the sample volume (between voltage probe leads). γ_H is a useful guide when one compares different materials. Usually, a low γ_H is associated with a sample with less defect density that contributes to the 1/f noise arising from the defect-mediated mobility fluctuation [18]. N can be calculated from carrier density $n=2\times10^{17}/cm^3$ and volume of the sample between two voltage leads ($30\times10^{-16}cm^3$). Our Si NW under test has N≈600. We obtained $\gamma_H \simeq 10^{-8}$ when we use the limiting value of the PSD for $V_{dc} \geq 0.2$ V.

For bulk crystalline Si, the noise has been studied extensively both in low-doped and degenerately doped crystals [15] as well as in films [19, 20]. In bulk Si wafers with low doping concentration, the value of γ_H lies in the range of 10^{-7} to 10^{-2} with the exact value being a sensitive function of impurity and defect process conditions [15, 17]. For the Si NW, we observed that the value can even be lower. We note, however, that in this size range, it has not been established that such a scaling of spectral power with 1/N truly holds as there can also be significant surface contributions. Thus, the use of γ_H, as a parameter for comparison is done with caution. The intrinsic contribution in a NW can be large because N is small. In a NW, if the γ_H is indeed low as observed, this will mitigate the increase in the intrinsic noise on size reduction. For even smaller devices with smaller diameter, less dopant and closer contacts, N can even be below 10.

In this report, we propose a likely scenario of suppression of the junction noise by V_{dc}. The noise at the M-S contact can arise in the depletion region where the SB forms. The traps in the depletion region can lead to substantial noise due to trapping-detrapping of carriers. Such a noise has been observed also in the depletion region of MOSFETs [7]. Flicker noise in sub-micron MOSFETs [7] have been investigated experimentally as well as theoretically, and it shows the existence of both $1/f^2$ and 1/f frequency components, where the $1/f^2$ component arises from charge exchange with traps in the oxide region. Application of the dc bias reduces the depletion width (d_{dw}). In an ideal SB, $d_{dw} \propto (\phi - V_{dc})^{1/2}$; for $V_{dc} \geq \phi$, $d_{dw} \rightarrow 0$. In such case, the trapping centres are occupied and cannot contribute to the trapping-detrapping process generated noise. This leads to severe suppression of the noise in the junction region. Another strong evidence that the noise at the junction arises from the trap states in the depletion region is the value of the exponent α. It has been shown that existence of trap states in the depletion region can lead to a power spectrum of the type $S_v(f) \propto 1/f^{\alpha}$ where α=2 [21]. We also found α≈2 for a very low dc bias, when the observed noise is mainly due to the junction noise. α rapidly reduces to ≈1 for high V_{dc}. The suggested mechanism for noise reduction with applied V_{dc} is the controlling of d_{dw} which can be a generic mechanism for an MSM device and thus has a general applicability for such junctions.

CONCLUSION

To summarize, we have measured the electrical noise in an MSM device consisting of a single stand of Si NW with a diameter of approximately 50 nm. The flicker noise as well as Nyquist noise was measured with ac excitation with a superimposed dc bias. On application of a dc bias, more than the Schottky barrier height reduces the contact resistance which leads to reduction of the Nyquist noise as well as the flicker noise ($\propto 1/f^{\alpha}$) in the MSM device along with a change of α from 2 to ≃1. The part of the noise suppressed by dc bias has been interpreted as arising due to trapping-detrapping noise in the depletion region at the interface. The residual noise has been has been linked to the noise in the single Si NW, which has the conventional 1/f spectral power density with an estimated Hooge parameter $\gamma_H \simeq 10^{-8}$.

ABBREVIATIONS

HRTEM:

High-resolution transmission electron microscopy.

DECLARATIONS

Acknowledgements

The authors thank Nanomission, Department of Science and Technology, Govt. of India for financial support as sponsored projects UNANST-II and Theme Unit of Excellence in Nanodevice Technology.

Authors' original submitted files for images

Below are the links to the authors' original submitted files for images.

11671_2013_1427_MOESM1_ESM.pdf Authors' original file for figure 1

11671_2013_1427_MOESM2_ESM.pdf Authors' original file for figure 2

11671_2013_1427_MOESM3_ESM.pdf Authors' original file for figure 3

11671_2013_1427_MOESM4_ESM.pdf Authors' original file for figure 4

11671_2013_1427_MOESM5_ESM.pdf Authors' original file for figure 5

Competing interests

The authors declare that they have no competing interests.

Authors' contributions

KD synthesized the Si NWs and fabricated the single NW device by nanolithography. SS did all the electrical measurements and the low-frequency noise measurements. SS performed the treatment and calculations on the experimental data and prepared the manuscript initially. AKR gave sufficient ideas and concepts to the whole work. All authors have read and approved the manuscript.

REFERENCES

1. Byon K, Tham D, Fischer JE, Johnson AT: Systematic study of contact annealing: ambipolar silicon nanowire transistor with improved performance. Appl Phys Lett 2007, 90: 143513/1–143513/3.
2. Goldberger J, Hochbaum AI, Fan R, Yang P: Silicon vertically integrated nanowire field effect transistors. Nano Lett 2006, 6: 973–977. 10.1021/nl060166j
3. Cui Y, Wei Q, Park H, Lieber CM: Nanowire nanosensors for highly sensitive and selective detection of biological and chemical species. Science 2001, 293: 1289–1292. 10.1126/science.1062711
4. Stern E, Klemic JF, Routenberg DA, Wyrembak PN, Turner-Evans DB, Hamilton AD, LaVan DA, Fahmy TM, Reed MA: Label-free immunodetection with CMOS-compatible semiconducting nanowires. Nature 2007, 445: 519–522. 10.1038/nature05498
5. Bae J, Kim H, Dang CH, Zhang Y, Choi YJ, Nurmikko A, Wang ZL, Zhang Xiao-Mei: Si nanowire metal-insulator-semiconductor photodetectors as efficient light harvesters. Nanotechnology 2010, 21: 095502/1–095502/5.
6. Paladino E, DArrigo A, Mastellone A, Falci G: Decoherence times of universal two-qubit gates in the presence of broad-band noise. New J Physics 2011, 13: 093037/1–093037/34.
7. Wei C, Zhou X, Singh N, Rustagi SC, Lo GQ, Kwong D-L, Xiong Yong-Zhong: Investigation of Low-frequency noise in silicon nanowire MOSFETs in the subthreshold region. IEEE Electron Device Lett 2009, 30: 668–671.
8. McWhorter AL: Semiconductor Surface Physics. Edited by Kingston RH. Philadelphia: University of Pennsylvania Press; 1957. 207–228 207–228
9. Peng KQ, Yan YJ, Gao SP, Zhu J: Synthesis of large-area silicon nanowire arrays via self-assembling nanoelectrochemistry. Adv Mater 2002, 14: 1164–1167. 10.1002/1521-4095(20020816)14:16<1164::AID-ADMA1164>3.0.CO;2-E
10. Chakravorty M, Naik J, Das K, Prewett PD, Raychaudhuri AK: Temperature dependent resistivity of platinum-carbon composite nanowires grown by focused ion beam on SiO2/Si substrate. Microelectronic Eng 2011, 88: 3360–3364. 10.1016/j.mee.2011.07.012
11. Norde H: A modified forward I-V plot for Schottky diodes with high series resistance. J Appl Phys 1979, 50: 5052–5053. 10.1063/1.325607
12. Scofield JH: ac method for measuring low-frequency resistance fluctuation spectra. Rev Sci Instrum 1987, 58: 985–993. 10.1063/1.1139587
13. Ghosh A, Kar S, Bid A, Raychaudhuri AK: A set-up for measurement of low frequency conductance fluctuation (noise) using digital signal processing techniques. e-print arXiv:cond-mat/0402130v1 1987, 58: 1–25.
14. Bora A, Raychaudhuri AK: Evolution of 1/fα noise during electromigration stressing of metal film: spectral signature of electromigration process. J Appl Phys 2006, 99: 113701/1–113701/7.
15. Raychaudhuri AK: Measurement of 1/f noise and its application in materials science. Curr Opin Solid State Mater Sci 2002, 6: 67–85. 10.1016/S1359-0286(02)00025-6
16. Van der Ziel A: Noise in Solid State Devices and Circuits. New York: Wiley Interscience; 1986.
17. Hooge FN: Discussion of recent experiments on 1/f noise. Physica 1976, 60: 130–144.
18. Dutta P, Horn PM: Low-frequency fluctuations in solids: 1/f noise. Rev Mod Phys 1981, 53: 497–516. 10.1103/RevModPhys.53.497
19. Li SB, Wu ZM, Jiang YD, Li W, Liao NM, Yu JS: Structure and 1/f noise of boron doped polymorphous silicon films.Nanotechnology 2008, 19: 085706/1–085706/6.
20. Li S, Jiang Y, Wu Z, Wu J, Ying Z, Wang Z, Li W, Salamo G: Origins of 1/f noise in nanostructure inclusion polymorphous silicon films. Nanoscale Res Lett 2011, 6: 281/1–281/6.
21. Rajan NK, Routenberg DA, Chen J, Reed MA: Temperature dependence of 1/f noise mechanisms in silicon nanowire biochemical field effect transistors. Appl Phys Lett 2010, 97: 243501/1–243501/3.

CHAPTER 11

HIGH-ENERGY SUB-NANOSECOND OPTICAL PULSE GENERATION WITH A SEMICONDUCTOR LASER DIODE FOR PULSED TOF LASER RANGING UTILIZING THE SINGLE PHOTON DETECTION APPROACH

Jaakko Huikari, Eugene Avrutin, Boris Ryvkin, Juha Kostamovaara

Keywords: Semiconductor laser diode, Gain switching, Laser radar, Single photon detection, Laser ranging, Solid-state scanner

INTRODUCTION

Recently, it has been shown that the single photon detection (single photon avalanche detector, SPAD) mode presents a very interesting option for the receiver of a pulsed time-of-flight (TOF) laser radar [1–4]. The main advantages of the single photon detection compared to linear detection (e.g. based on avalanche photo detection, APD) are the sensitivity and the possibility of realizing the SPAD or even an array of SPADs on a standard CMOS integrated circuit (IC) technology. The sensitivity is strongly wavelength dependent and limited by the photon detection probability (PDP). The PDP is typically 5–2 % in the wavelength range of 800–900 nm meaning that approximately 50 photons per pulse are needed for a valid detection. This is about an order of magnitude less than what is needed in a typical linear receiver [4]. In addition, no separate analogue amplifier channel is needed which simplifies the receiver considerably. A SPAD is essentially a reverse biased p-n junction with a bias voltage higher than the breakdown voltage of the junction. The detected photon introduces a fast voltage change over the junction with a magnitude of approximately $V_{bias} - V_{br}$. Thus, a digital level signal is immediately achieved with a proper bias. The noise of the detector is being produced by the random detections of thermally generated charge-carriers within the depleted region of the diode, and also by the photons due to the background radiation (from the Sun) falling into the field of view of the receiver. It is important also to note that the timing jitter of a CMOS based SPAD is quite low, typically only of about 50–100 ps (FWHM) [5, 6].

The possibility of realizing a sensitive and low jitter SPAD detector or a SPAD array in a standard CMOS technology does not only simplify the radar construction, but makes it possible to suggest completely new laser radar configurations. For example, realizing a miniaturized solid-state scanner based on the pulsed time-of-flight approach and the appealing concept of focal plane scanning becomes possible. From the miniaturization point of view, this approach would benefit from the availability of semiconductor laser source producing highly energetic and short pulses with a characteristic pulse length on the scale of the jitter of the SPAD (~100 ps). Radiometric calculations show that a pulse energy of ~1 nJ is needed for non-cooperative targets within a range of tens of meters from the emitter/receiver, with the size of optics of ~5 cm^2 [4].

The achievable single shot precision of such a laser radar working in the single photon detection mode is set by the optical pulse width and the single photon detector response jitter. With a pulse width of 100 ps, a single shot distance measurement precision of the order of a few centimeters is available (67 ps is equivalent to 1 cm). Thus, such a configuration brings forth a prospect of constructing a very compact, high speed and accurate TOF-laser radar and possibly also solid-state 2D and 3D imagers.

Citation: Huikari, J., Avrutin, E., Ryvkin, B. et al. High-energy sub-nanosecond optical pulse generation with a semiconductor laser diode for pulsed TOF laser ranging utilizing the single photon detection approach, doi:10.1007/s10043-016-0189-7.

In the earlier work, it was shown that highly intensive and fast optical pulses can be produced with semiconductor lasers utilizing a large equivalent spot size, or the ratio (d_a/Γ_a) of the active layer thickness d_a to the confinement factor Γ_a [7–10]. Rate equation based simulations showed that the use of a large equivalent spot size of >2 μm in the laser diode construction results in enhanced gain switching and single (after pulse-free) optical pulses with an energy in the range of a few nanojoules and a duration around 100 ps as is shown in details in the later sections of this paper. Moreover, a driving pulse with a peak current of <10 A and a pulse width of ~1 ns only is needed; such pulse parameters are readily available with MOS-switch based pulsing electronics [12]. As a result, not only the SPAD based receiver, but also the semiconductor laser diode based transmitter can be miniaturized, which is highly desirable in the anticipated 3-D measuring applications.

The goal of this paper is to present and compare measured results from bulk and quantum well laser diodes utilizing the "enhanced gain switching" principle and aiming at the above mentioned performance parameters. The comparison is carried out with regard to the parameters, e.g. the temperature characteristics, which are relevant especially for the laser radar system level design. Section 2 discusses briefly the laser designs studied, and the measured results are given in Sect. 3. Some system level results are given in Sect. 4, and finally conclusions are made in Sect. 5.

LASER STRUCTURES

All lasers used in this work were of the asymmetric waveguide type investigated in our earlier work [7, 8, 10] with the refractive index contrast between the optical confinement layer (OCL) and the p-cladding substantially stronger than that between the OCL and the n-cladding, and the active layer position shifted strongly from the centre of the OCL towards the p-cladding. With most of the transverse mode power localized within the OCL, the lasers were of the broad asymmetric as opposed to Narrow Asymmetric waveguide type [13]. As shown previously [7, 8,13], lasers of this type combine the large effective spot size crucial for enhanced gain-switching, built-in single transverse mode operation, low internal losses maintained at high currents, good injection efficiency, and good fabrication tolerance. Two types of laser structures were investigated, with either quantum well (QW) or bulk active layers. The QW lasers were very similar to those used in [11]. They had an active layer in the form of five thin (4 nm thick each) GaAs/ AlxGa1-xAs ($x \approx 0.3$) QWs corresponding to a total da = 20 nm; the operating wavelength of these lasers was 808 nm. The value $da/\Gamma a \approx 3$ μm was achieved by shifting the active layer towards the p-cladding beyond the position of peak modal intensity. The injection efficiency of the QW lasers was measured as ≈0.75, and the internal loss, as ≈1.5 cm^{-1}. The bulk lasers had a waveguide structure similar to that of QW lasers and the same $d_a/\Gamma_a \approx 3$ μm, but with a broader waveguide layer and a GaAs active layer 80 nm thick; the operating wavelength was ≈870 nm.

EXPERIMENTAL

The time domain measurements were performed to determine the relation between the drive current pulse and the optical output pulse for the bulk and quantum well semiconductor laser designs, and specifically to determine their relative positions in the time domain.

The laser diodes' drive current pulse was determined by measuring the voltage drop across a damping resistor in series with the laser diode. Optical pulse power was determined by combining the results of laser pulse time domain measurements and laser pulse train average power measurements. In the time domain measurement, the optical energy was collected with a lens pair and guided, via graded index optical fiber, to the input of a 24 GHz OE converter, whereas pulse train average power was measured with a calibrated photodiode detector.

Shown in Figs. 1 and 2 are the time domain measurement results of bulk and quantum well laser designs, respectively. Figures display separate time domain measurement results, note the drive current amplitude increasing from right to left, on the same frame for a convenient assessment. Results show driving current pulses with a amplitudes of an ampere scale and a duration of the order of 1 ns producing high speed, ~100 ps, laser pulses, with the optical power reaching up to 10 W with a bulk laser diode, and slightly less with a quantum well LD.

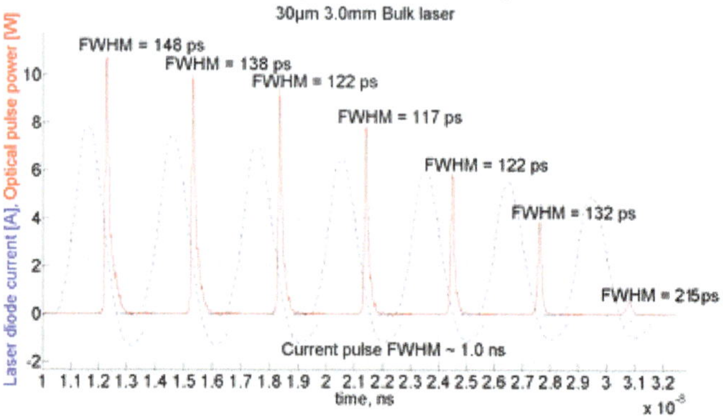

Figure. 1 Time domain measurement results of a 30 μm 3.0 mm (stripe width and cavity length) bulk laser

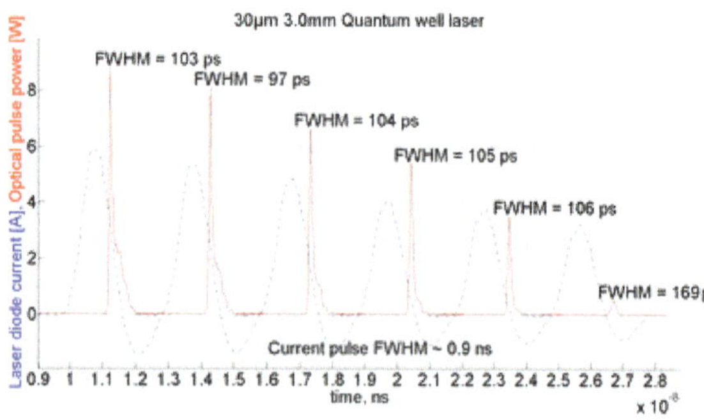

Figure. 2 Time domain measurement results of a 30 μm 3.0 mm (stripe width and cavity length) quantum well laser

To determine and compare the temperature dependence of the output from the 30 μm wide, 3.0 mm long bulk and quantum well lasers, the time domain measurements were carried out at three different operation temperatures. The temperature performance result plots (Figs. 3, 4) show the laser pulse peak value as a function of the drive current pulse peak value at the temperatures of 0, 30 and 60 °C. During the measurements, the laser transmitter temperature was controlled using a thermal oven. The transmitter heating up due to laser drive current pulsing is assumed to be negligible at the relatively low 10 kHz pulsing frequency used in the measurements.

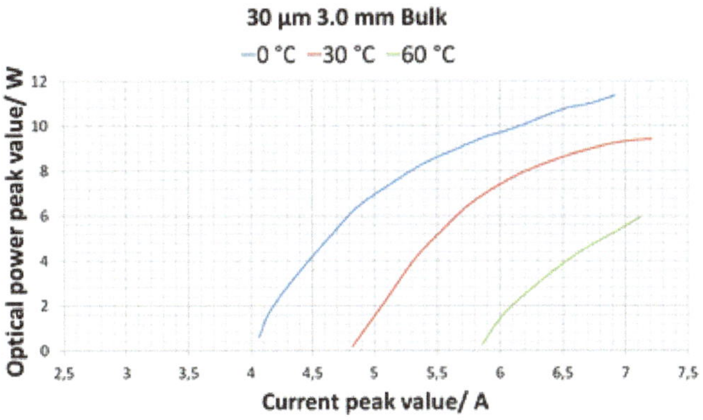

Figure. 3 The optical power peak values vs. drive current peak values of the 30 µm 3.0 mm bulk laser in three different temperatures

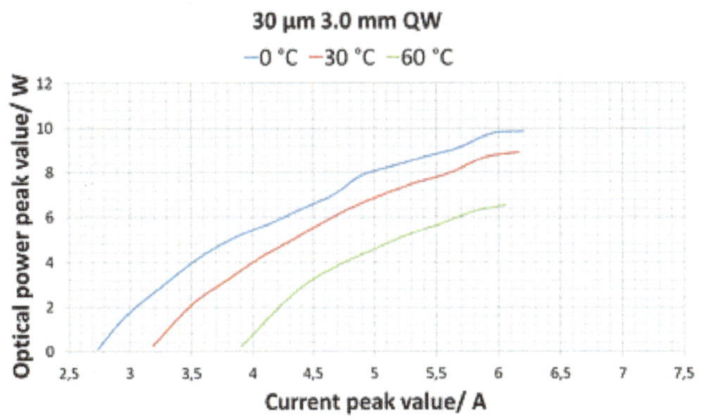

Figure. 4 The optical power peak values vs. drive current peak values of the 30 µm 3.0 mm quantum well laser in three different temperatures.

The bulk laser diode's measurement result in Fig. 3 and that of a quantum well laser diode in Fig. 4 show the achievable optical peak power being similar for both laser diode designs. However, the quantum well laser's threshold current is significantly lower (mainly because of the larger thickness of the bulk active layer). The figures also show the temperature increase causing more dramatic effect in optical output power the closer the laser's driving current peak value is to the lasing current threshold value. For instance, in the case of the bulk laser, at the drive current peak value of 6 A, 60 °C degrees temperature increase almost switches off the laser, while in the case of the quantum well laser, the optical output power peak value decreases by ~35 % only. Preliminary calculations show however that the main reason for this is likely to be simply the much smaller total active layer thickness d_a of the quantum well laser compared to that of the bulk lasers, rather than the properties of the QW material directly caused by size quantization. This means that a bulk laser with a substantially decreased d_a may potentially offer temperature stability comparable to that of the QW lasers with the same d_a/Γ_a.

The laser beam spatial sweep measurements were performed to define and compare the spatial distribution (in the slow axis direction/plane) of the laser pulse energy, and also to measure the spatial homogeneity of the time domain behavior of emission. The measurement was done using a laboratory laser radar system, the target being a 2 mm wide black rubber strip (reflectance ~3 %) which was swept across the focused laser beam with a width of approximately 28 mm, at a distance of 24.5 m from the laser radar system. The horizontal transition step of the strip between successive measuring points was also 2 mm. An illustration of the strip target spatial sweep setup is shown in the Fig. 5. At every measurement point, a measuring time of 30 s was used to accumulate a histogram of laser pulse flight times with a time interval measurement unit.

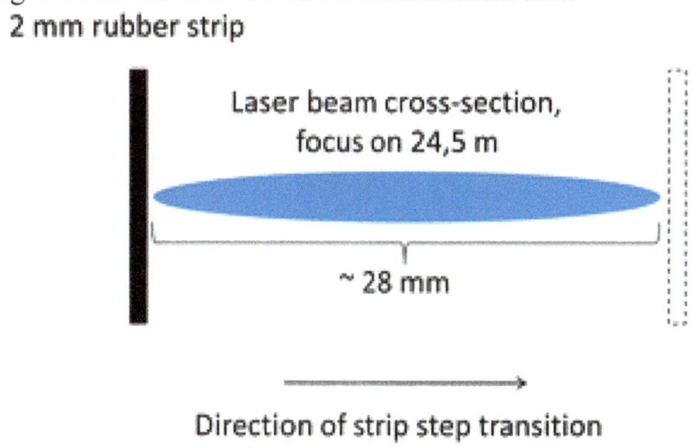

Figure. 5 The strip target spatial sweep over the laser beam at 24.5 m

The measured detection rate of the laser radar system at both laser designs was ~1 % indicating that the detector operating mode was in the single photon mode. At a detection rate higher than ~10 %, signal distortion begins to occur due to photon pile-up, since detection is a result of several photons incident on the detector [14]. Comparing the results in Figs. 6 and 7, we note that the quantum well laser diode shows more evenly distributed spatial beam energy than the bulk laser. The bulk laser diode distribution shows two energy peaks, the narrower of which one has a higher energy. Noticeable variation in the time-domain shapes of the measured hit distributions is also visible in both of the laser diode designs.

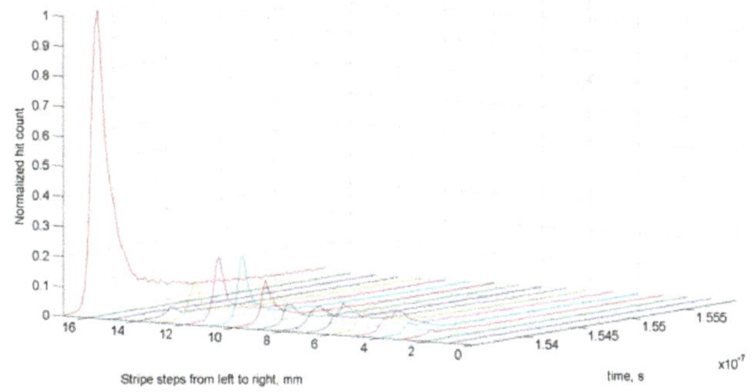

Figure. 6 Time of flight histogram representation of spatial and time-domain energy distribution of the 30 µm 3.0 mm bulk laser

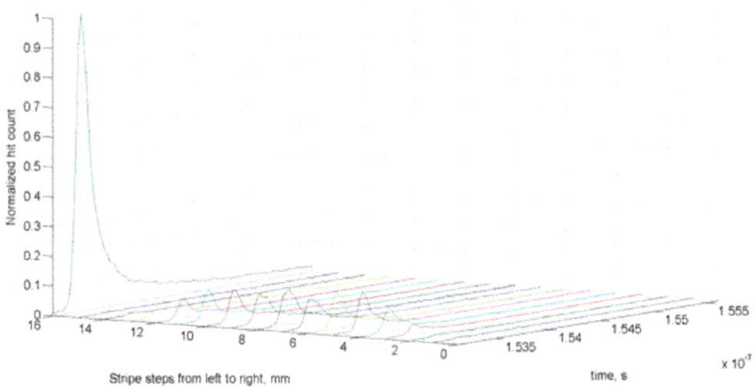

Figure. 7 Time of flight histogram representation of spatial and time-domain energy distribution of the 30 µm 3.0 mm quantum well laser

The temperature behavior of the 30 μm wide, 3.0 mm long quantum well laser diode is thus more appealing from the practical application point of view, and thus it was selected for further studies. These included the 2-D (wavelength versus time) studies of the optical output with a monochromator/streak set-up. Also, a more accurate measurement of the time-domain shape of the optical pulse was expected with the streak camera measurement.

The triggering signal of the streak camera sweep unit was produced from the laser diode current pulse which has significantly smaller jitter with respect to the optical pulse than that measured from the driver input. Using this approach, a FWHM of the jitter between current pulse and streak camera sweep triggering pulse was ~40 ps.

A 3D plot of the streak camera measurement results is shown in Fig. 8.

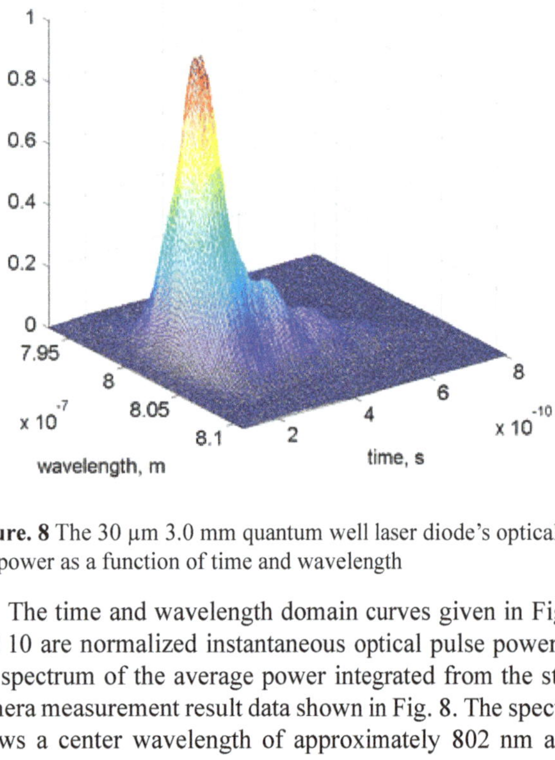

Figure. 8 The 30 μm 3.0 mm quantum well laser diode's optical output power as a function of time and wavelength

The time and wavelength domain curves given in Figs. 9 and 10 are normalized instantaneous optical pulse power and the spectrum of the average power integrated from the streak camera measurement result data shown in Fig. 8. The spectrum shows a center wavelength of approximately 802 nm and a FWHM of 4 nm, as can be seen in Fig. 10.

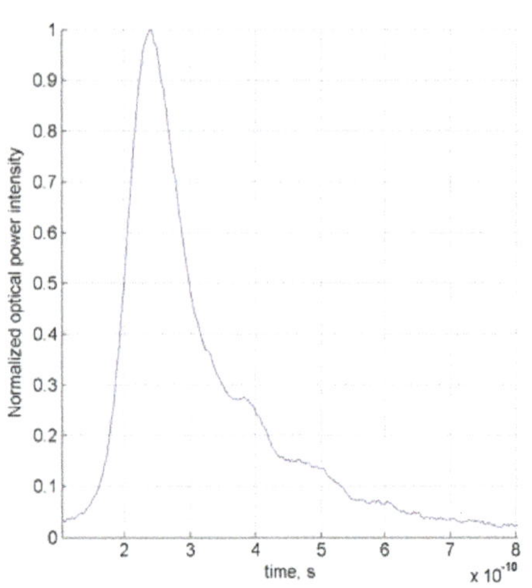

Figure. 9 Time domain integration of streak camera measurement result of the 30 μm 3.0 mm quantum well laser diode

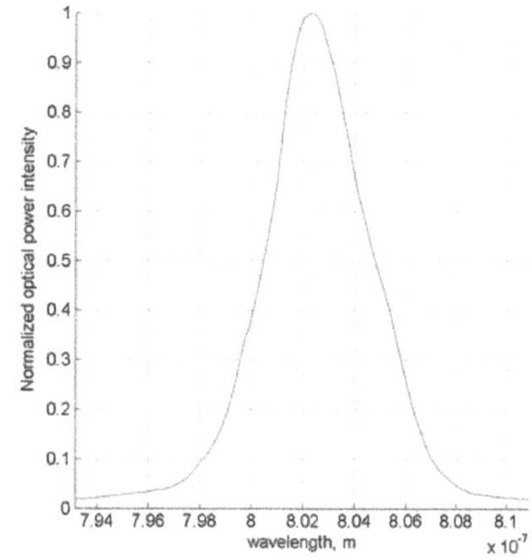

Figure. 10 Wavelength domain integration of streak camera measurement result of the 30 μm 3.0 mm quantum well laser diode

A far field distribution measurement was performed to determine the QW laser diode's beam divergence and thus to estimate the size of the required transmitter optics. The laser diode transmitter was attached on a turn table with two perpendicular orientations and the emitted average power was measured with an average optical power meter, applying a 2 mm entrance slit, 1 m away from the laser diode. From the results given in Fig. 11, the fast and slow axis FWHM of 15° and 4°, respectively, can be seen. From the practical point of view, the small wide axis value of 15° is highly desirable since it keeps the size of the transmitter optics reasonable while providing a high optical energy collection efficiency.

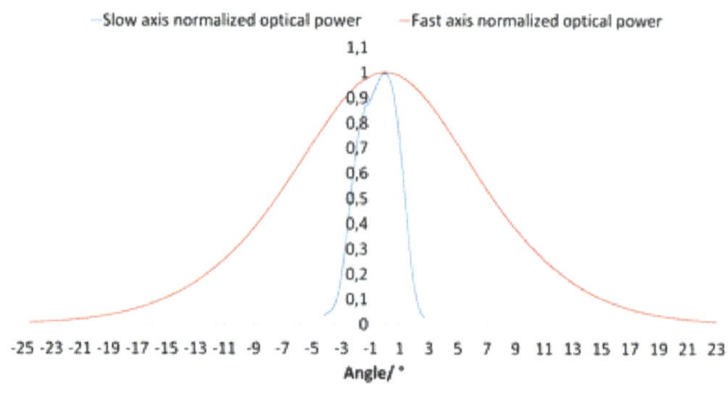

Figure. 11 Fast and slow axis beam divergence of the 30 μm 3.0 mm quantum well laser diode

LASER RADAR EXPERIMENTS

Some preliminary system level measurements were done to get an idea of overall performance of the single photon detection based laser radar. The transmitter consisted of the 30 μm wide, 3.0 mm long quantum well laser diode, and the receiver used a single ~20 μm CMOS SPAD element with a receiver aperture of 18 mm. The system's performance parameters of interest were the achievable detection rate and the single shot precision. The target was a black rubber surface (reflectance ~3 %) at a distance of ~25 m from the radar. Transmitted laser pulse counts were 10, 20, 30, 40, 50, 100, 200, 400 and 800 and each pulse count measurement was repeated ten times. The applied pulsing frequency was 100 kHz. Measurements were performed in normal indoor lightning conditions.

Measurement results presented in Fig. 12 show TOF histograms of the above mentioned pulse counts. The hit counts in individual measurements were 2, 8, 10, 17, 17, 34, 74, 101 and 182, and thus the detection rate varies between 20 and 40 %. The cumulative shape of the results distribution resembles a convolution of the optical pulse and the SPAD time-domain response which is becoming increasingly pronounced with increasing pulse counts. Most of the hits are concentrated around the expected flight time, corresponding to the distance to the target, and show also the expected diffusion tail characteristic of CMOS SPADs [5]. The single shot precision based of the time-of-flight (TOF) histograms is ~20 mm (FWMH).

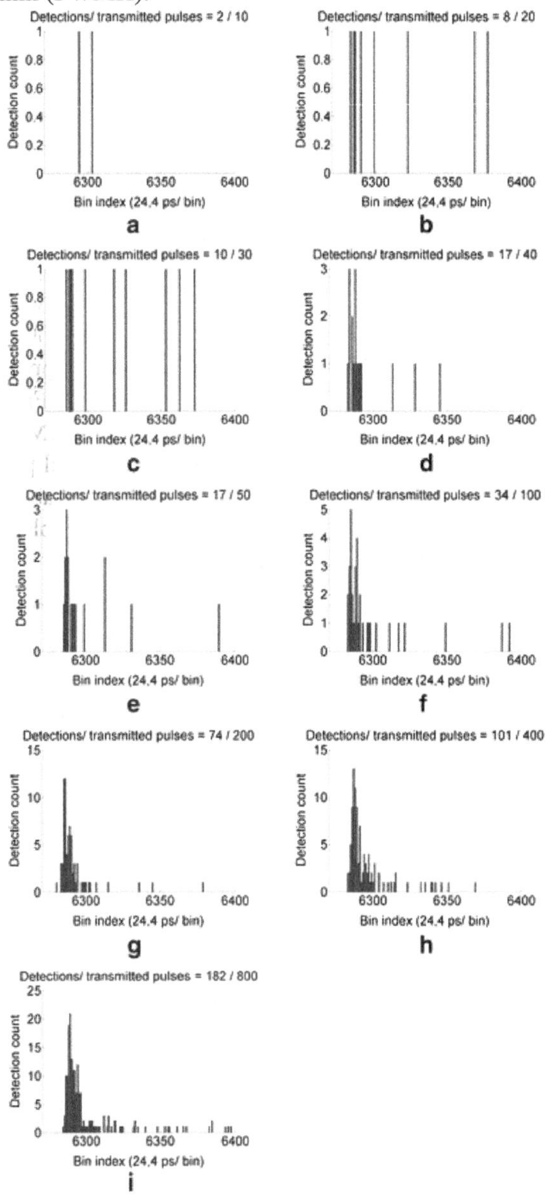

Figure. 12 Time of flight histograms with varying number of transmitted total pulse counts

CONCLUSIONS

It is shown experimentally that high energy, short (~100 ps) optical pulses can be produced with semiconductor lasers with a large equivalent spot size, or the ratio (d_a/Γ_a) of the active layer thickness d_a to the confinement factor Γ_a. QW and bulk laser diodes with $d_a/\Gamma_a \approx 3$ μm and stripe width/cavity length of 30 μm/3 mm were realized, and achieved a pulse energy of the order of ~1 nJ and a pulse length of about 100 ps, when pumped by a current pulse with an amplitude of 6–8 A and a duration of 1 ns. The far-field characteristics were also studied with a view to establishing the time-homogeneity and energy distribution. The wide axis value of the far-field distribution was measured to be 15° for a QW laser diode, which makes it possible to achieve high optical energy collection and low divergence with a low F number of the transmitter optics.

It was also demonstrated that an energy level of ~1 nJ is quite sufficient for radar measurements for distances up to tens of meters to non-cooperative targets with a receiver aperture of <20 mm. For example, a black rubber target with a reflectance of ~3 % gave a detection rate of 20–40 % at a distance of 25 m. We conclude that bulk and quantum well laser diodes utilizing the "enhanced gain switching" principle based on the large equivalent spot size are quite promising candidates for the role of a transmitter in single photon detection based laser radar devices and applications, since they allow considerable miniaturization in the transmitter part of the laser radar.

REFERENCES

1. Niclass, C., Soga, M., Matsubara, H., Ogawa, M., Kagami, M.: A 0.18-μm CMOS SoC for a 100-m-Range 10-Frame/s 200 × 96-pixel time-of-flight depth sensor. IEEE J. Solid State Circuits **49**(1), 315–330 (2014)
2. Ito, K., Niclass, C., Aoyagi, I., Matsubara, H., Soga, M., Kato, S., Maeda, M., Kagami, M.: System design and performance characterization of a MEMS-based laser scanning time-of-flight sensor based on a 256 × 64-pixel single-photon imager. IEEE Photonics J. Vol. 5, Issue 2, 15p., 2013
3. Perenzoni, D., Gasparini, L., Massari, N., Stoppa, D.: Depth-range extension with folding technique for SPAD-based TOF LIDAR systems. Proceedings of the IEEE SENSORS Conference, pp. 622–624 (2014)
4. Kostamovaara, J., Huikari, J., Hallman, L., Nissinen, I., Nissinen, J., Rapakko, H., Avrutin, E., Ryvkin, B.: On laser ranging based on high-speed/energy laser diode pulses and single-photon detection techniques. IEEE Photonics J. 7(2), 1–15 (2015)
5. Pancheri, L., Stoppa, D.: Low-Noise CMOS single-photon avalanche diodes with 32 ns dead time. Proceedings of the 37th European Solid State Device Research Conference, ESSDERC, pp. 362–365 (2007)
6. Veerappan, C., Charbon, E.: CMOS SPAD Based on photo-carrier diffusion achieving PDP >40% from 440 to 580 nm at 4 V excess bias. IEEE Photonics Technol. Lett. **27**(23), 2445–2448 (2015)ADS
7. Ryvkin, B.S., Avrutin, E.A., Kostamovaara, J.: Asymmetric waveguide laser diode for high-power optical pulse generation by gain switching. J. Lightw. Technol. **27**(12), 2125–2131 (2009)ADS
8. Ryvkin, B.S., Avrutin, E.A., Kostamovaara, J.: Quantum well laser with an extremely large active layer width to optical confinement factor ratio for high-energy single picosecond pulse generation by gain switching. Semicond. Semicond. Sci. Technol. **26**(4), 045010 (2011). **(pp. 1–4)**ADS
9. Lanz, B., Ryvkin, B.S., Avrutin, E.A., Kostamovaara, J.: Performance improvement by a saturable absorber in gain-switched asymmetric-waveguide laser diodes. Opt.

Express**21**(24), 29780–29791 (2013)ADS

10. Avrutin, E.A., Ryvkin, B.S., Kostamovaara, J.T., Kuksenkov, D.V.: Strongly asymmetric waveguide laser diodes for high brightness picosecond optical pulses generation by gain switching at GHz repetition rates. Semicond. Sci. Technol. **30**(5), 055006 (2015)ADS
11. Huikari, J., Avrutin, E., Member, I.E.E.E., Ryvkin, B., Nissinen, J., Kostamovaara, J., Senior Member, IEEE: High-energy picosecond pulse generation by gain switching in asymmetric waveguide structure multiple quantum well lasers. IEEE J. Select. Topics Quantum Electron. **21**(6), 1–6 (2015)
12. Hallman, L., Huikari, J., Kostamovaara, J.: A high-speed/power laser transmitter for single photon imaging applications. Proceedings of the IEEE SENSORS Conference, pp. 1157–1160 (2014)
13. Ryvkin, B.S., Avrutin, E.A., Kostamovaara, J.: Narrow versus broad asymmetric waveguides for single-mode high-power laser diode. J. Appl. Phys. **113**, 013104 (2013)ADS
14. Becker, W.: Advanced time-correlated single photon counting techniques. Springer, Germany (2005)

CHAPTER 12 CIRCULAR POLARIZATION IN A NON-MAGNETIC RESONANT TUNNELING DEVICE

Lara F dos Santos, Yara Galvão Gobato, Márcio D Teodoro, Victor Lopez-Richard, Gilmar E Marques, Maria JSP Brasil , Milan Orlita, Jan Kunc, Duncan K Maude, Mohamed Henini, Robert J Airey

INTRODUCTION

The understanding of the physics governing the dynamics of spin-polarized carriers in semiconductor structures is a fundamental issue for the development of new spintronic devices. In the past years, several systems have been proposed for spin-based devices, including magnetic metal/semiconductor junctions, all metallic devices, and all semiconductor systems [1, 2, 3, 4, 5, 6, 7, 8, 9, 10]. However, the change of the polarization requires the use of an applied external magnetic field to change the contact magnetization. For some device applications, it would be interesting to have devices where the spin character of the injected or detected electrons could be voltage selected. One possible approach to achieve this goal is based on resonant tunneling diodes (RTDs) because the spin character of the carriers in the structure could be voltage controlled [11, 12, 13, 14, 15].

In this work, we have investigated the polarization-resolved photoluminescence (PL) from different regions in a non-magnetic asymmetric *n*-type RTD with a GaAs quantum well (QW) and AlAs and AlGaAs barriers. This asymmetry was used to increase the accumulation of charge of carriers in the QW. Under applied bias, electrons tunnel through the double-barrier structure creating a two-dimensional electron gas (2DEG) in the QW and at the accumulation layers next to the barriers which densities and *g*-factors are bias voltage dependent. The spin-dependent tunneling of carriers was studied by analyzing the current-voltage characteristics ($I(V)$) and the right (σ^+) and left (σ^-) circular polarized PL from the contact layers and from the QW under magnetic fields up to 19 T. High magnetic fields were used in order to increase the spin-related effects in our non-magnetic RTD. The main goal of the present work is to investigate the fundamental physics of spin-related effects in our structures, but this is an essential step for analyzing the feasibility of using RTD structures for spintronic devices in the future.

We have observed small oscillations on the QW circular polarization degree as a function of the applied voltage with values up to -70% at 19 T. We have also observed optical emission from spin-polarized 2DEG and two-dimensional hole gas (2DHG) in the GaAs contact layers next to emitter and collector barriers. Under applied bias voltage, polarized carriers from contact layer tunnel through the double-barrier region and contribute to the spin polarization of carriers in the QW. The circular polarization of the QW emission seems to depend on various other points, including the *g*-factors of the different layers, the spin-polarization of injected carriers from the contact region, the density of carriers along the structure, and the Rashba and Zeeman effects.

Our device was grown by molecular beam epitaxy on a n^+ (001) GaAs substrate. The double-barrier region consists of 2 μm *n*-GaAs (1×10^{18} cm^{-3}), 0.1 μm *n*-GaAs (1×10^{17} cm^{-3}), 51 Å undoped GaAs spacer, 40 Å AlAs barrier, 50.9 Å GaAs QW, 42 Å $Al_{0.4}Ga_{0.6}As$ barrier, 51 Å GaAs spacer, 0.1 μm *n*-GaAs (1×10^{17} cm^{-3}), and 0.51 μm *n*-GaAs (1×10^{18} cm^{-3}). Circular mesas of 400 μm diameter were processed with annular AuGe contacts which allow optical measurements. Micro-PL measurements were performed at 4 K under magnetic fields up 19 T parallel to the tunnel current. The measurements were performed by using optical fibers and a Si CCD system coupled with a Jobin-Yvon spectrometer. A linearly polarized beam from an Ar^+ laser (all lines) was used for optical excitation. Therefore, photogenerated carriers in the structure do not present any preferential spin polarization. The right (σ^+) and left (σ^-) circularly polarized emissions were

Citation: Lara F dos Santos, Yara Galvão GobatoEmail author, Márcio D Teodoro, Victor Lopez-Richard, Gilmar E Marques, Maria JSP Brasil, Milan Orlita, Jan Kunc, Duncan K Maude, Mohamed Henini and Robert J Airey, Circular polarization in a non-magnetic resonant tunneling device, DOI: 10.1186/1556-276X-6-101.

selected with appropriate optics and by reversing the current in the electromagnet.

Figure 1a shows a schematic band diagram of our device under forward bias voltage and light excitation. Under applied bias, a pseudo-triangular QW is created next to the emitter barrier. Electrons which occupy the quasi-bound states in the triangular QW form a 2DEG. Resonant tunneling can occur between 2DEG states in this triangular well and resonant states in the double-barrier structure (labeled e1). Photogenerated holes can also occupy the quasi-bound states in the triangular QW next to the top contact (collector barrier) and form a 2DHG. Therefore, resonant tunneling can also occur between 2DHG states and hole resonant states (hh1, lh1, and etc.) in the QW.

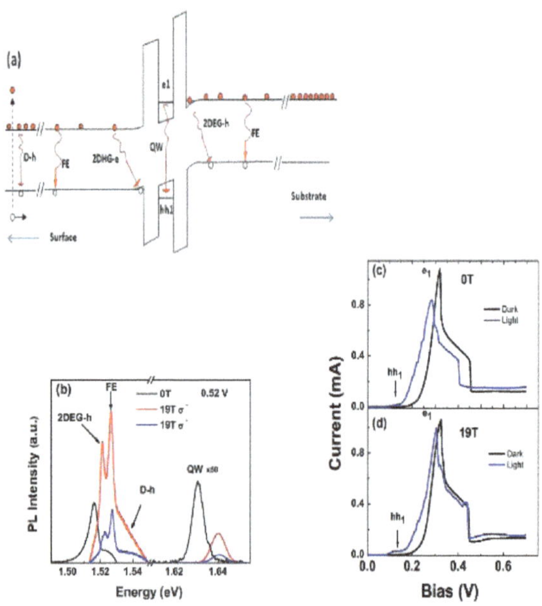

Figure 1 Schematic band diagram of our device under forward bias, light excitation, and magnetic field parallel to the tunnel current (a). Typical PL emission from contact layers and QW region under 0.52 V and 19 T (b). Current voltage characteristics curves for 0 and 19 T (c,d).

Under applied bias, photo-created holes can tunnel (resonantly or non-resonantly) and recombine with tunneling electrons into the QW and contact layers. The PL intensity from the QW is, in first approximation, proportional to the product of hole and electron densities. Therefore, it is very sensitive to the variation of charge density in the QW which can be voltage controlled in resonant tunneling devices. As a consequence, the PL intensity is, in general, correlated to the $I(V)$ characteristic curves. In our experimental conditions, the optical emission from the QW was not detected under zero bias voltage, which indicates that the optical generation of carriers inside the QW is negligible.

Figure 1b shows typical polarization-resolved PL spectra for our device at 0.52 V. In general, the GaAs contact emission includes several bands: the free-exciton (FE) transition from the undoped space-layer, the recombination between photogenerated holes and donor related electrons from the *n*-doped GaAs layers (D-H), and the indirect recombination between free holes (electrons) and confined electrons (holes) localized at the 2DEG (or 2DHG) formed at the accumulation layer next to the barriers (2DEG-h or 2DHG-e emissions). In particular, only the 2DEG-h space-indirect emission was recently observed for *p-i-n* RTDs [13]. However, its voltage dependence and its contribution to the spin polarization of carriers in the QW were not investigated. In order to have more information about the contribution of both spin-polarized 2DEG and 2DHG to circular polarization degree of the QW emission, we performed a detailed measurement of the PL emissions as a function of the applied bias in our resonant tunneling structure.

Figure 1c shows the current voltage characteristics curve ($I(V)$) under dark and under light excitation at zero magnetic field. We have observed one electron resonant peak at $V_1 = 0.32$ V which was associated with the resonant tunneling through the first confined electron state **e1** in the QW. Under light excitation, this resonant peak (**e1**) shifts to lower voltages. This shift is an evidence that the hole charge density has increased in the double-barrier region. Actually, an increase of the hole density in the QW reduces the total charge accumulated in the QW and shifts the electron resonant peak to lower voltages. The hole density at the accumulation layer also increases when we increase the applied voltage which results in the formation of a 2DHG next to the collector barrier (at the surface side, as illustrated in Figure 1a). As mentioned earlier, the energy position of confined levels in the 2DHG and QW can be voltage controlled. The hole resonant tunneling condition can be obtained by the alignment between the confined levels at the accumulation layer (2DHG) and QW. This effect is evidenced by the observation of an additional structure at 0.125 V in the $I(V)$ curve under light excitation which was associated with the first heavy hole resonance (hh1) (Figure 1c). This additional structure hh1 is better defined under magnetic field (Figure 1d). In addition, we have observed that photocurrent under low voltages is markedly larger than the electron current in the dark, which indicates that the holes actually become the effective majority carrier under this voltage condition.

Figure 2a presents the voltage dependence of the QW PL at 0 T. The PL intensity increases in the electron resonance region (**e1**) and decreases after resonant tunneling condition. Therefore, the QW PL intensity presents a good correlation with the electron resonance which is due to the important increase of electron carrier density in the QW under resonant condition. As mentioned earlier, the PL intensity is proportional to the product of the hole and electron densities and, therefore, it is very sensitive to the variation of charge density in the QW (due to accumulated holes or electrons) which results in a modulation of the PL intensity near the resonant voltages. Figure2b presents the voltage dependence of the QW PL intensity under 19 T for both σ^+ and σ^- polarizations. Under magnetic field, confined levels in the QW and contact layers split into spin-up and spin-down Zeeman states and the optical recombination can occur with well-defined selection rules giving information about the spin polarization of carriers in the structure. We have also observed a good correlation between the PL intensity and $I(V)$ characteristic curve for both polarization σ^+ and σ^-. We have also observed that the QW emission is highly σ-polarized. The voltage dependence of the polarized degree of this emission will be discussed later in this manuscript.

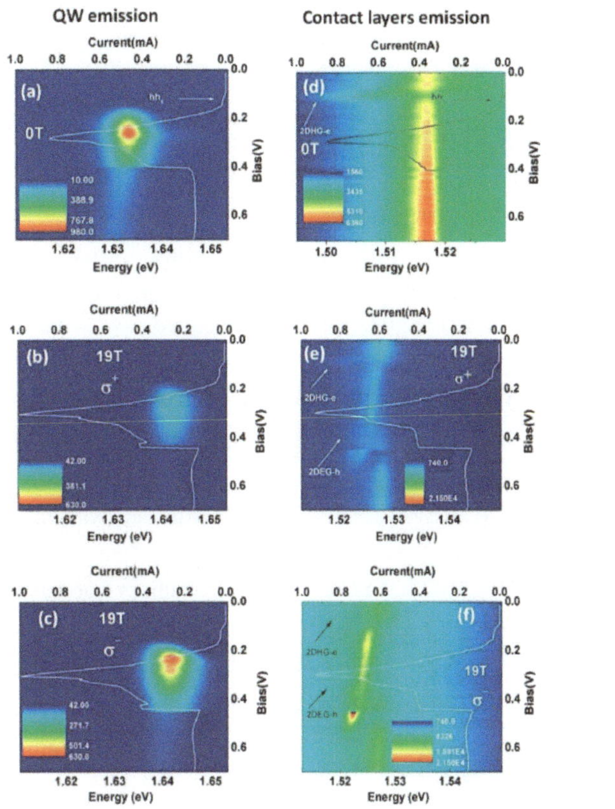

Figure 2 Voltage dependence of PL for QW (a) and contact layers (d) under zero magnetic field and polarization resolved PL for QW (b,c) and contact layers (e,f) under 19 T.

Figure 2d-f presents the voltage dependence of the PL from the GaAs contact layers. As discussed earlier, this emission includes several bands: the FE transition, the recombination between photogenerated holes and donor-related electrons (broad band) and the voltage-dependent peaks which were associated with the indirect recombination between free electrons (holes) and the 2DHG (2DEG) at the accumulation layers (Figure 1a). The 2DHG-e emission is observed for low bias voltage, before the onset of hole resonant tunneling condition while the peak attributed to the 2DEG-h recombination is only observed under applied magnetic field, for bias voltages in the range of $0.1 < V < 0.5$ V (Figure 2e,f).

We observed that an increase the applied bias results in a decrease of 2DHG-e emission intensity and in an increase of FE emission intensity. When we increase the applied voltage, photogenerated holes are swept away from the 2DHG by the increased electric field, and excitons are predominantly formed in the GaAs layer reducing the intensity of this 2DHG-e indirect transition. At hole resonant tunneling condition, holes tunnel through the QW which results in a reduction of hole density accumulated in the 2DHG which can also reduce the PL intensity of 2DHG-e emission. An abrupt transfer of FE emission to 2D electron - 3D hole emission was previously observed with increasing magnetic field (perpendicular to the 2DEG plane) for integer and fractional filling factors ($\nu < 2$) on high quality modulation doped GaAs/AlGaAs heterojunction (HJ) [16, 17, 18, 19]. Actually, it was observed that for filling factors $\nu < 2$ the FE PL intensity decreases and a new lower energy PL line abruptly appears and gain intensity at expense of the exciton PL. This abrupt transfer was explained by a phenomenological dynamical model which considers an exciton dissociation near the magnetized 2DEG. In this model, the dissociation rate depends on exciton dynamics in two well potentials that is formed by FE near the HJ interface and 3D hole interacting with the 2DEG [18]. Our system is, however, more complex than a simple HJ structure. A variation of the applied voltage in the RTD results not only in strong variations of the carrier densities at the accumulation layers, and therefore variation of the filling factor of the 2D gases in the structure, but it also directly alters the electric field along the structure, and consequently, the potential profile $V(z)$ at the accumulation layers. A complete analysis of the results requires detailed calculations, but we point out some general points which are consistent with our interpretation. The 2DHG-e peak is only observed before the onset of hole-tunneling, as for larger voltages the relatively small reservoir of photo-created holes (2DHG) accumulated at the top barrier interface must be mainly depleted. The 2DEG-h transition is observed at the electron resonant tunneling condition. Its intensity (Figure 2f) initially increases with bias voltage, which is consistent with increasing densities of tunneling holes and electrons accumulated at the 2DEG, but it shows some reduction at the **e1** resonance peak, as at this condition the density of the 2DEG should be somehow reduced. At about 0.45 V, when the $I(V)$ characteristic curve shows an abrupt current reduction, the 2DEG-h shows an abrupt increase of intensity which is consistent with a sudden increase of electron density accumulated at the 2DEG. For large voltages (>0.5 V), the 2DEG-h emission tends to vanish, which may be associated with a reduced efficiency on the localization of holes around the 2DEG due to the significantly large electric field or to reaching a critical density of electrons at the 2DEG. As mentioned earlier, on the previous works on GaAs/GaAlAs HJs [16, 17, 18, 19], the h-2DEG was only observed for magnetic fields larger than a critical value that corresponded to the filling factor $\nu = 2$, which was pointed out as the limit case at which holes are still localized near the 2DEG. In our measurements, we maintained the magnetic field constant at 19 T. However, with increasing bias voltages, the density of electrons accumulated at the 2DEG should increase, and therefore, its filling factor should also increase. Therefore, it is possible that the condition $\nu \geq 2$ is attained for an applied bias voltage of about 0.5 V, resulting in the fading out of the 2DEG-h transition.

The observed 2DEG-h emission presents a high circular polarization degree with abrupt energy discontinuities after the electron resonance. This effect can be explained considering the increase of electron density in the accumulation-layer after the electron resonance and by changes in the overlap between the 2D electron and 3D hole wave functions induced by magnetic field which affects the 2DEG-h radiative recombination lifetimes of photo-excited holes [18, 19].

Figure 3 shows the voltage dependence of the excitonic spin splitting and circular polarization degree from the QW PL under 19 T. The circular polarization degree was calculated from the following relation: $(I_{\sigma+} - I_{\sigma-})/(I_{\sigma+} + I_{\sigma-})$; where $I_{\sigma+}$ ($I_{\sigma-}$) are the integrated intensity of the right (left) circular polarization. We have observed small oscillations on the voltage dependence of the polarization degree. The QW spin-splitting presents a small variation with applied bias probably due to the Rashba and Zeeman splitting tuning of hole levels by the effective electric field [12]. It was shown earlier that the electronic structure of RTDs are affected by the variation of the effective field in the double-barrier region and by modulation of Rashba SO and screening effects induced by hole charge buildup in the QW which results in a voltage modulation of spin-splitting [12]. However, we have observed that the circular polarization of the QW emission does not follow the measured spin-splitting energy of this emission. Therefore, it cannot be attributed to a simple thermal occupation effect of the QW excitonic states, which have a rather small effective *g*-factors. On the other hand, we observe that when we have a maximum in the excitonic spin-splitting we observed a minimum in the polarization degree. It seems that the excitonic spin splitting tends to change the sign of polarization degree of carriers in the QW. This effect could be explained if the *g*-factors of electrons and holes present opposite signs [20]. Under electron resonant condition the sign of polarization degree tends to be defined by the sign of *g*-factor of minority carriers (holes). In addition, we observe that, under higher voltages, the QW and contact layer emissions present similar values of polarization degree which indicates that carriers tunnel to the QW with a polarization degree previously defined in the contact layers. However, the quantitative voltage dependence on the QW polarization degree seems to be rather complex and probably involves other effects such as the alignment of the spin-split QW levels at the resonant condition, the spin polarization of electrons and holes in contact layers prior to their tunneling into the QW, assuming that they maintain their spin polarization during the tunneling process.

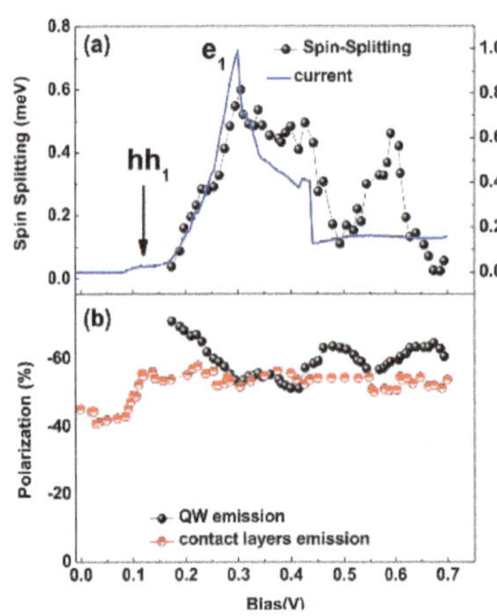

Figure 3 Voltage dependence of spin-splitting from QW emission (a) and circular polarization degree of contact layers and QW at 19 T (b).

In conclusion, we have observed small oscillations on the polarization degree from the QW as a function of the voltage. We have evidence of highly polarized 2DEG and 2DHG in the RTD which can contribute to the polarization degree of carriers in the QW. The voltage dependence of the 2DEG-h emission under magnetic field presents some anomalies which can be explained by the voltage dependence of tunneling dynamics of carriers in the structure. Our results imply that the double-barrier structure creates a polarized two-dimensional gas with a strongly enhanced *g*-factor, which can act as a spin-polarized source of injected carriers in the structure. However, the circular polarization of carriers in the double-barrier region seems also to depend on various other points, including the *g*-factors of the different layers, the spin-polarization of carriers in the contact region, the density of carriers along the structure, and the Rashba and Zeeman effects in the valence band.

DECLARATIONS

Acknowledgements

The financial support from FAPESP, CAPES, CNPq, and U.K. Engineering and Physical Sciences Research Council and EuroMagNET II is gratefully acknowledged.

Authors' original submitted files for images

Below are the links to the authors' original submitted files for images.

11671_2010_36_MOESM1_ESM.pdf Authors' original file for figure 1

11671_2010_36_MOESM2_ESM.pdf Authors' original file for figure 2

11671_2010_36_MOESM3_ESM.pdf Authors' original file for figure 3

Competing Interests

The authors declare that they have no competing interests.

Authors' contributions

LFS prepared figures and participated in the analyses of the data. YGG conceived of the study, carried out the PL and transport measurements, analyzed the data and wrote the paper. MDT prepared figures. VLR, GEM and MJSPB participated in the draft of the manuscript. MO and JK participate in the photoluminescence alignment and measurements. DKM is responsable for the transport setup. MH grown the RTD sample and RJA processed our RTD.

REFERENCES

1. Hanbicki A, Van Erve OMJ, Magno R, Kioseoglou G, Li CH, Jonker BT, Itskos G, Mallory R, Yasar M, Petrou A: Analysis of the transport process providing spin injection through an Fe/AlGaAs Schottky barrier. *Appl Phys Lett* 2003, 82: 4092. 10.1063/1.1580631
2. Jiang X, Wang R, Shelby RM, Macfarlane RM, Bank SR, Harris JS, Parkin SSP: Highly spin-polarized room-temperature tunnel injector for semiconductor spintronics using MgO(100). *Phys Rev Lett* 2005, 94: 056601. 10.1103/PhysRevLett.94.056601
3. Motsnyi VF, Van Dorpe P, Van Roy W, Goovaerts E, Safarov VI, Borghs G, De Boeck J: Optical investigation of electrical spin injection into semiconductors. *Phys Rev B* 2003, 68: 245319. 10.1103/PhysRevB.68.245319
4. Fiederling R, Keim M, Reuscher G, Ossau W, Schmidt G, Waag A, Molenkamp LW: Injection and detection of a spin-polarized current in a light-emitting diode. *Nature* 1999, 402: 787. 10.1038/45502
5. Ohno Y, Young DK, Beschoten B, Matsukura F, Ohno H, Awschalom DD: Electrical spin injection in a ferromagnetic

semiconductor heterostructureD. *Nature* 1999, 402: 790. 10.1038/45509

6. Oestreich MJ, Hübner MJ, Hägele D, Klar PJ, Heimbrodt W, Rühle WW, Ashenford DE, Lunn B: Spin injection into semiconductors. *Appl Phys Lett* 1999, 74: 1251. 10.1063/1.123515
7. Jonker BT, Park YD, Bennett BR, Cheong HD, Kioseoglou G, Petrou A: Robust electrical spin injection into a semiconductor heterostructure. *Phys Rev B* 2000, 62: 8180. 10.1103/PhysRevB.62.8180
8. Braden JG, Parker JS, Xiong P, Chun SH, Samarth N: Direct measurement of the spin polarization of the magnetic semiconductor (Ga,Mn)As. *Phys Rev Lett* 2003, 91: 056602. 10.1103/PhysRevLett.91.056602
9. Mattana R, George JM, Jaffres H, Van Dau FN, Fert A, Lepine B, Guivarc'h A, Jezequel G: Electrical detection of spin accumulation in a p-type GaAs quantum well. *Phys Rev Lett* 2003, 90: 166601. 10.1103/PhysRevLett.90.166601
10. Gruber Th, Keim M, Fiederling R, Reuscher G, Ossau W, Schmidt G, Molenkamp LW, Waag A: Electron spin manipulation using semimagnetic resonant tunneling diodes. *Appl Phys Lett* 2001, 78: 1101. 10.1063/1.1350600
11. Slobodskyy A, Gould C, Slobodskyy T, Becker CR, Schmidt G, Molenkamp LW: Voltage-controlled spin selection in a magnetic resonant tunneling diode. *Phys Rev Lett* 2003, 90: 246601. 10.1103/PhysRevLett.90.246601
12. de Carvalho HB, Galvão Gobato Y, Brasil MJSP, Lopez-Richard V, Marques GE, Camps I, Henini M, Eaves L, Hill G: Voltage-controlled hole spin injection in nonmagnetic GaAs/AlAs resonant tunneling structures. *Phys Rev B* 2006, 73: 155317. 10.1103/PhysRevB.73.155317
13. de Carvalho HB, Brasil MJSP, Galvão Gobato Y, Marques GE, Galeti HVA, Henini M, Hill G: Circular polarization from a nonmagnetic p-i-n resonant tunneling diode. *Appl Phys Lett* 2007, 90: 62120. 10.1063/1.2472522
14. dos Santos LF, Galvão Gobato Y, Marques GE, Brasil MJSP, Henini M, Airey R: Light controlled spin polarization in asymmetric n-type resonant tunneling diode. *Appl Phys Lett* 2007, 91: 073520. 10.1063/1.2772662
15. dos Santos LF, Galvão Gobato Y, Marques GE, Brasil MJSP, Henini M, Airey R: Polarization resolved luminescence in asymmetric n-type GaAs/AlGaAs resonant tunneling diodes. *Appl Phys Lett* 2008, 92: 143505. 10.1063/1.2908867
16. Ashkinadze BM, Voznyy V, Cohen E, Ron A, Umansky V: Condensation of bulk excitons on a magnetized two-dimensional electron gas in modulation-doped heterojunctions. *Phys Rev B* 2002, 65: 073311. 10.1103/PhysRevB.65.073311
17. van der Meulen HP, Sarkar D, Calleja JM, Hey R, Friedland KJ, Ploog K: Free versus localized hole magnetophotoluminescence in semiconductor heterojunctions near integer filling factors. *Phys Rev B* 2004, 70: 155314. 10.1103/PhysRevB.70.155314
18. Ashkinadze BM, Linder E, Cohen E, Rudenkov VV, Christianen PCM, Maan JC, Pfeiffer LN: Exciton to two-dimensional electron-hole photoluminescence transitions driven by the quantum Hall effect in photoexcited heterojunctions. *Phys Rev B*2005, 72: 075332. 10.1103/PhysRevB.72.075332
19. Ashkinadze BM, Cohen E, Rudenkov VV, Christianen PCM, Maan JC, Pfeiffer LN: Transition from two-dimensional electron-hole to geminate-exciton photoluminescence in GaAs/$Al_xGa_{1-x}As$ heterostructures under a high in-plane magnetic field. *Phys Rev B* 2007, 76: 075344. 10.1103/PhysRevB.76.075344
20. Snelling MJ, Blackwood E, MacDonagh CJ, Harley RT, Foxon CTB: Exciton, heavy-hole, and electron g- factors in type-I GaAs/AlxGa1-xAs quantum-wells. *Phys Rev B* 1992, 45: 3922. 10.1103/PhysRevB.45.3922

CHAPTER 13 SOLAR ENERGY FOR ELECTRICITY AND FUELS

Olle Inganäs and Villy Sundström

INTRODUCTION

Renewable energies are still dominated by bio-energy and hydro-energy. This technology is well developed and more energy can be produced but are inherently limited (Fig. 1). Vegetation exists on some 80 million km^2 with an average net bio-energy production of around 0.1 W/m^2. Higher production would require active nitrogen. Hydro-energy is even more limited. Photovoltaic electricity from solar cells has undergone a rapid development and is rapidly being used both in minor private systems as well as in large-scale installations connected to the national grids. Wind energy has also reach maturity and undergoes presently a rapid worldwide implementation. The main challenge with the renewable energy is the intermittency requiring major storage or large-scale integration.

This article focuses on solar energy, identifying the need for breakthroughs in more efficient ways to produce (i) electricity from more powerful and cost efficient solar cells (ii) the possibility of direct conversion of solar energy into fluid (such as ethanol or methanol) or gas forms (methane of hydrogen), and (iii) the need for producing significantly higher bioconversion including breeding of special plants and genetic engineering of cyanobacteria. The annual flow of energy from the sun dwarfs all other non-renewable energy flows and stocks, and is several orders of magnitude above what humankind needs (Fig. 1). Very major secondary flows of solar energy are concentrated by the thermal engine of the atmosphere, supplying flow of water for hydropower, and flow of air for wind power. From inside the Earth, geothermal heat is delivered from nuclear processes. All these are very small by comparison to the solar influx. The solar influx over land on a horizontal surface varies considerably between different regions and with the diurnal and annual cycle. The most favorable areas are in the subtropics where values can reach almost 300 W/m2 in annual average, while in parts of northern Europe values can be as low as around 50 W/m2 with very low values during winter. The diurnal variation of the solar energy influx matches somewhat but not perfectly the activity patterns of societies. There is therefore a great need to store solar derived energy as electrical or chemical energy, to be used at some later time.

PHOTOVOLTAIC

Current Scientific Status

Photovoltaic devices generate electrical power upon illumination (Fig. 2). The electrical current is generated as

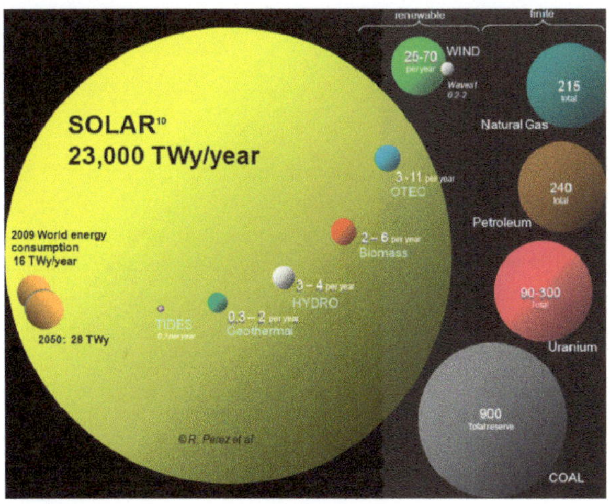

Figure. 1 A comparison of finite and renewable planetary energy re-

Citation: Inganäs O, Sundström V. Solar energy for electricity and fuels. Ambio. 2016;45(Suppl 1):15-23. doi:10.1007/s13280-015-0729-6.

serves measured in TW/years. Total recoverable reserves are shown in TW_y for the finite resources, and yearly potential for the renewables. OTEC = Ocean thermal energy conversion. One TW_y is 8760 TW_h. This figure does not include shale gas or shale oil nor the energy available in methane hydrate. Source: Perez and Perez (2009).

photons from sunlight are absorbed and charges are generated in a semiconductor material. Semiconductors such as silicon absorb a large fraction of sunlight, but below the absorption edge (band gap) of the semiconductor, no absorption occurs. This controls the photocurrent and thus the electrical power of the devices, in combination with the voltage which is influenced by very basic physics and materials science parameters. The science of photovoltaic is a mature field. Novel photovoltaic materials arrive, and scientists later learn how to accommodate these novel phenomena to the scientific terminology of photovoltaics. The thermodynamic limitations analyzed by Shockley and Queisser (S–Q) in (1961) resulted in a maximum efficiency of slightly over 33 % in a single-junction device using a band gap of 1.1–1.4 eV. These limitations are due to heat losses from high energy photons above the single-band gap, and from loss of photons not absorbed below the band gap. It is also based on the equilibrium flux of photons to the photovoltaic device from the sun, and the flow of photons to the universe from the photovoltaic device. This is a description that is valid for well established materials, like crystalline and polycrystalline silicon-based photovoltaic, but also for amorphous inorganic semiconductors in thin film format, as well as newer organics in wet or dry state conditions, and new hybrid organic–inorganic materials. Though the science is mature, the implications of this science have not always been fully acknowledged. The recent improvement, to 28.8 % power conversion efficiency (PCE), of high performance gallium arsenide (GaAs) single-junction solar cells, has been accomplished by photonic engineering, to manipulate the balance between the incoming and outgoing

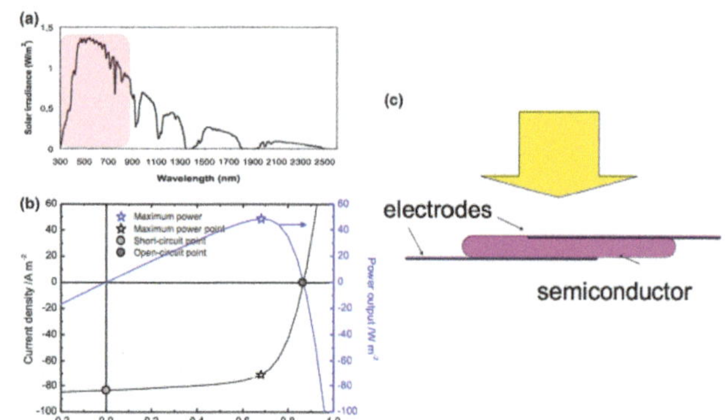

Figure. 2 a The solar radiation arriving at the surface of the earth is distributed over different wavelengths of light, with tails extending far out in the infrared and invisible region. With semiconductor materials, light with energy higher/wavelengths shorter than the band gap of the material can be absorbed (shadow in graph), but none below the band gap. b The semiconductor is contacted with electrodes and exposed to solar light. Photocurrent J_{sc} from the devices depends on absorption and charge generation in the active material; the photovoltage V_{oc} basically depends on the materials. c The current–voltage curve shows both these parameters, and the device is used to deliver electricity at the maximum power delivery point, indicated with a star.

photon flux, with the reward coming as voltage from the cell (Miller et al. 2012). Improved energy conversion is found in multi-junction solar cells, or tandem solar cells, where the heat and transmission losses are minimized by use of two, three, or more different materials of different band gaps, and connecting these devices in electrical series. These are very high performance solar cells, with power conversion efficiencies up to 44.7 % under the direct sun spectrum at very high concentrations. These solar cells are mostly used in space, or under concentrated solar light on earth. The delicate growth of several compound semiconductors on top of each other to create these structures results in a higher material and processing cost. Therefore, these solar cells are mostly relevant together with concentrating optical elements, and thus require high performance, low-cost concentrators, which only work well in direct sunlight. Such concentrator systems are today becoming more widely available and applications in the multi MW range have been realized.

Attempts to use the part of the solar spectrum found in the infrared and far infrared, carrying 50 % of the solar energy, for photovoltaic energy conversion is of relevance, in particular in combination with tandems. For single-band gap devices, the optimum band gap is 1.1–1.4 eV, and the lower energy part of the solar spectrum is sacrificed.

Current Technological Developments

Developments of silicon photovoltaic markets and technology

The goal of 0.4 €/W_p for photovoltaic (PV) modules is close to appear in the silicon photovoltaic market, where the wholesale price is now 0.5 €/W_p, and where the financial difficulties of most companies indicates that variable costs in their production are c. 0.4 €/W_p. The dramatic reductions of silicon PV sales prices (to 20 % during the last 5 years) are mainly due to market pressure resulting from oversupply, increased production volumes leading to cost reduction by scale, and to some degree also due to technological advancement. Extrapolating that pace of technological improvement in the near future, short-term market and technology analysis indicate that by year 2017 the production cost of 0.3 €/W_p will be reached. The learning curve of silicon PV may, however, decelerate as sales prices go so low that new technology cannot be acquired and financed, or by reaching the technology limits given by the requirement for crystalline layers with – due to the indirect band gap of silicon—large thickness. However, already today photovoltaic electricity generation cost can be as low as 12 €cents/KWh in countries like France or Germany and 8 €cents/KWh in the south of Europe. Within the next decade solar electricity from photovoltaics is likely to become the cheapest form of renewable energy in most parts of the world. Silicon is an established photovoltaic material, it is widely available, non-toxic and it is likely to dominate the market for the foreseeable future. However, production cost of electricity is not the only element, as the electricity system cost represents typically 40–60 % of the final cost. This major cost reduction of silicon PV puts significant pressure on all alternative technologies like thin film inorganic photovoltaics (e.g., a-Si, CdTe, CIGS) which lost in market share during recent years, but improved somewhat in record performance numbers. All PV technologies in the future will have to approach higher conversion efficiencies to reduce the contribution of all other module components. Whether there is a place for low efficiency, low-cost solar cells to compete with silicon photovoltaic is now more doubtful than 5 years back, as the cost of silicon has dropped. The combination of silicon PV with compound semiconductors (thin film III-V compounds or nanowires) in hybrid tandem photovoltaic promises big improvements of power conversion efficiencies, and may be one of the most attractive pathways for rapid improvement in PCE. Also these approaches can benefit from the advances in silicon photovoltaic like lower material costs and enhanced manufacturing technologies.

Hybrid organic/inorganic solar cells

High performance dye sensitized solar cells, where an organic compound in contact with a high band gap inorganic semiconductor (e.g., TiO_2, ZnO, ZrO_2) generates charge upon photon absorption, still require a redox liquid electrolyte for best performance. Alternative redox design using cobalt compounds has improved efficiencies to 13 %. Solid state versions of the dye sensitized solar cells have lower PCE. A new inorganic absorber, a perovskite based on lead halogen compound first used for dye sensitized solar cells, has rapidly demonstrated that this compound can be used at its best (18 % power conversion efficiency) in thin film geometries (Jeon Jeon et al. 2015), as optical absorption and electronic transport is sufficient in this material; the advantages of the dye sensitized cells are thereby removed for this absorber.

Solid state organic solar cells

Development of organic donor/acceptor-based photovoltaic (OPV), where donor and acceptor are typically organic molecules or polymers, is accelerating, with best efficiencies now at 12 % (Heliatek 2012; Mitsubishi 2013). The photovoltage from these materials is controlled by a charge transfer state generated at the interface between the donor/acceptor (Vandewal et al. 2009). Though the basic physics of these photovoltaic materials is now somewhat mature, there is still a large room for improvement by chemical design. With improvement of 1 % unit per year, and limits to PCE still above 20 %, there is reason to expect 15–20 % PCE as an outcome of novel materials. As the active materials in OPVs are carbon based, these are generally very scalable materials. The production methods for OPV include printing from solvents or evaporation of molecules through vacuum. Both are amenable for large area deposition, with somewhat different potential speed of production and capital costs of equipment. Printing is the faster and cheaper, and rapidly scalable to produce large areas of photovoltaic modules on the global scale. For both OPV and dye sensitized cells, they operate well under low lights, high temperatures and considerably improve the light harvesting compared to silicon devices. The extra yield of almost a 1/3 from an organic-based module will give the same amount of kWh from a 15 % organic module as compared to a 20 % silicon module. In the comparison of early organic photovoltaic systems, which have no market share yet, issues of lifetime and module efficiency are largely unresolved. Whether they will be sufficient to compete with mature silicon photovoltaic is not clear, and they will most probably not enter into power plants but rather be used in building integration.

Needs for breakthroughs

Scalability of semiconductor materials

Silicon- or carbon-based materials, which can be scaled up, deliver 25 % (Si) and 12–13 % (carbon based) power conversion efficiency in best lab devices. Several of the competing thin film technologies suffer from a dependence on rare or toxic materials which reduces their potential to play a significant role when photovoltaics will be used in large-scale systems. Therefore, the use of earth abundant materials or efficient recycling concepts will be important for large-scale deployment of the technologies. Removing materials limitations due to less abundant elements (for active materials e.g., indium, gallium, tellurium, for current collectors silver) by substitution with more abundant elements (e.g., zinc, sulfur or copper, aluminum) will be a general and critical target.

Decreasing the energy payback time of solar cells

Measures and means needed

- Reducing the energy input in photovoltaic modules production. Low-temperature processes, substrates with low embodied energy, repeated use of substrates for growth of high quality materials/devices, closure of the materials cycle during the life cycle of devices;
- Extending the technical lifetime of high performance photovoltaic by improved encapsulation and protection.

Materials design

- Designing the electronic properties of carbon-based materials by understanding and controlling the path from single molecular structures via nano-morphology to film growth. A prime example is bulk heterojunctions in organic solid state solar cells which need a nanometer control of morphology in the key active layer.

Areas where scientific and materials breakthroughs may be possible and of potential importance

- Multiple-band gap tandem solar cells incorporating scalable materials;
- Light concentration through far field optics for multiple-band gap materials;
- Nanophotonics in dielectrics and semiconductors for light-matter coupling, light trapping/ light management;
- Plasmonics for light-matter coupling, to direct energy to the semiconductor structure in photovoltaic devices;
- Combination with solar fuel generation, e.g., hydrogen generation by photovoltaic layers with catalyst electrodes.

Identification of research needs of the coming decades

Nanophotonic strategies for light trapping in thin structures

The well-established Yablonovitch limit for light trapping in the limit of geometrical optics is not relevant when arriving at thin film structures that must be analyzed with the wave optics approach. This goes for thin film devices with extremely thin absorbers, and new strategies for optical compression of solar light into such thin absorbers have been proposed. Considerable experimental developments are necessary to exploit such strategies, shaping, and integrating thin film absorbers in a benign optical environment.

Transparent electrode materials, not relying on elements of low abundance

- Carbon based or abundant metal oxide transparent conductors for electrodes;
- New combinations of device/module assembly to generate lower currents and higher voltages for decrease of losses in photovoltaic modules.

Beyond S–Q limits?

The perennial quest for ways of moving beyond the Shockley–Queisser limits for single- and multiple-band gap absorbers in photovoltaic devices may require new approaches for generating electricity. When developing hybrid approaches based on high performance semiconductors, the added energy conversion with novel materials and structures may include up-conversion. Without a band gap alternative physical mechanisms must be found. Current developments in thermoelectrics points to new combinations of photovoltaics and thermoelectrics; optical antennas and rectennas show interesting avenues for creating absorbing structures somewhere between geometric antennas and molecular/elemental semiconductors with defined band gaps.

Solar Fuels: Artificial Photosynthesis

Direct conversion of solar energy into a fuel mimicking the catalytic processes of photosynthesis would be a way to solve the problem of storing solar energy. A direct process would minimize the number of steps needed from light to fuel, and therefore potentially allow for high conversion efficiencies. The raw material for the (storage) fuel is an important key aspect for the solar fuels. The desirable raw material must be essentially inexhaustible, cheap, and widely available. Most scientists target water as the raw material. Processes in which water is split (oxidized) into its constituents by solar energy can become large contributors to shift away from fossil fuels on a global scale. The target fuel is another important issue. Many scientists target hydrogen as the solar fuel. When water is used

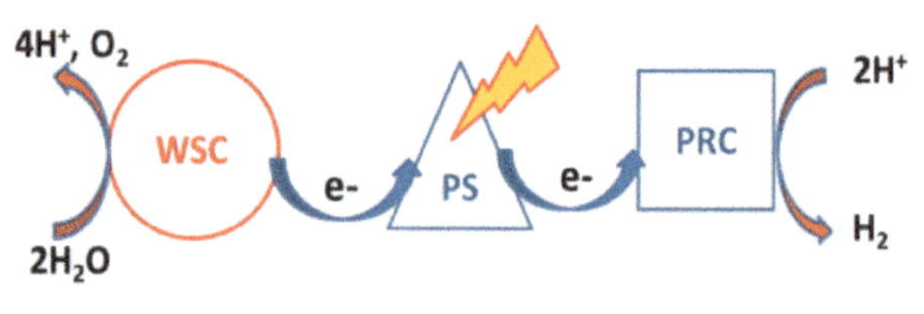

Figure. 3 Schematic picture of an artificial photosynthesis assembly consisting of a light collecting photosensitizer (PS), a proton reducing catalyst (PRC), and a water splitting catalyst (WSC)

as starting material this is natural for scientific reasons. However, the transition to a hydrogen-based economy is not easy. The hydrogen technologies have their own hurdles to overcome and technological issues to tackle before hydrogen can become a worldwide spread fuel. An alternative is to use CO_2 itself as a second raw material (together with water) to create a carbon-based solar fuel. This might have lower technological problems on a shorter term but will not remove CO_2 from the energy cycle (see Fig. 1). In addition, some of the science involved in developing methods for fuel production is most likely more difficult than for hydrogen production. The envisaged key components of an assembly for production of a solar fuel (H_2) using water as raw material are illustrated in Fig. 3 (Hammarstro¨m and HammesSchiffer 2009; Cogdell et al. 2013). A photo-sensitizer (PS) absorbs the light from the sun and delivers energized electrons to the proton reducing catalysts (PRC) that generates the fuel, H_2. The electrons delivered by the sensitizer are replenished by oxidizing (splitting) water with the help of a water splitting catalyst (WSC). In an operating device the WSC and PRC functions would most likely be driven by different sensitizers and physically separated and located on different photo-electrodes in contact with water.

Predicted power conversion efficiencies: Direct processes for solar fuel

Today, no complete device for (direct) solar fuel generation exists, but sensitizers and catalysts that perform the individual reactions have been developed. Nevertheless, a maximum theoretical Shockley–Queisser (S–Q) efficiency of *30 % for (direct) solar fuel generation (hydrogen and oxygen from water) can be estimated for a single-band gap sensitized process and assuming a realistic value (*0.15 V) for the over-potential (Hanna and Nozik 2006; Blankenship et al. 2011). It is foreseeable that the time curve for the development of solar fuel generation devices, with respect both to efficiency and cost, will be faster than for the different PV technologies, because the development to some extent will benefit from progress already made in neighboring fields. The most important being the scientific level in light management, electron and proton transfer, material and nano-science, biochemical, and natural photosynthesis. Technological advances in characterization methods, such as time resolved spectroscopy of many kinds, synchrotrons and new laser techniques, are also vital. Sensitizers and catalysts available today frequently contain elements that are scarce, expensive, or toxic. For large-scale commercial implementation of solar fuels produced by artificial photosynthesis technologies materials based on abundant and environmentally benign elements have to be developed, which most likely will take an additional 10 years or more. This restriction is analogous to PV technology limitations.

In comparison, for solar fuel generation relying on a combination of PV and electrolysis, the thermodynamic limits are given by the S–Q treatment in combination with conditions for electrolysis. For the case of photoelectrolysis, with production of H_2 and O_2, the over potentials for oxidation of water to oxygen and reduction of protons to hydrogen must be added to the thermodynamic potential of 1.23 V for dissociation. Driven by solar cells, this requires more than one cell, added in series, to generate sufficient voltage for electrolysis. With electrolysis having a quantum efficiency of 80–90 %, and a single-band gap cell of 33 % power conversion efficiency, the maximum efficiency from photon to hydrogen may be limited to &30 %. The "arti- ficial leaf" construction of Nocera (2012), comprising a triple junction amorphous silicon photovoltaic cell interfaced with hydrogen- and oxygen-evolving catalysts made from a ternary alloy (NiMoZn) and a cobalt_phosphate cluster (Co-OEC), respectively, is an example of where this technology stands today. The reported overall efficiency of this device is 2.5 %.

Methods to produce direct solar fuels

At present there is no direct process that works on any technological scale. The large potential in the direct processes make them nevertheless attractive future technologies and their development is expected to be require heavy research initiatives in several fields. As mentioned above and illustrated in Fig. 3, the key steps for direct solar fuel production mimicking photosynthesis are two catalytic reactions, oxidation of water and reduction of a substrate to produce the fuel. If the substrate is protons a relatively well established catalytic process yields the hydrogen fuel. A carbon-based fuel, for example, an alcohol, might be easier to use, but is probably more difficult to make since it would involve less established catalytic processes. A highly critical aspect is that the catalysts should be made from earth abundant metals like cobalt, manganese, iron and nickel, while scarce elements like iridium, indium, palladium, platinum, rhenium, and ruthenium are not available in amounts large enough to allow development of processes and materials of a scale sufficient to replace fossil fuels. Molecular and semiconductor nanostructure processes. The light-driven catalyst can be molecular (Hammarstro¨m and Hammes-Schiffer 2009; Sundstro¨m 2009; Nocera and Guldi 2009; Kamat and Bisquert 2013) or non-molecular (Cook et al. 2010; Poddutoori et al. 2011; Vagnini et al. 2013; Joya et al. 2013). The physical limitations are equal and the scientific problems are of equal magnitude. The development of catalysts that can assist the light-driven oxidation of water is the main research problem. The capture of solar energy and the formation of hydrogen are easier to achieve. Systems combining the two reactions (light-driven water oxidation and hydrogen formation) have the highest potential with respect to solar energy to fuel conversion of all systems envisioned today (Fig. 3). The catalysts and maybe the entire system for artificial photosynthesis can be entirely molecular in nature. They can thereby be varied through small, deliberate synthetic modifications to improve and fine-tune their properties. They are also amenable to studies with high-level molecular or kinetic spectroscopy. Thereby, the catalytic process can be followed and understood to a very detailed level. A special angle to this research involves biomimetic approaches. Here, knowledge and chemical principles from extremely efficient enzymes are applied in entirely synthetic systems for artificial photosynthesis. A successful scientific example involves the development of di-iron catalysts for the reduction of protons to hydrogen. These are designed from deep knowledge about the structure and function of so called hydrogenase enzymes and there are recent examples of very efficient catalysts in the category (Tamagnini et al. 2007). Another example involves mimics of the natural photosynthetic reaction centres, with both catalysis and light harvesting in future artificial photosynthesis devices (Magnuson et al. 2009).

In non-molecular systems the light-driven catalysis occurs on metal surfaces, semiconductors or nanostructured carbon-based materials while the catalysts involved for water splitting are often cores of metal oxides, sometimes doped with other metals. A severe limitation is that systems many times are based on catalysts made from scarce and expensive metals. Another disadvantage when compared to the molecular systems is that it is much more difficult to study the mechanism for the reactions involved. An important advantage is that many non-molecular systems are seemingly sturdier against degradation and inactivation while most molecular systems studied to date are unstable and easily break during illumination. It is not obvious that this situation will always prevail when more functional systems have been better characterized. An analogy is the Gra¨tzel cell where the individual parts are fairly unstable but the integrated system is very stable. A rapid development involves ideas where molecular and non-molecular systems are mixed. They are sometimes known as hybrid systems and the science is very broad and collaborative, involving several scientific fields. Here, the solar energy capture system is semiconductor based or made from some other nanotechnology while the chemistry is carried out by linked molecular catalysts. These catalysts are then made from abundant materials like cobalt, iron, manganese, or nickel similar to the "purely" molecular systems described above. Often catalysis is driven by light via incorporation of a photo-sensitizer between the semiconductor and the catalyst. It is not unlikely that these mixed hybrid systems will become dominant in research and maybe technology since they combine advantages from both fields. Thermochemical cycles. A totally different technology is the employment of thermal processes (Pagliaro et al. 2010) for solar fuels production, which involve generation of very high temperatures in closed environments to split water into its constituents directly. This results in a solar fuel when the high temperature is achieved in a reaction vessel in a solar tower by concentration of solar energy in a heliostat. The technology represents interesting engineering and physical science and is very demanding technically involving very high temperatures and huge systems like heliostats. They are mainly suitable to very sunny locations.

Direct solar fuel needs for breakthroughs

Efficient photo-sensitizers and photocatalysts

For large-scale implementation of direct solar fuel technologies, water will necessarily be the ultimate electron source for reducing the fuel-producing substrate, protons, CO_2, etc. New and highly efficient catalysts for water oxidation and fuel generation are urgently needed, as well as sensitizers based on abundant elements:

- Photo-sensitizers frequently contain noble metals (ruthenium), which have to be replaced with cheap and abundant elements. Iron is one example but dyes based on Fe and similar transition metals have unfavorable properties for solar energy conversion, necessitating scientific breakthroughs. At the same time fully organic dyes are appearing. Thus, it can be expected that sensitizers suitable for most applications and for large-scale implementation are likely to appear in not too distant future.

- To achieve highly efficient catalysts for water oxidation and hydrogen (liquid fuel) generation is a considerably more difficult and complex problem. Catalysts may be either molecular complex or solid state (e.g., metal oxide) based. Understanding the mechanisms of O-O and H-H bond formation is key to the development of efficient catalysts. Most of today's catalysts have low efficiencies and are frequently based on scarce or noble metals, making them unsuitable for large-scale commercial implementation. Here, scientific breakthroughs are needed, both concerning efficiency and developing new catalysts based on abundant and cheap elements. Proof-of-concept catalysts with high efficiencies, based on metals like ruthenium, are likely to appear within the next 10 years, but catalysts suitable for commercial applications most likely will take considerably more time.

Liquid fuels

Liquid solar fuels would facilitate the use of existing infrastructure for the utilization of the fuel. By combining the protons and electrons released as a result of water oxidation with CO_2 in a catalytic process, alcohols or other liquid fuels could be synthesized. This is a scientifically challenging task and there are today no good catalysts for CO_2 reduction. Another problem is that the concentration of CO_2 in the atmosphere is low, necessitating an energyconsuming concentration step, or coupling of the solar fuel production to CO_2-emissions from e.g., a fossil-fuel power plant or a CCS plant (see MacElroy 2016).

Functional devices for direct solar fuel production

No complete device for (direct) solar fuel production exists. New concepts to assemble sensitizers and catalysts into a functional device producing a fuel through lightdriven water splitting have to be developed. One possibility that is considered within the field is to attach sensitizer catalyst assemblies to a nanostructured electrode surface and placing these electrodes in separate compartments, one for fuel (hydrogen) generation and the other for oxygen evolution. Here, suitable electrode systems have to be developed—methods for sensitizer-catalyst assembly attachment to the electrode must be developed, and electrochemical properties of catalysts and electrode materials matched. Considering that this development is still in its scientific infancy, it will probably take more than 10 years until efficient proof-of-concept systems have been developed and considerably more time until commercially suitable devices are available.

Scalability of photo-sensitizers and catalysts

Like for PV, large-scale implementation of solar fuel technology requires cheap and earth abundant materials. Today's molecular or semiconductor photo-sensitizers and photocatalysts often contain scarce, noble (and expensive), or toxic metals, which obviously need to be replaced by more abundant and benign elements. This may take a considerable amount of time and effort, since the photophysical, photochemical, and catalytic properties of materials based on the abundant elements are frequently very different from those of the metals used today. It will require extensive development work to achieve requested properties and efficiencies of the abundant materials.

ACKNOWLEDGMENTS

VS would like to thank S. Styring for useful discussions and for providing parts of the text on solar fuels.

REFERENCES

1. Blankenship, R.E., D.M. Tiede, J. Barber, G.W. Brudvig, G. Fleming, M. Ghirardi, M.R. Gunner, W. Junge, et al. 2011. Comparing photosynthetic and photovoltaic efficiencies and recognizing the potential for improvement. Science 332: 805–809.
2. Cogdell, R.J., A.T. Gardiner, P.I. Molina, and L. Cronin. 2013. The use and misuse of photosynthesis in the quest for novel methods to harness solar energy to make fuel. Philosophical Transactions of the Royal Society A 371: 20110603.
3. Cook, T.R., D.K. Dogutan, S.Y. Reece, Y. Surendranath, T.S. Teets, and D.G. Nocera. 2010. Solar energy supply and storage for the legacy and nonlegacy worlds. Chemical Reviews 110: 6474–6502.
4. Hammarstro¨m, L., and S. Hammes-Schiffer. 2009. Artificial photosynthesis and solar fuels. Accounts of Chemical Research 42: 1859–1860.
5. Hanna, M.C., and A.J. Nozik. 2006. Solar conversion efficiency of photovoltaic and photoelectrolysis cells with carrier multiplication absorbers. Journal of Applied Physics 100: 074510.
6. Heliatek. 2012. Retrieved January 10, 2015 from www.heliatek.com. Jeon, N.J., J.H. Noh, W.S. Yang, Y.C. Kim, S. Ryu, J. Seo, and S.I. Seok. 2015. Compositional engineering of perovskite materials for high-performance solar cells. Nature 517: 476–480.
7. Joya, K.S., Y.F. Joya, K. Ocakoglu, and R. van de Krol. 2013. Watersplitting catalysis and solar fuel devices: Artificial leaves on the move. Angewandte Chemie International Edition 52: 10426–10437.
8. Kamat, P., and J. Bisquert. 2013. Solar fuels. Photocatalytic hydrogen generation. Journal of Physical Chemistry C 117: 14873–14875.
9. Magnuson, A., O. Johansson, T. Polivka, M. Anderlund, P. Lindblad, S. Styring, S. Ott, R. Lomoth, et al. 2009. Biomimetic and microbial approaches to solar fuel generation. Accounts of Chemical Research 42: 1899–1909.
10. Miller, O.D., E. Yablonovitch, and S.R. Kurtz. 2012. Strong internal and external luminescence as solar cells approach the Shockley– Queisser limit. IEEE Journal of Photovoltaics 2: 303–311.
11. Mitsubishi. 2013. Retrieved January 10, 2015 from http://www.mkagaku.co.jp/english/r_td/strategy/technology/topics/opv/.
12. Nocera, D.G. 2012. The artificial leaf. Accounts of Chemical Research 45: 767–776.
13. Nocera, D., and D. Guldi. (ed.). 2009. Themed issue: Renewable energy, Chemical Society Reviews. 38:

1–300.

14. Pagliaro, M., A.G. Konstandopoulos, R. Ciriminna, and G. Palmisano. 2010. Solar hydrogen: Fuel of the near future. Energy & Environmental Science 3: 279–287.
15. Perez, R., and M. Perez. 2009. A fundamental look at energy reserves for the planet. IEA SHC Solar Update. Retrieved from: http:// www.asrc.cestm.albany.edu/perez/Kit/pdf/a-fundamental-lookat%20the-planetary-energy-reserves.pdf.
16. Poddutoori, P., D.T. Co, A.P.S. Samuel, C.H. Kim, M.T. Vagnini, and M.R. Wasielewski. 2011. Photoinitiated multistep charge separation in ferrocene–zinc porphyrin–diiron hydrogenase model complex triads. Energy & Environmental Science 4: 244–2450.
17. Shockley, W., and H.J. Queisser. 1961. Detailed balance limit of efficiency of p–n junction solar cells. Journal of Applied Physics 32: 510.
18. Sundstro¨m, V. (ed.). 2009. Solar energy conversion Dalton Transactions. Themed Issue volume 45.
19. Tamagnini, P., E. Leita, P. Oliveira, D. Ferreira, F. Pinto, D. Harris, T. Heidorn, and P. Lindblad. 2007. Cyanobacterial hydrogenases: Diversity, regulation and applications. FEMS Microbiology Reviews 31: 692–720.
20. Vagnini, M.T., M.W. Mara, M.R. Harpham, J. Huang, M.L. Shelby, L.X. Chen, and M.R. Wasielewski. 2013. Interrogating the photogenerated Ir(IV) state of a water oxidation catalyst using ultrafast optical and X-ray absorption spectroscopy. Chemical Science 4: 3863–3873.
21. Vandewal, K., K. Tvingstedt, A. Gadisa, O. Ingana¨s, and J.V. Manca. 2009. On the origin of the open-circuit voltage of polymer– fullerene solar cells. Nature Materials 8: 904–909.

CHAPTER 14

WAVEGUIDE DESIGN OPTIMIZATION FOR LONG WAVELENGTH SEMICONDUCTOR LASERS WITH LOW THRESHOLD CURRENT AND SMALL BEAM DIVERGENCE

Abdulrahman Al-Muhanna, Abdullah Alharbi, Abdelmajid Salhi

Keywords: Semiconductor Laser, Far-Field, Simulation

INTRODUCTION

Antimonide materials have become increasingly important for a number of electronic and optoelectronic devices. In particular antimonide-based semiconductor laser diodes emitting in the 2 - 5 μm range have attracted considerable attention for a variety of applications: trace gas sensing, military countermeasures, pollution monitoring and molecular spectroscopy [1,2]. In the 2 - 5 μm wavelengths range, the best results, with low threshold currents and high optical powers, were achieved by type I GaInAsSb-AlGaAsSb diode lasers grown on GaSb substrates [3-9]. Continuous wave (cw) output powers as high as 1W were demonstrated at room temperature at $\lambda \sim 2$ and 2.5 μm [6,7]. Beyond 2.75 μm quantum cascade lasers based on InAs/AlSb also show promising results [8,9]. Cw operation of diode lasers, using type I GaInAsSb quantum wells with AlGaInAsSb barriers and emitting at 3.36 μm, has recently been achieved [10]. The results for type I lasers emitting between 2 and 3 μm, mentioned above, are based on a broadened-waveguide structure. The laser structure consists of a wide waveguide with two separate confinement layers (SCL), typically 375 nm in width enclosing the GaInAsSb quantum wells, and utilize 90% aluminum content cladding layers [3,6,7]. The wide-waveguide design produces a small overlap of the optical mode with the cladding layers resulting in low internal losses, α_i, and high modal gain as a result of the high quantum well confinement (Γ_{QWs}) [3,11,12]. The light coupling efficiency into an optical system of finite aperture is strongly dependent on the beam divergence in the fast axis which in the described design has a typical value of full width at half maximum (FWHM) of 70°. This large value makes it difficult to collect the laser power with standard optics. Therefore, to reduce VFF, new epitaxial structures were adopted.

The new structures allow for a broader optical mode inside the waveguide without altering the confinement factor of the quantum wells and free carrier absorption in order to maintain the same characteristics. Normally, device characteristics are expected to degrade with a large optical field due to the increase of free-carrier absorption, particularly in the p-doped cladding. We take the additional precaution of tailoring the doping profile in the p-cladding layer in order to maintain low internal losses. In this paper we propose new epitaxial structures for long wavelength GaSb-based lasers which provide a small vertical beam divergence and low threshold-current density. The first new design includes a narrow optical waveguide and smaller refractive index step between the cladding and the waveguide layers. The insertion of a V-Shaped Weaker Waveguide (VSWW) in the n-cladding layer leads to considerable improvement of the vertical far-field FWHM. A value as low as 35° is achieved in the new design nearly a 50% reduction of the far-field obtained with the conventional design (67°). The threshold current density was kept low to a value of 190 A/cm2 for 1000 × 100 μm^2 devices by a careful adjustment of doping in the highly absorbing p-doped cladding layer. In the second design, instead of inserting a VSWW in the n-cladding layer, two layers with low refractive index with respect to the cladding layers are inserted between n and p-type cladding and the optical waveguide layers. This approach leads also to a reduction of the VFF to 34° keeping the threshold current density to a reasonable value of 210 A/cm^2.

Citation: A. Al-Muhanna, A. Alharbi and A. Salhi, "Waveguide Design Optimization for Long Wavelength Semiconductor Lasers with Low Threshold Current and Small Beam Divergence," Journal of Modern Physics, Vol. 2 No. 4, 2011, pp. 225-230. doi: 10.4236/jmp.2011.24031.

STRUCTURE DESIGN AND SIMULATION

The base-line structure used in the modeling consists of three 10-nm-thick $Ga_{0.65}In_{0.35}As_{0.11}Sb_{0.89}$ QWs, emitting at 1 = 2.38 μm, separated by 35-nm-thick $Al_{0.25}Ga_{0.75}As_{0.02}Sb_{0.98}$ barriers and enclosed between 375-nm–thick Al0.25Ga0.75As0.02Sb0.98 confining layers. The waveguide is surrounded by two $Al_{0.9}Ga_{0.1}As_{0.07}Sb_{0.93}$ n-type (2 × 1018 cm^{-3}, Te) and p-type (5 × 1018 cm^{-3}, Be) cladding layers. A 0.25 μm p^{+} -GaSb is used as contact layer. The p-doping level of the first 0.2 μm near the active zone was decreased to 5 × 1017 cm^{-3} in order to reduce the free carrier absorption. Further details describing the laser structure can be found in [3]. The modeling of the near field, far-field and confinement factors of the optical modes was performed using FIMMWAVE from Photon Design. The threshold current density of the laser is performed by a HAROLD [13]. The laser software solves self-consistently the Poisson equation, the current continuity equations, the carrier's capture-escape balance equations, and the photon rate equation. In addition, it solves the vertical and longitudinal wave and Schrödinger's equations. It calculates also the optical gain using a parabolic band approximation. The carrier transport through the multi-quantum well structure is included in the model. The model allows also the calculation of the distribution of electron and hole concentrations, for both confined and unconfined carriers. The calibration of the laser model was described in our previous work using the conventional laser structure described previously [12].

The model calibration leads to an Auger coefficient of 3.2 × 10^{-28} cm^{6} s^{-1}, a hole cross section in the quantum well of $\sigma_{p\text{-}QW}$ = 80 × 10^{-18} cm^{2} . The hole cross section in the p-type cladding layers was fixed to $\sigma_{p\text{-}Clad}$ = 46 × 10^{-18} cm^{2} as found experimentally in [11]. Figure 1 shows the refractive index profile of the conventional laser structure as structure S_1. In this structure the combination of a high refractive index step between the cladding and the waveguide layers of Δn = 0.29 (correspond to an aluminum content of 90% in the cladding layers and 25% in the waveguide) and a broad-waveguide leads to a small overlap of the optical mode with the cladding layers and a high value for the overlap of the optical mode with the quantum wells. The variation of Γ_{QWs} and VFF as a function of the waveguide thickness is presented in Figure 2. The VFF has a value as high as 67° and Γ_{QWs} = 4.4% for conventional laser structure S_1 with a total waveguide thickness of 850 nm. When the refractive index step is decreased in structure S_1 to Δn = 0.166 (obtained by reducing the aluminum content to 50% in the cladding layers) the VFF and Γ_{QWs} decreases to 53° and 3.75%, respectively (Figure 2). Decreasing the waveguide thickness is accompanied by an increase of the threshold current density as a result of the increase of the free carrier losses. This behavior is illustrated in Figure 3 in which we show the threshold current density and the free carrier losses versus waveguide thickness for the conventional structure with aluminum content of 50% and 90% (Δn = 0.166 and 0.29, respectively). For both structures, decreasing the waveguide thickness is accompanied by an increase of free carrier loss due to a higher optical mode overlap with the p-type cladding layer. Structures with a smaller index step between the cladding and waveguide always possess higher free-carrier losses regardless of the waveguide thickness as shown in Figure 3. For the structure with Δn = 0.29 (S_1), the threshold current density for 1000 × 100 μm^{2} laser is 135 A/cm^{2} (waveguide thickness of 850 nm) remains constant until the waveguide thickness reaches 650 nm. This is due to the fact that the increase in free carrier absorption is compensated by an increase of Γ_{QWs} (see Figure 2). Below a waveguide thickness of 650 nm the threshold begins to increase and reaches a value of 160 A/cm^{2} for a waveguide thickness of 400 nm.

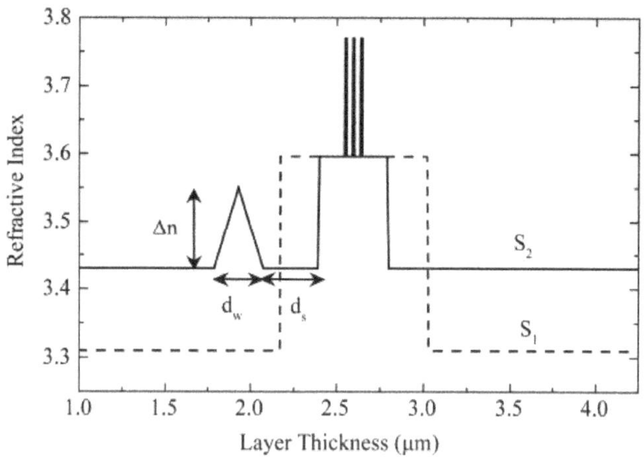

Figure 1: Refractive index profile of the conventional laser structure S1 (dashed line) and the new laser structure S2 (solid line).

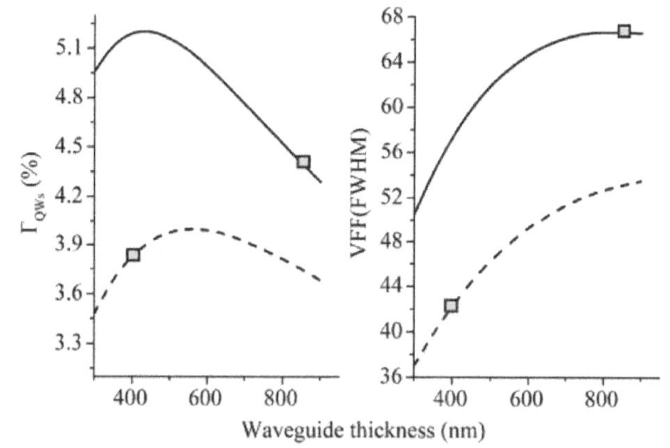

Figure 2: Quantum well confinement factor, ΓQWs, and vertical far-field, VFF, as a function of the waveguide thickness for a conventional rectangular waveguide for two aluminum compositions in the AlGaAsSb cladding layer (50 and 90% indicated in the figure by dashed line and solid line, respectively).

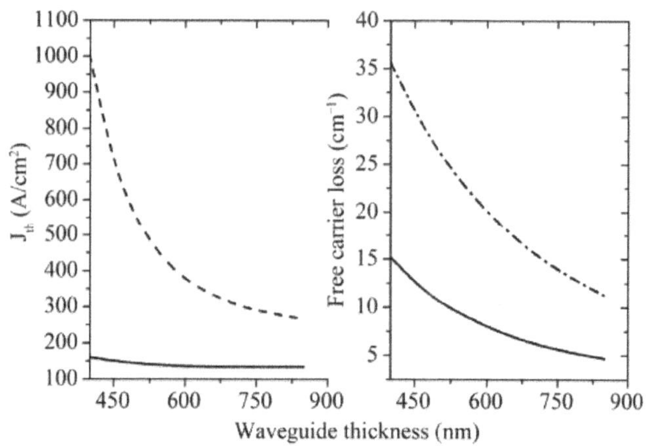

Figure 3: Threshold current density, Jth, and free carrier loss as a

function of waveguide thickness for a conventional rectangular waveguide for two aluminum content in the AlGaAsSb cladding layer (50% and 90% indicated in the figure by dashed line and solid line, respectively).

160 A/cm^2 for a waveguide thickness of 400 nm. The situation is different in the case of the structure with $\Delta n = 0.166$. The increase of Γ_{QWs} while decreasing the waveguide thickness is not sufficient to compensate the rapid increase of free carrier absorption as the optical field expands into cladding layers. Jth increases from 245 A/cm^2 for a waveguide thickness of 850 nm to 1.01 kA/cm^2 for a waveguide thickness of 400 nm The VFF is reduced to only 42°, Γ_{QWs} is reduced to 3.83%, 13% lower than that for the conventional structure S_1. The drawback for this design is the extremely large value of Jth which can be decreased by optimizing the doping profile in the p-type cladding layer. In the conventional laser structure S_1 the doping level in the p-type cladding layer is 5 × 1018 cm–3 except the first 0.2 μm, which has high overlap with the optical field, where p = 5 × 1017 cm^{-3}. This doping profile provides a low internal loss of ~4.7 cm^{-1} which increases to 35.6 cm^{-1} (corresponding to Jth = 1010 A/cm^2) for the situation where the waveguide thickness is 400 nm and $\Delta n = 0.166$. Adopting the following doping profile: the first 0.5 μm of the p-type cladding layer close to the waveguide is doped p = 2 × 1017 followed by 0.2 μm with a linear doping ramp from 2×10^{17} to 3×10^{18} cm^{-3} and the remaining p-type cladding remains constant, p = 3×10^{18} cm^{-3}, leading to a decrease in the threshold current density to ~190 A/cm^2. Further reduction of the VFF can be accomplished by inserting a VSWW into the n-cladding layer [14]. The refractive index profile of the new structure (S_2) is shown in Figure 1. The VSWW represents a weaker waveguide compared to the waveguide core. The incorporation of the VSWW into the n-cladding layer allows an expansion of the optical near field in the vertical direction and hence decreases the corresponding VFF. In order to optimize the VFF, three parameters are modified separately: The separation, d_s, between VSWW and the waveguide core, the thickness of VSWW layer, d_w, and the aluminum content in the middle of VSWW. Figure 4 shows the variation of Γ_{QWs} and VFF as a function of the separation ds for a VSWW thickness d_w of 350 nm and for fixed aluminum content in the middle of the VSWW of 15% ($\Delta n = 0.26$). When d_s is varied from 0.5 μm to 1.9 μm, Γ_{QWs} increases from 3.07% to 3.76%, however the VFF decreases and reaches a minimum of 35° for $d_s = 1.3$ μm. The predicted value is lower than the lowest reported value for antimonide lasers (44°) [15]. Increasing the separation d_s leads to an expansion of near field and subsequently a decrease in the VFF. After the separation reaches the optimum value any further increase of d_s weakens the effect of the VSWW on the near field and hence increases again the VFF. The effect of the VSWW thickness d_w on Γ_{QWs} and VFF is depicted in Figure 5 for a fixed separation ds = 1.3 μm and 15% aluminum content in the centre of the VSWW.

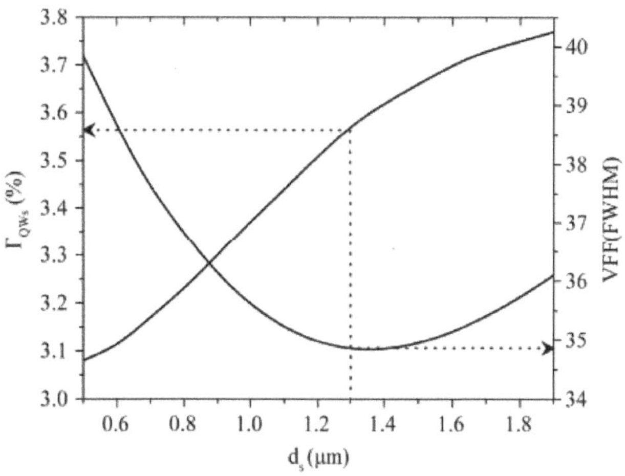

Figure 4: Quantum well confinement factor, Γ_{QWs}, and vertical far-field, VFF, as a function of ds (separation between the V-shaped and rectangular waveguides) for a 350 nmthick V-shaped weaker waveguide. The aluminum content in the centre of the V-shaped weaker waveguide is fixed to 15% ($\Delta n = 0.26$).

Increasing d_w from 0 to 500 nm is accompanied by a decrease of the VFF from 42° to 30°. For $d_w = 350$ nm, Γ_{QWs} decreases only by a factor of 7%. The corresponding VFF is as low as 35° which represents ~50% reduction from that of the conventional structure S_1 (Figure 6). A further increase of d_w, strengthen VSWW waveguide effect and will result in a broader near field and dramatic decrease in ΓQWs which will lead to an increase of Jth. For 500 nm d_w, Γ_{QWs} reaches a minimum value of 2.84%, 26% lower than that obtained without a VSWW. The insertion of the VSWW allows a reduction of the VFF without affecting the threshold current; in fact our simulation shows that the threshold current density remains the same, 190 A/cm^2. This is expected because the slight decrease of Γ_{QWs} is compensated by a decrease in the free carrier absorption losses since the optical mode expands toward the n-type cladding layer which leads to a decrease of the optical overlap with the p-type cladding layer. When we studied the effect of the aluminum content in the center of VSWW, similar behavior was observed. For fixed d_w and d_s, decreasing the aluminum content in the centre of VSWW from 50% to 10% results in reduction of VFF and slight decrease in Γ_{QWs}. However, low value of the Al-content will result in a stronger VSWW waveguide and therefore a dramatic decrease in Γ_{QWs}. Another alternative design to reduce VFF and maintaining low threshold current density is the use of a depressed cladding structure design. Starting from the structure with narrow optical waveguide (400 nm) and with a low refractive index step between the cladding and the optical waveguide ($\Delta n = 0.166$) corresponding to Al content of 25% and 50% in the waveguide and cladding layers, respectively.

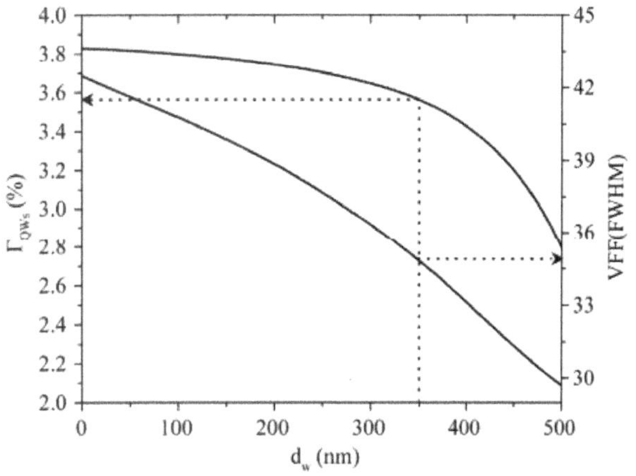

Figure 5: Quantum well confinement factor, Γ_{QWs}, and vertical far-field, VFF, as a function of the thickness, d_w, of V-Shaped weaker waveguide. The aluminum content in the centre of the V-shaped weaker waveguide and d_s are fixed to 15% ($\Delta n = 0.26$) and 1.3 µm, respectively.

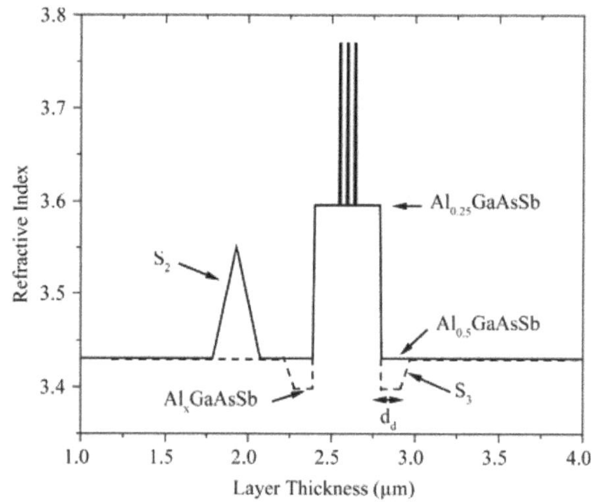

Figure 7: Refractive index profile of the structure S_2 with VSWW (solid line) and the depressed cladding structure S_3 (dashed line).

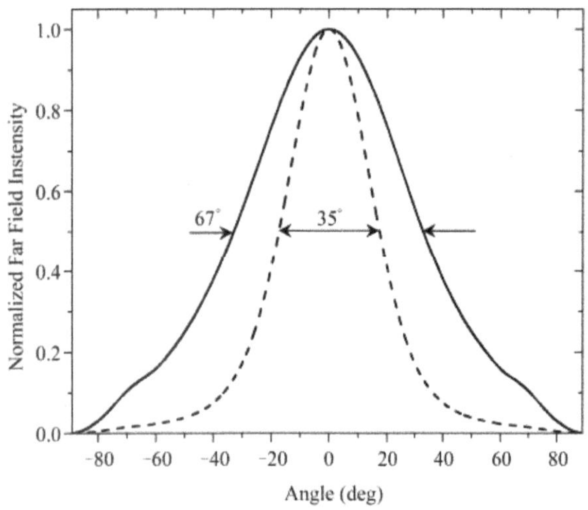

Figure 6: Vertical far-field for the conventional (solid line) and new laser design (dashed line).

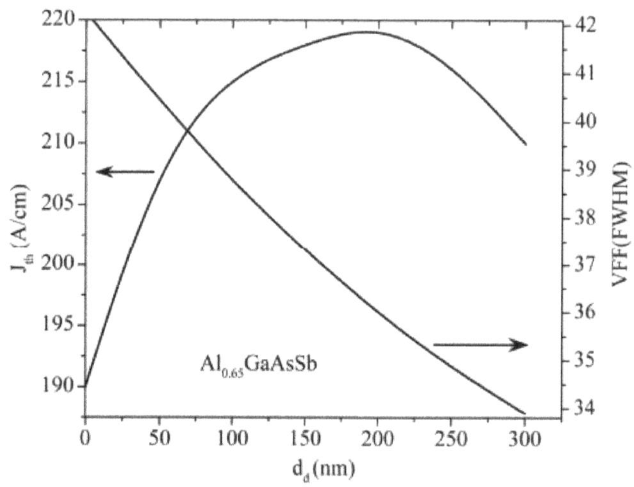

Figure 8: Threshold current density, J_{th}, and vertical Farfield of the depressed cladding structure S_3.

Two thin layers with Al content higher than 50% are inserted between the waveguide and the claddings. These two thin layers have lower refractive index than the cladding layers and their thicknesses (d_d) will alter the distribution of the near field and consequently VFF. The refractive index of the new Structure S_3 and the previous structure with VSWW (S_2) are shown in Figure 7. The threshold current density and VFF as a function of d_d is shown in Figure 8. The VFF decreases from 42° to 34° when the thin layers thickness, d_d, increases from zero to 300 nm which is comparable to the value obtained by the structure with VSWW. The Al content of these layers is fixed to 65% in this case.

As the thicknesses of these layers are increased, the optical mode is depressed and the near field becomes broader resulting in a reduction of the VFF. When increasing d_d, the threshold current density increases from 190 A/cm^2 and reaches a maximum of 219 A/cm^2 for d_d = 200 nm, then it starts to decrease again. This behavior is linked directly to the quantum wells optical confinement factor which is not reported here. The quantum well optical confinement factor starts to decrease from 3.83% until it reaches a minimum of 3.69% then increases again to 3.72% for d_d = 300 nm which gives a threshold current density of 210 A/cm^2 . This value is comparable to the value obtained with the structure having a VSWW in the n-cladding layer. A further increase of the thin layers will reduce VFF, however we have to limit the thickness d_d to 300 nm in order to reduce the effect of this layer on thermal conductivity with the increase of the Al content in these layers compared to the structure with VSWW.

CONCLUSIONS

We have reported the design of long-wavelength GaSbbased quantum well lasers with a small beam divergence while maintaining a low threshold current density. A narrow vertical far-field as low as 35° can be achieved using two approaches. First design has a combination of a thin optical waveguide, reduced refractive index step between the confining and cladding layers, and a V-Shaped weaker waveguide in the n-cladding layer. The threshold current density remains at a low value of 190 A/cm^2 by tapering the doping profile in the p-type cladding layer. Second design uses depressed cladding structure by inserting two thin layers with lower refractive index between the waveguide and the cladding layers resulting in low VFF and comparable threshold current density. These two designs significantly increase the light-coupling efficiency into an optical system and also have the advantage of having lower aluminum content in the structure (50% instead of 90% in the AlGaAsSb cladding layers).

REFERENCES

1. M. Mattiello, M. Niklès, S. Schilt, L. Thévenaz, A. Salhi, D. Barat, A. Vicet, Y. Rouillard, R. Werner and J. Koeth, "Novel Helmholtz-based Photoacoustic Sensor for Trace Gas Detection at Ppm Level Using GaInAsSb/GaAlAsSb DFB Lasers," Spectrochimica Acta A, Vol. 63, No. 5, 2006, pp. 952-958. doi:10.1016/j.saa.2005.11.006
2. S. Kassi, M. Chenevier, L. Gianfrani, A. Salhi, Y. Rouillard, A. Ouvrard and D. Romanini, "Looking into the Volcano with a Mid IR DFB Diode Laser and Cavity Enhanced Absorption Spectroscopy," Optics Express, Vol. 14, No. 23, 2006, pp. 11442-11452. doi:10.1364/OE.14.011442
3. A. Salhi, Y. Rouillard. J. Angellier and M. Garcia, "Very-Low-Threshold 2.4-μm GaInAsSb–AlGaAsSb Laser Diodes Operating at Room Temperature in the Continuous-Wave Regime," IEEE Photonics Technology Letters, Vol. 16, No. 5, 2004, pp. 2424-2426. doi:10.1109/LPT.2004.835623
4. M. Rattunde, C. Mermelstein, J. Schmitz, R. Kiefer, W. Pletschen, M. Walther and J. Wagner, "Comprehensive Modeling of the Electro-Optical-Thermal Behavior of (Algain)(Assb)-Based 2.0 μm Diode Lasers," Applied Physics Letters, Vol. 80, No. 22, 2002, pp. 4085-4087. doi:10.1063/1.1481979
5. C. Lin, M. Grau, O. Dier and M. -C. Amann, "Low Threshold Room-Temperature Continuous-Wave Operation of 2.24 - 3.04 μm GaInAsSb/AlGaAsSb Quantumwell Lasers," Applied Physics Letters, Vol. 84, No. 25, 2004, pp. 5088-5090. doi:10.1063/1.1760218
6. D. Z. Garbuzov, R. U. Martinelli, H. Lee, P. K. York, R. J. Menna, J. C. Connolly and S. Y. Narayan, "Ultralowloss Broadened-Waveguide High-Power 2 μm AlGaAsSb/InGaAsSb/GaSb Separate-Confinement Quantum Well Lasers," Applied Physics Letters, Vol. 69, No. 7, 1996, pp. 2006-2008. doi:10.1063/1.116861
7. J. G. Kim, L. Shterengas and G. L. Belenky, "HighPower Room-Temperature Continuous Wave Operation of 2.7 and 2.8 μm In(Al)GaAsSb/GaSb Diode Lasers," Applied Physics Letters, Vol. 83, No. 10, 2003, pp. 1926- 1928. doi:10.1063/1.1605245
8. J. Devenson, O. Cathabard, R. Tessier and A. N. Baranov, "InAs/AlSb Quantum Cascade Lasers Emitting at 2.75-2.97 μm," Applied Physics Letters, Vol. 91, No. 25, 2007. doi:10.1063/1.2825284
9. J. Devenson, O. Cathabard, R. Tessier and A. N. Baranov, "High Temperature Operation of λ~3.3 μm Quantum Cascade Lasers," Applied Physics Letters, Vol. 91, 2007. doi:10.1063/1.2794414
10. L. Shterengas, G. L. Belenky, T. Hosoda, G. Kipshidze and S. Suchalkin "Continuous Wave Operation of Diode Lasers at 3.36 μm at 12°C," Applied Physics Letters, Vol. 93, 2008.
11. M. Rattunde, J. Schmitz, R. Kiefer and J. Wagner, "Comprehensive Analysis of the Internal Losses in 2.0 μm (AlGaIn)(AsSb) Quantum-Well Diode Lasers," Applied Physics Letters, Vol. 84, No. 23, 2004, pp. 4750- 4752. doi:10.1063/1.1760216
12. A. Salhi and A. Al-Muhanna, "Self Consistent Analysis of Quantum Well Number Effects on the Performance of 2.3 μm GaSb-Based Quantum Well Laser Diodes," IEEE Journal of Selected Topics in Quantum Electronics, Vol. 15, No. 3, 2009, pp. 918-924. doi:10.1109/JSTQE.2008.2012000
13. FIMMWAVE and HAROLD by Photon Design http://www.photond.com
14. B. Qiu, S. S. McDougall, X. Liu, G. Bacchin and J. H. Marsh, "Design and Fabrication of Low Beam Divergence and High Kink Free Power Lasers," IEEE Journal of Quantum Electronics, Vol. 41, No. 9, 2005, pp. 1124- 1130. doi:10.1109/JQE.2005.853359
15. M. T. Kelemen, J. Weber, M. Rattunde, G. Kaufel, J. Schmitz, R. Moritz, M. Mikulla and J. Wagner, "High Power 1.9 μm Diode Laser Arrays with Reduced Far-Field Angle," IEEE Photonics Technology Letters, Vol. 18, No. 4, 2006, pp. 628-630. doi:10.1109/LPT.2006.870146

CHAPTER 15

COMBINED ELECTROMAGNETIC AND DRIFT DIFFUSION MODELS FOR MICROWAVE SEMICONDUCTOR DEVICE

Samir Labiod, Saida Latreche, Mourad Bella, Christian Gontrand

Keywords: Drift-Diffusion Model, GUMMEL's Method, Backward Euler, MAXWELL's Equation, 3-D FDTD Method

INTRODUCTION

Propagation and radiative effects become more and more important for the integrated circuit domain, today submicron semiconductor are operated under high frequencies. This is particularly important for the characterization of interconnected structures loaded at digital and drivers. Indeed, it is well known that electromagnetic compatibility (EMC) and signal integrity (SI) are strongly affected by the geometry of interconnects and by the possibly complex nonlinear/dynamic behavior of the electronic devices collocated at their terminations.

In this paper, the semiconductor analysis is based on the time-domain drift-diffusion model (DDM) in conjunction with electromagnetic model (Maxwell's equations) [1,2]. Where the active device model (AD model) correspond to the Poisson equation and the carrier transport ones, obtained by splitting the Boltzmann transport equation (BTE) into its first two moments.

In first part of this wok, we develop a modeling and a simulation for a PIN diode.

For this we give a simple discretization scheme for the DDM model using a finite different algorithm to achieve the space-domain integration, whereas Euler backward algorithm is adopted to accomplish time-domain integration. The resulting equations can be written as implicit sparse matrix systems. In this approach, the three equations are coupled together through the whole process of computation. The well known Gummel's iteration is most commonly used to solve the coupled problem from the active device model at each time step, whereas each matrix system is solved using an efficient direct UMFPACK method.

In a second part, Finite-difference time-domain (FDTD) method is used to solve electromagnetic equations, where it can be coupled to an efficient solution algorithm by incorporating a "lumped-device" [3]. The design simulation is employed to 3-D structure of the printed circuit including PIN diode (lumped device). The Berenger Perfectly Matched Layer condition (PML) is accomplished for the FDTD computation. The innovation of this condition is that electromagnetic waves of arbitrary incidence, polarization, and frequency are matched at the boundary and subsequently absorbed in the PML layer.

ACTIVE DEVICE MODEL

The active device model (AD model) used is based on the moments of Boltzmann's transport equations obtained by the integration over the momentum space [4]. This model formulates the problem using three dependent variables V, n, and p.

The Poisson's equation can be employed for the potential within the diode together with the charge conservation for electrons and holes.

$$\nabla^2 V = -\frac{q}{\varepsilon}\left(p(x)-n(x)+Dop(x)\right) \tag{1}$$

where $Dop = N_A - N_D$ and accounts for the net ionized impurity concentration.

The electron and hole current continuity equation for semiconductor devices can be written as

Citation: S. Labiod, S. Latreche, M. Bella and C. Gontrand, "Combined Electromagnetic and Drift Diffusion Models for Microwave Semiconductor Device," Journal of Electromagnetic Analysis and Applications, Vol. 3 No. 10, 2011, pp. 423-429. doi: 10.4236/jemaa.2011.310067.

$$\frac{\partial n}{\partial t}-\frac{1}{q}\nabla \boldsymbol{J}_n=-q\cdot r_{SRH} \tag{2}$$

$$\frac{\partial n}{\partial t}+\frac{1}{q}\nabla \boldsymbol{J}_p=-q\cdot r_{SRH} \tag{3}$$

where r_{SRH} is the Shockley-Read-Hall recombination.

$$r_{SRH}=\frac{n\cdot p-n_i^2}{\tau_p\cdot(n+n_i)+\tau_n\cdot(p+n_i)} \tag{4}$$

J_n, J_p can be described by a drift-diffusion model:

$$\boldsymbol{J}_n=-q\cdot n\cdot\mu_n\cdot\nabla V+q\cdot D_n\cdot\nabla n \tag{5}$$

$$\boldsymbol{J}_p=-q\cdot n\cdot\mu_p\cdot\nabla V+q\cdot D_n\cdot\nabla p \tag{6}$$

NUMERICAL APPROACH

Discretization Scheme

The discretization uses a first and second order of (1) in 1-D-finite difference mesh, leads to have

$$V^k(i+1)+V^k(i-1)-2\cdot V^k(i)=\frac{q\cdot h^2}{\varepsilon}\left[n^k(i)-p^k(i)-Dop(i)\right] \tag{7}$$

Euler implicit method seeks to approximate the derivatives in (2) and (3) with regard to the discrete solutions points defined by spatial and temporal cells [5,6]. The electron and hole continuity equation may be discretized in implicit form as follows

$$a_1(i)\cdot n^{k+1}(i-1)+a_2(i)\cdot n^{k+1}(i+1)+\left[a_3(i)+a_4(i)+1\right]\cdot n^{k+1}(i)=\frac{\tau\cdot u_T\cdot h^2\cdot\mu_0}{\Delta t\cdot L_D^2\cdot\mu_n}\left[R(i)-n^k(i)\right] \tag{8}$$

$$a_5(i)\cdot p^{k+1}(i-1)+a_6(i)\cdot p^{k+1}(i+1)+\left[a_7(i)+a_8(i)+1\right]\cdot p^{k+1}(i)=\frac{\tau\cdot u_T\cdot h^2\cdot\mu_0}{\Delta t\cdot L_D^2\cdot\mu_n}\left[R(i)-p^k(i)\right] \tag{9}$$

The variables $a_{1\text{-}8}$ are matrix coefficient which can be found in [Mirzavand], and the indices i describe spatial discretization, k corresponds to the time progression. The mesh size h is limited by the Debye length and Δt by the dielectric relaxation time The corresponding Matrix systems of (7), (8) and (9) can be written as:

$$A\cdot V^k=Y^k \tag{10}$$

$$B\cdot n^{k+1}=n^k \tag{11}$$

$$C\cdot p^{k+1}=p^k \tag{12}$$

where

$$Y^k=\frac{q\cdot h^2}{\varepsilon}\begin{pmatrix} n^k(1)-p^k(1)-Dop(1)\\ \vdots \\ n^k(\mathrm{i})-p^k(\mathrm{i})-Dop(\mathrm{i})\end{pmatrix}$$

A, B and C are the sparse matrix.

Numerical Whole

In order to derive the iteration procedure, at each time step we relate Gummel's method to direct UMFPACK method which is well known to converge quadratically [7].

The Equations (10), (11) and (12) are solved in a decoupled manner. Poisson equation is solved at all grid points, followed by electron continuity equation, and then by hole continuity one. Each matrix system is solved using an efficient direct numerical UMFPACK method [8,9].

The whole numerical procedure to calculate the final solution at each time step using Gummel's iterations can eventually be summarized as follows:

Step 1: Initialize carriers (Evaluate n^0 and p^0)

Step 2: Solve (10) (Evaluate V^0)

For k from 1 to final time iterations Begin Gummel iterations Step 3: Solve (10), (11) and (12) using UMFPACK method Step 4: Test convergence End Gummel iterations Step 5: Set $n^k \leftarrow n^{k+1}$ and $p^k \leftarrow p^{k+1}$

End time iterations.

ELECTROMAGNETIC WAVE APPROXIMATION

The electromagnetic wave propagation can be completely characterized by solving Maxwell's Equations [10]. These equations are first-order linear the field vectors at any point in the space at any time. The current density distribution is obtained from the solution of the driftdiffusion model [11,12].

$$\nabla\times\boldsymbol{H}=\varepsilon\frac{\mathrm{d}\boldsymbol{E}}{\mathrm{d}t}+\boldsymbol{J}_{media}+\boldsymbol{J}_{lumped} \tag{13}$$

$$\nabla\times\boldsymbol{E}=-\mu\frac{\mathrm{d}\boldsymbol{H}}{\mathrm{d}t} \tag{14}$$

Here $\boldsymbol{J}_{media}$ account for the contribution of the current flowing along the distributed media, whereas $\boldsymbol{J}_{lumped}$ includes the contributions of lumped elements. Such a contribution comes from the solution of related device model's equations.

We consider a locally uniform medium in the vicinity of an x-directed lumped element [13]. A schematic of the coupling between the solvers is given in Figure 1. The entire computational domain is discretized with a FDTD mesh sizes Δx, Δy, Δz along the three Cartesian coordinates [14].

Then the electric field of the lumped element may be defined as [15]:

$$E_x^{t+1}\left(i+\frac{1}{2},j,k\right)=E_x^{t}\left(i+\frac{1}{2},j,k\right)$$
$$+\frac{\Delta t}{\varepsilon\cdot\Delta y}\left[H_z^{t+\frac{1}{2}}\left(i+\frac{1}{2},j+\frac{1}{2},k\right)-H_z^{t+\frac{1}{2}}\left(i+\frac{1}{2},j-\frac{1}{2},k\right)\right]$$
$$-\frac{\Delta t}{\varepsilon\cdot\Delta z}\left[H_y^{t+\frac{1}{2}}\left(i+\frac{1}{2},j,k+\frac{1}{2}\right)-H_y^{t+\frac{1}{2}}\left(i+\frac{1}{2},j,k-\frac{1}{2}\right)\right]$$
$$-\frac{\Delta t}{\varepsilon}\left[J_{media,x}^{t+\frac{1}{2}}\left(i+\frac{1}{2},j,k\right)\right.$$
$$\left.+\frac{1}{2}\left(J_{lumped,x}^{t+1}\left(i+\frac{1}{2},j,k\right)+J_{lumped,x}^{t}\left(i+\frac{1}{2},j,k\right)\right)\right] \tag{15}$$

where, J_{lumped} is an impressed current density through which the PIN diode will be incorporated. Such a contribution comes from the time domain solution of the related device model's equations. The active device is oriented in the x-direction, so the current density is given by:

$$J_{lumped}=\frac{I_{Diode\ Current}}{\Delta z\cdot\Delta y}$$

To preserve the computational efficiency of the "leapfrog" solution scheme, we assume that the device equations are solved at even time-steps; this permit the averaging of current density contribution between the two times steps k and k + 1.

The time step Δt is bounded by courant stability criterion [16]

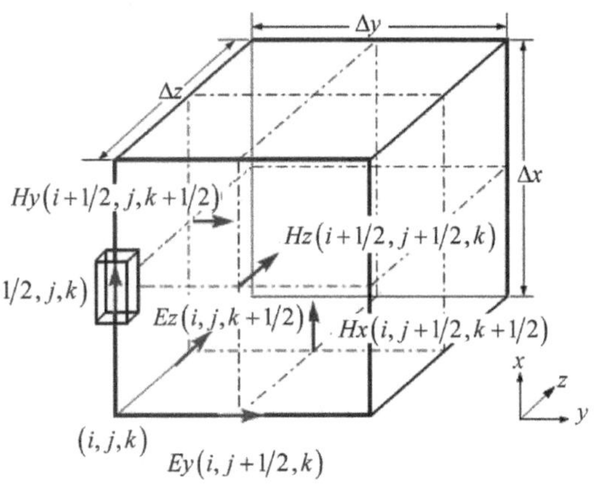

Figure 1: Example of a figure caption.

$$\Delta t\leq\frac{1}{C\sqrt{\frac{1}{(\Delta x)^2}+\frac{1}{(\Delta y)^2}}} \tag{16}$$

To simulate the infinite space surrounding the structure, absorbing boundaries conditions are considered. At interconnections, metal surfaces behave, to a good approximation, as perfect electric conductors (PEC). Therefore, we set the tangential component of the electric field to zero on the metal boundaries.

In the cell (i + 1/2, j), we solve drift-diffusion model and Maxwell's equations, where Equation (15) is adopted to combine between the two model to update the electric and magnetic fields. This algorithm can be efficiently carried out in the tree steps [15,17]:

For k from 0 to final time Step 1: update magnetic field components H using (14)

Step 2: solve semiconductor device using (10), (11), (12) and (15)

Step 3: update electric field components E using (13)

End time iterations

RESULTS AND DISCUSSIONS

Stationary Simulation Results

The PIN diode considered in this simulation is 2.5 μm length. Here, a uniform mesh that covers the 1-D cross section of the diode is used with 500 nodes.

The P and N junctions are doped as Gaussian function with a peak density of 5.1017 At $^{\bullet}$ cm^{-3} at the two contacts.

The device is biased with a direct potential of Vp = 0.7 v, the DC parameter distributions are obtained by solving the drift-diffusion model.

Figure 2 represents the norm of residual error using Newton's iterations and direct UMFPACK methods respectively.

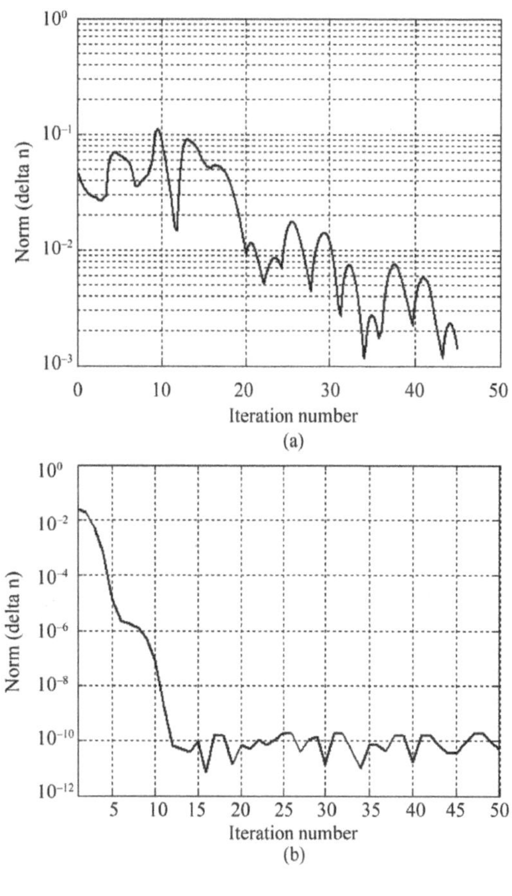

Figure 2: Norm of residual versus Gummel's iterations for 1-D p-i-n device, (a) with Newton iteration, (b) with direct UMFPACK method.

When the UMFPACK method is considered, the CPU time is reduced by 97.34% with a maximum numerical dispersion error of 10 - 11 for this 1D problem. We show clearly that the combined "Gummel-UMFPACK" iteration has significantly better convergence characteristics than the combined "Gummel-Newton raphson" iteration.

Carriers density distributions and potential profile of the device for an applied bias of Vp = 0.7 v is presented in Figure 3.

The static I-V characteristic of PIN diode, as computed for different net doping by means of the numerical model, is shown in Figure 4. Current characteristics are calculated directly by substituting the numerical results in the discretized form of (5) and (6), and then integrate it over the anode contact. The calculated results present a good agreement with those obtained in [15].

Time Domain Results

At initial time t = 0, the PIN diode is forward biased with the potential Vp = 0.7. As the DC simulation time in-

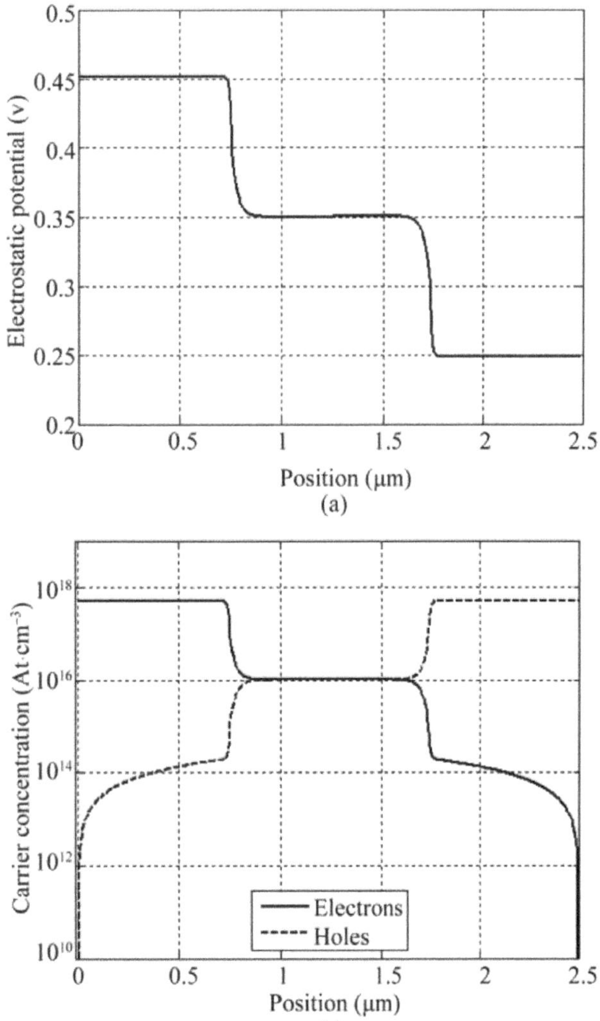

Figure 3: Sample DC results obtained using the proposed algorithm, (a) potential distribution; (b) Carrier density distribution.

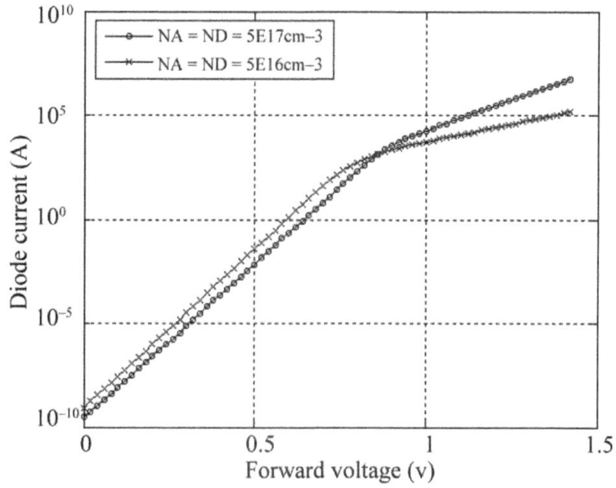

Figure 4. I-V characteristics of the PIN diode, at different doping concentrations.

dependent (the electromagnetic model is not included), where the time step is $\Delta t = 0.05$ ps .

When a PIN diode is forward biased, holes and electrons are injected from the P and N regions into the intrinsic region. These carriers not recombine immediately. Instead, a finite quantity of charge is stored and results in lowering of the resistivity of the intrinsic region as shown in Figure 5.

Figure 6 represents the variation of the output current versus the time for different values of the potential Vp. We can show clearly that the transient solution converge towards results obtained in the steady state solution (view Figure 4).

Frequency Domain

The FDTD simulation is performed with uniform grids of space steps Δx = 0.2 mm, Δy = 0.2 mm and Δz = 0.15 mm. The computation domain is divided into a grid of 100 × 50 × 24 and performed for 4000 time steps, It's located in an unbounded medium, requiring absorbing boundaries placed at a sufficient distance from the circuit to avoid nonphysical reflections.

Figure 7 represents the proposed circuit. The excitation pulse is a Gaussian one with a maximum frequency as 50 GHz. An internal resistor (50 ohm) corresponds at the 1st port and a matched load (50 ohm) is connected to the 2nd port.

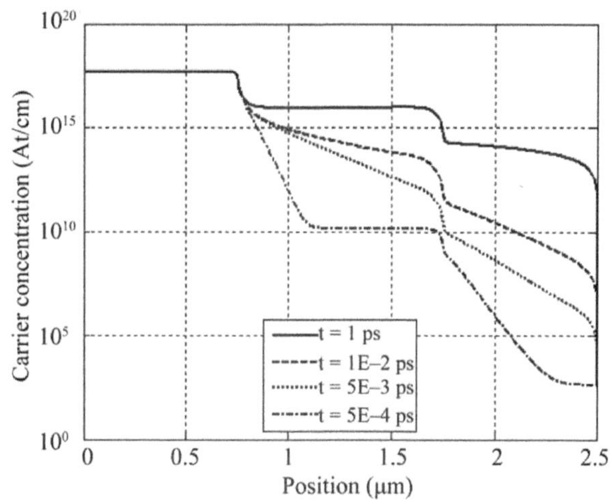

Figure 5: Electrons distribution versus time.

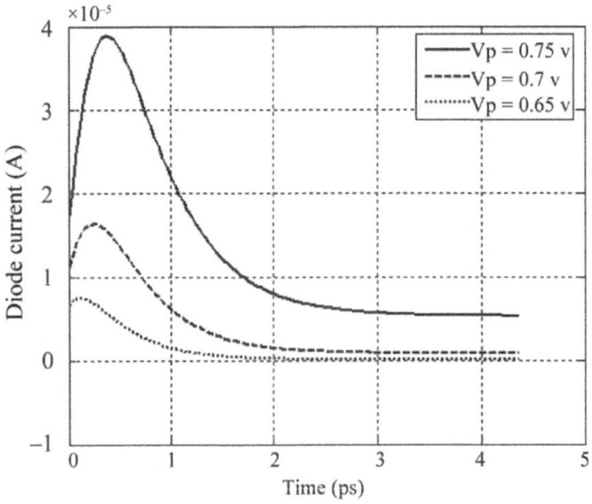

Figure 6: Diode current versus time.

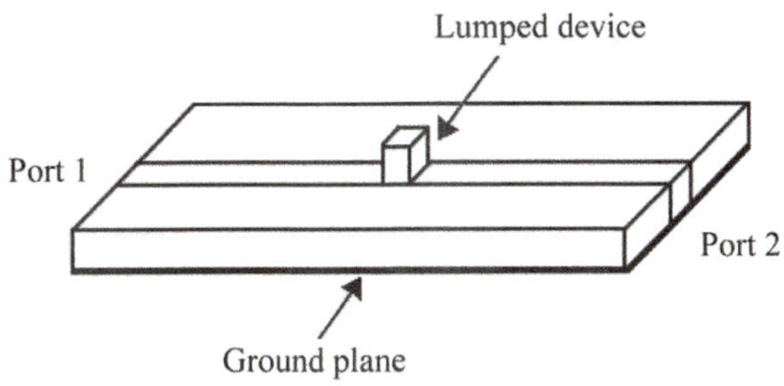

Figure 7: The simulated structure.

Voltage sources and load resistor have been described by analytical models, where the surfaces of the two ports and the ground plane are related by:

$$E(i,j,k) = R_{load} \cdot H(i,j,k)$$

R_{load} = 50 ohm.

The PIN diode is described by numerical solutions obtained in time domain simulation.

Figure 8 shows the special distribution of electric field Ex(x, y, t) beneath the structure at t = 0.882 ps.

The S-parameters of the PIN diode with 50 frequency points is linearly distributed in the frequency—band (1 KHz - 50 GHz). They are calculated and compared with those obtained using finite element software as SILVACOTCAD, which are illustrated in Figures 9 and 10. Time responses for the anode and cathode contacts are calculated using implicit Backward Euler algorithm. A good fit between the proposed model and the finite element simulator (SILVACO-TCAD) is observed.

The insertion loss for the proposed structure is shown in Figure 11, where the real and imaginary parts are presented.

To obtain frequency domain comportment, Fourier Transform algorithm is used in output and input time responses, [18] present the details on how such analysis is performed using FFT.

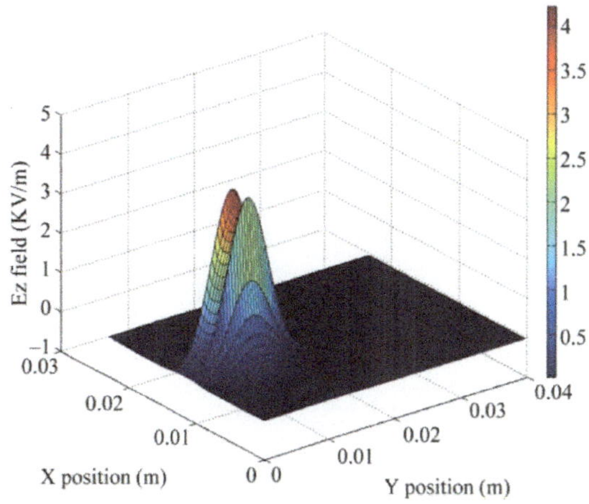

Figure 8: Electric field distribution for 250 time steps.

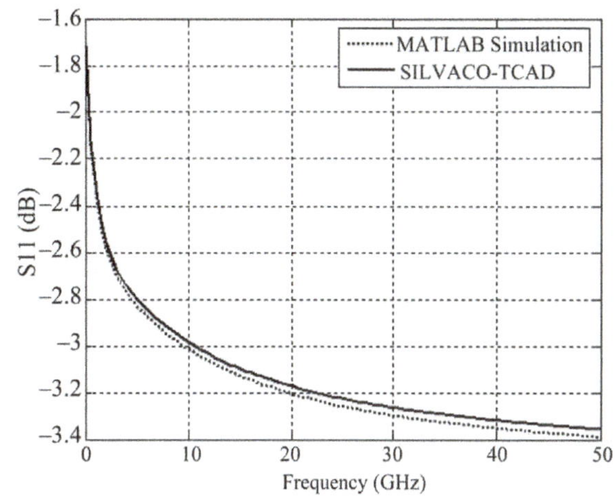

Figure 9: Set of calculated return loss for the PIN diode.

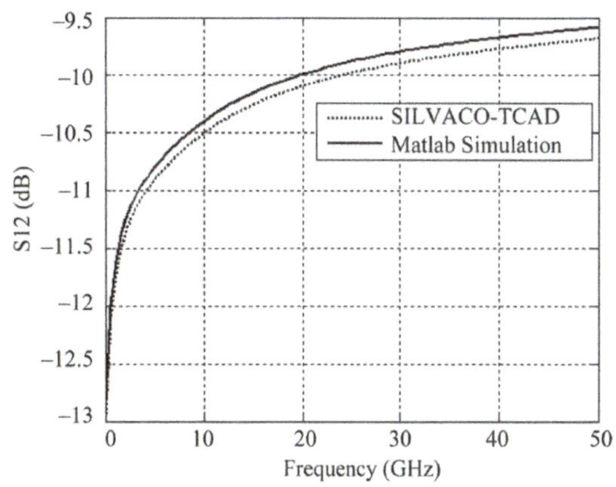

Figure 10: Set of calculated insertion loss for the PIN diode.

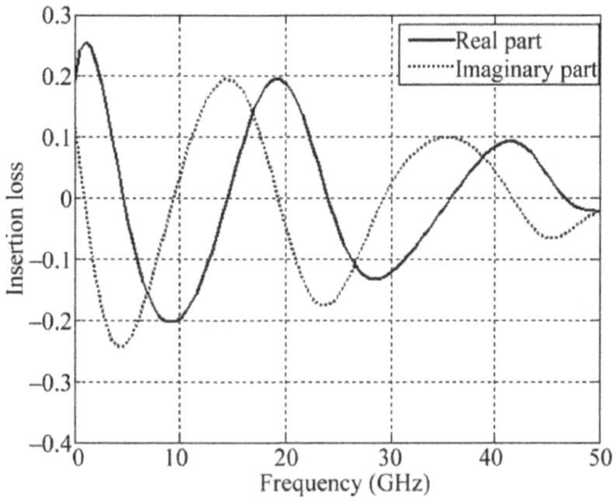

Figure 11: Set of calculated insertion loss for proposed structure.

CONCLUSIONS

In this paper, we have presented an efficient time-domain electromagnetic wave simulator for modeling a PIN diode considered in a micro-wave circuit. The features distributed analysis both the active semiconductor devices and passive interconnection domains.

The mathematical model of the proposed device consists of a set of nonlinear second-order partial differential equations which are solved on uniform mesh using implicit finite difference technique. The descretization of drift-diffusion model is established using backward Euler scheme. The implicit system obtained is solved using the well known Gummel's scheme related with an efficient Direct UMFPACK method.

Using this approach, the proposed algorithm reaches over 95.34% reduction in CPU time. The modeling results are qualitatively in agreement with theoretical concepts. Ended, the transient solution converges towards the steady state solution which confirms the validity of the proposed algorithm.

In other hand, we have presented an efficient timedomain electromagnetic algorithm for modeling a PIN diode used for mm-wave applications. This model combines both accurate modeling of transport phenomena in PIN diode and electromagnetic wave effects present around the device at high frequencies. The "leapfrog" scheme of 3-D FDTD method is used for the simulation of the propose circuit. Results obtained are in agreement with those obtained by other simulator software like SILVACO-TCAD.

Our conclusion is that, for the following works, we essentially will use the "FDTD" simulator for two essential reasons: the first one: this simulator is also dedicated to other devices analysis. The other reason is we will go on to develop our time domain simulator using CrankNicolson scheme, taking care of the no totally resolved boundary problems, nowadays.

Nomenclature

q: Electronic charge.
n_i: Intrinsic carrier concentration.
n: Electron density.
p: Hole density.
V: Electrostatic potential.
V_p: Anode voltage.
μ_T: Thermal voltage.
μ_n, μ_p: Electron and hole mobility.
D_n, D_p: Electron and hole diffusion coefficient.
τ: Dielectric relaxation time.
τ_n, τ_p: Carrier lifetimes of electron and hole.
ε: Dielectric constant.
C: Propagation speed in the medium.
L_D: Debye Length.
E: Electric field.
H: Magnetic field.

REFERENCES

1. K. J. Willis, J. S. Ayubi-Moak, S. C. Hagness and I. Knezevic, "Absorbing Global Modeling of Carrier-Field Dynamics in Semiconductors Using EMC-FDTD," Journal of Computational Electronics, Vol. 8. No. 2, 2009, pp. 153- 171. doi:10.1007/s10825-009-0280-4
2. M. Sirbu, B. Sebastien, P. Lepaul and F. Aniel, "Coupling 3-D Maxwell's and Boltzmann's Equations for Analyzing a Terahertz Photoconductive Switch," IEEE Transactions on Microwave Theory and Techniques, Vol. 53, No. 9, 2005, pp. 2991-2998.doi:10.1109/TMTT.2005.854228
3. W. Sui, D. Christensen and C. Durney, "Extending the Two-Dimensional FD-TD Method to Hybrid Electromagnetic Systems with Active and Passive Lumped Elements," IEEE Transactions on Microwave Theory and Techniques, Vol. 40, No. 4, 1992, pp. 724-730.doi:10.1109/22.127522 [Paper reference:1]
4. K. M. Sze and K. K. Ng, "Physics of Semiconductor Devices," 3rd Edition, Wiley, Hoboken, 2007, pp. 40-63. [Paper reference:1]
5. A. Quarteroni, R. Sacco and F. Saleri, "Numerical Mathematics," Springer, Berlin, 2000.
6. R. Mirzavand, A. Abdipour and G. Moradi, "Full-Wave Semiconductor Devices Simulation Using ADI-FDTD Method," Progress in Electromagnetic Research M, Vol. 11, 2010, pp. 191-202. doi:10.2528/PIERM10010604
7. S. K. Khaitan, J. D. McCalley and Q. Chen, "Multifrontal Solver for Online Power System Time-Domain Simulation," IEEE Transactions on Power Systems, Vol. 23, No. 4, 2008, pp. 1727-1737. doi:10.1109/TPWRS.2008.2004828
8. A. Aste and R. Vahldieck, "Time-Domain Simulation of the Full Hydrodynamic Model," International Journal of Numerical Modeling, Vol. 16, No. 12, 2003, pp. 161-174.doi:10.1002/jnm.491
9. M. Movahhedi and A. Abdipour, "Efficient Numerical MeThods for Simulation of High-Frequency Active Device," IEEE Transactions on Microwave Theory and Techniques, Vol. 54, 2006, pp. 2636-2645. doi:10.1109/TMTT.2006.872937
10. J. E. Marsden, L. Sirovich and S. S. Antman, "Computational Electromagnetics," Springer, Berlin,

2005. [Paper reference:1]

11. J. Mix, J. Dixon, Z. Popovic and M. Piket-May, "Incorporating Non-Linear Lumped Element in FDTD: The Equivalent Source Method," International Journal of Numerical Modelling: Electronic Networks, Devices, Vol. 12, No. 1-2, 1999, pp. 157-170.doi:10.1002/(SICI)1099-1204(199901/04)12:1/2<157::AID-JNM323>3.0.CO;2-V

12. S. G. Talocia, I. S. Stievano and F. G. Canavero, "Hybridization of FDTD and Device Behavioral-Modeling Techniques," IEEE Transactions on Electromagnetic Compatibility, Vol. 45, No. 1, 2003, pp. 31-42. doi:10.1109/TEMC.2002.808035

13. H. Chuang and L. Kuo, "3-D FDTD Design Analysis of a 2.4-Ghz Polarization-Diversity Printed Dipole Antenna with Integrated Balun and Polarization-Switching Circuit for WLAN and Wireless Communication Applications," IEEE Transactions on Microwave Theory and Techniques, Vol. 51, No. 2, 2003, pp. 374-381. doi:10.1109/TMTT.2002.807838 [Paper reference:1]

14. M. N. O. Sadiku, "Numerical Techniques in Electromagnetics," CRC Press, Boca Raton, 2001. [Paper reference:1]

15. P. Ciamolini, L. Roselli and G. Stopponi, "Mixed-Mode Circuit Simulation with Full-Wave Analysis of Interconnections," IEEE Transactions on Electron Devices, Vol. 44, No. 11, 1997, pp. 2098-2105. doi:10.1109/16.641390 [Paper reference:2]

16. F. Zheng, Z. Chen and J. Zhang, "Toward the Development of a Three-Dimensional Unconditionally Stable FiniteDifference Time-Domain Method," IEEE Transactions on Microwave Theory and Techniques, Vol. 48, No. 9, 2000, pp. 1550-1558. doi:10.1109/22.869007 [Paper reference:1]

17. O. Gonzalez, J. A. Pereda, A. Herrera and A. Vegas, "An Extension of the Lumped-Network FDTD Method to Linear Two-Port Lumped Circuits," IEEE Transactions on Microwave Theory and Techniques, Vol. 54, No. 7, 2006, pp. 3045-3051.doi:10.1109/TMTT.2006.877058

18. X. Zhang and K. K. Mei, "Time-Domain Finite Difference Approach to the Calculation of the Frequency-Dependent Characteristics of Microstrip Discontinuities," IEEE Transactions on Microwave Theory and Technique, Vol. 36, No. 12, 1988, pp. 1775-1787.

CHAPTER 16

EVALUATION OF FATIGUE LIFE OF SEMICONDUCTOR POWER DEVICE

Kazunori Shinohara, Qiang Yu

Keywords: Power Electronics, Power Modules, Reliability, Damage, Simulation Technology, Solder Joint Fatigue, Crack Propagation, Mechanical Stress

INTRODUCTION

Nowadays, increasing attention is being paid to environmental problems such as global warming. In this regard, hybrid cars attract interest as fuel-efficient transportation. The growth of the fuel-efficient transportation market is anticipated. In hybrid cars and electrical cars, electric power is used as the driving energy source. In order to drive a motor efficiently, direct current must be converted to alternating current by an inverter. The electrical device used in the inverter is called the power device [1,2]. Currently, the power devices in a car is used as an auxiliary for the internal-combustion engine. In order to use a power device as the driving energy source in a car such as a hybrid or an electrical car, there must be a strong flow of electrical current in the device. On the other hand, the available space in a car inverter is limited. The power device must be compact. This compactness cannot be avoided with the use of high power density, a higher degree of power device reliability is required than can be found in the literature on the subject.

In the methods of evaluating fatigue in power devices found in the literature, we see that a stress test cycle is applied [3]. Bond wire heel crack failures have been observed in experiments. In evaluating the fatigue of a power device, the thermal distribution is closely related to the fatigue life. The thermal distribution at the solder junction and in the power device is computed using a combination of electrical and heat analysis [4,5]. The fatigue life cycle has also been predicted using thermal stress analysis [6-11]. In an experiment, the fatigue life of a power device has been evaluated by a thermal cycle test [12], in which the temperature is raised to a specified point and then lowered to room temperature. This raising and lowering constitutes one cycle, and such cycles are repeated until failures appear in the power device.

The temperature distribution in a power device tends to become more uniform when here is a low degree of electrical loss under a low power current density. Therefore, by using a thermal cycle test, the fatigue life cycle can be evaluated. Increased compactness of power devices is a great need in our time. A power device is also used under a high current density caused by the high current. Because heat increases at the local solder junction of power devices, the temperature distribution tends to become non uniform. This may explain the difference between the evaluation of the fatigue life cycle based on a thermal cycle test as opposed to one based on the power cycle.

In this study, a method of evaluating fatigue based on the power cycle is presented, using a combination of electrical and heat stress analysis [13]. The discretization procedure is based on the finite element method [14]. The propagation of cracks is modeled in the solder junction, since this cracking is the main cause of power device failure. Crack propagation thus becomes an important factor in the evaluation of the fatigue life cycle. By comparing the results of the thermal cycle test with those of the power cycle test, we obtain new insight into the evaluation of fatigue life.

POWER DEVICE STRUCTURE

The structures of a power device are shown in Figure 1. [17]. These consist of multilayered structures (aluminum wires, a

Citation: K. Shinohara and Q. Yu, "Evaluation of Fatigue Life of Semiconductor Power Device," Engineering, Vol. 2 No. 12, 2010, pp. 1006-1018. doi: 10.4236/eng.2010.212127.

copper plate, an aluminum nitride plate, a silicon plate, and an aluminum plate (heat sink)). The silicon is soldered to the copper plate. Solder is an alloy metal that is composed of tin, silver, and copper. The aluminum nitride plate is attached to the aluminum plate through the copper plate. Semiconductor devices can be subject to significant mechanical stress resulting from a mismatch between the CTE of the different layers. The mismatch between the CTE of the silicon and the copper plate is much greater than that between the other layers. Solder is most suitable for use because it has low elasticity, low yield strength, and a relatively high ductility.

To prevent expansion, heat needs to be dissipated through the use of a sandwich-like structure. Solder degradation increases the silicon resistance. Cyclical temperature changes result from the creep strain and plastic strain that are produced by the power cycle, which in turn produces cracking in the solder due to fatigue. It is necessary for the structure to release heat from the silicon. The contact area of the aluminum nitride plate is larger than that of the laminated structure between the silicon and the heat sink plate. In Figure 1, boundaries (i) and (ii) are on the collector side and emitter side, respectively. The current flows from the collector side (i) to the emitter side (ii) through aluminum wires. The current on boundary (i) is the collector current. The difference in voltage between (i) and (ii) is the collector-emitter voltage.

THEORY

Electrical Analysis

In Maxwell's equations, a steady direct current is as follows:

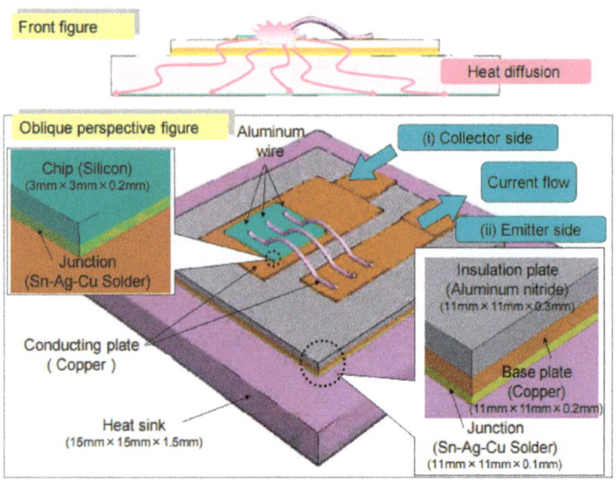

Figure 1: Power device structure (Wire type).

$$I = \int_S J \cdot n dS \tag{1}$$

The variables S, n, and J indicate the surface, outward normal, and current density, respectively. The current density, J, is as follows:

$$J = \sigma \cdot E \tag{2}$$

where E and σ represent the electric field strength and the electrical conductivity, respectively. The dissipation of the electrical energy, P, by the current in the conductivity is as follows:

$$P = J \cdot E \tag{3}$$

In this study, it is assumed that the dissipated electrical energy is transformed into heat energy.

Heat Analysis

Heat is generated by the dissipation of electrical energy. For a thermal analysis, the equation is as follows:

$$C\frac{\partial T}{\partial t} = \lambda \Delta T \tag{4}$$

where T, t, C, λ and Δ represent the temperature, time, specific heat, coefficient of thermal conductivity, and Laplacian, respectively. Using the linear expansion coefficient α, the increment of the thermal strain is as follows:

$$\Delta\varepsilon_\theta = \alpha \Delta T \tag{5}$$

Stress Analysis

Strain consists mainly of elastic strain (ε_e), plastic strain (ε_p) and creep strain (ε_c), which is time dependent. In this study, the total strain consists of the elastic strain, plastic strain, creep strain, and thermal strain (ε_θ).

$$\Delta\varepsilon = \Delta\varepsilon_e + \Delta\varepsilon_p + \Delta\varepsilon_c + \Delta\varepsilon_\theta \tag{6}$$

In the case that the external stress does not affect this strain, the strain is zero ($\Delta\varepsilon = 0$).

$$\Delta\varepsilon_\theta = -\Delta\varepsilon_e - \Delta\varepsilon_p - \Delta\varepsilon_c \tag{7}$$

Thermal strain that is caused by material expansion is converted to elastic strain, plastic strain, and creep stain. A body is deformed due to a force exerted upon its surface between materials. In a state of elasticity, a body can return to its initial condition. However, the initial condition of the body cannot fully recover its plastic behavior, and the strain remains in the body. In creep behavior, the body is deformed in relation to the amount of time that passes. Creep strain is calculated as follows:

$$\frac{d\varepsilon}{dt} = A\left(\sigma_m^2\right)^n \tag{8}$$

where A and n are the parameters shown in Table 5. The variable σ_m represents the von Mises stress as follows:

$$\sigma_m^2 = \frac{(\sigma_1 - \sigma_2)^2 + (\sigma_2 - \sigma_3)^2 + (\sigma_3 - \sigma_1)^2}{2} \tag{9}$$

where the variables $\sigma_1 \sim \sigma_3$ represent the principal stresses.

Coffin-Mansion Law

The fatigue life cycle N is related to the strain range, $\Delta\varepsilon$. Coffin and Manson presented an approximate line correlating N and $\Delta\varepsilon$ using a double logarithmic graph. This relationship is the Coffin-Manson Law, which is expressed as follows:

$$\Delta\varepsilon \cdot N^{\alpha} = C \tag{10}$$

where C and α are the constant of proportion and the fatigue coefficient, respectively. In this study, N is defined as follows [15]:

$$N = 1000\left(\frac{\Delta\varepsilon}{0.01}\right)^{-1.24} \tag{11}$$

An isothermal mechanical fatigue test is performed using a BGA test piece, in order to investigate the fatigue characteristics of Sn-Ag-Cu lead-free solder and the relationships between the number of cycles and the beginning of fatigue cracking and the strain range. Equation (11) is derived from the Coffin-Mansion Law by using these plots in Figure 2.

The plastic strain range and creep strain range affect the fatigue life cycle. In this study, the strain range $\Delta\varepsilon$ is defined as follows:

$$\Delta\varepsilon = \frac{|\Delta\varepsilon_p| + |\Delta\varepsilon_c|}{2}, \quad ,(12)$$

where $\Delta\varepsilon_p$ and $\Delta\varepsilon_c$ represent the ranges of plastic strain and creep strain, respectively.

Cumulative Damage Rule (Miner's Rule)

In a real environment, steady repetitive loads are rare, as there are usually complex changes in the load duration and amplitude. It is therefore difficult to quantitatively estimate material damage. The cumulative damage rule is generally applied in order to estimate damage approximately. This rule is an evaluation method based on damage superposition. The damage ratio for the strain range experienced in one cycle is assumed to be 1/N. The fatigue damage is derived from Equation (10) as follows:

$$\frac{1}{N} = \left(\frac{\Delta\varepsilon}{C}\right)^{\frac{1}{\alpha}} \tag{13}$$

The damage D caused by loading n times is assumed to be

$$D = \frac{n}{N}. \tag{14}$$

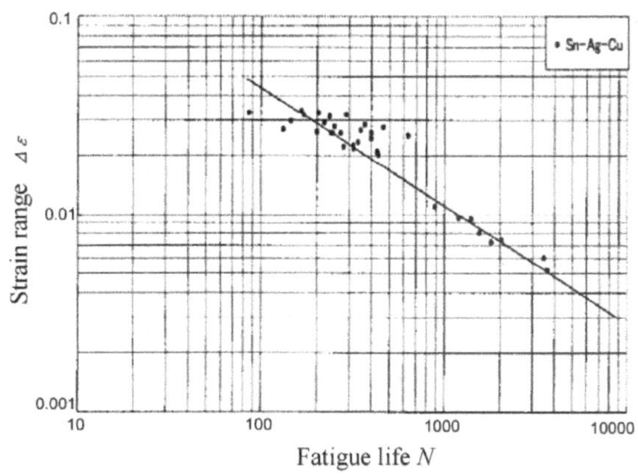

Figure 2: S-N curve of solder [15].

These loads are applied under the same conditions. The cumulative damage caused by different loads is assumed to be as follows:

$$D = \sum_i \left\{\frac{n_i}{N_i}\right\} \tag{15}$$

These loads are applied under different conditions. This relationship is also called Miner's rule. The index i represents the phase of the crack's progress, as shown in **Figure 3**. The phase indicates a point in time that is defined by arbitrarily dividing the time span. If the value of the damage reaches D = 1, a crack occurs in the material. In the case of i = 0 (the initial condition), the material displays no damage. The fatigue life cycle is defined as N_1 = N (Equation (11)). After the phase index i = 0, additional damage is caused in the phase i = 1. Therefore, the equation is as follows:

$$n_i = N_i(1 - D) \tag{16}$$

The quantity in the parenthesis, 1-D, indicates the ratio of the damage not yet caused. The variable N_i is obtained by the Coffin-Manson Law (Equation (11)).

ALGORITHM

The analysis algorithm, which is shown in Figure 4, consists of five steps. In the first step, the boundary conditions and the material constants are set in the calculation model. In the second step, the combined electrical-thermal analysis is performed iteratively. The algorithm determines the temperature data at each time step for all nodes. The power cycle from the start time to the end time is obtained, as shown in Figure 8. In the third step, the combined thermal-stress analysis is performed in order to obtain the strain values (for plastic and creep strain). In the third step, the algorithm requires the temperature distribution data. In the fourth step, the number of cycles required to determine the fatigue life is calculated. The fourth step also consists of five stages, which can be described as follows.

First Stage

In this study, it is assumed that the initial crack occurs at the point of maximum strain. Subsequently, the solder domain replaces the crack domain by a threshold at every phase i.

The threshold is determined by the Manson-Coffin Law and cumulative damage rule. In order to calculate the crack length, one line is used to determine how the crack progresses. In the computational domain, 3D domains are used for cracks. To determine a crack length (1D), the crack domain should be projected on an axis (one line). Therefore, an axis (one line) is set along the direction of the crack progress after checking the 3D crack domain, as shown in Figure 24 of the phase I = 9.

Second Stage

The fatigue life cycle is calculated using the Coffin-Manson Law. The strain range (Equation. (12)) in the Coffin-Manson law is calculated by the finite element method.

Third Stage

The damage at the node point (I) (the maximum strain) in Figure 3 is defined as D = 1. Therefore, the initial crack occurs at the maximum strain point. In each phase, i, the fatigue life cycle, N, at the maximum strain is defined as the variable in Equation (11). By using N, the fatigue life cycle at the other node point is obtained.

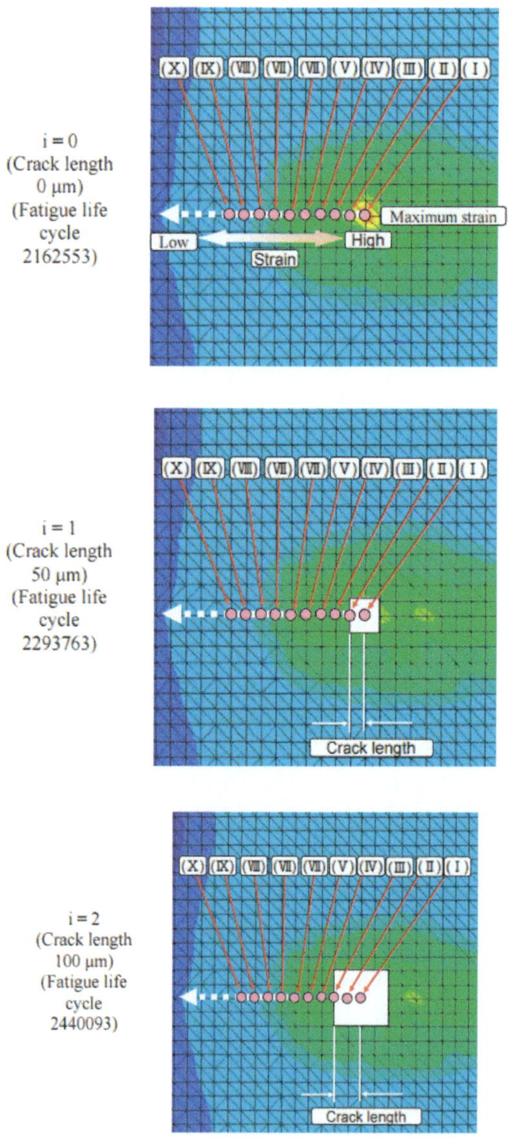

Figure 3: Crack progress.

Table 1: Fatigue life cycle in phase I = 0.

Coordinate	Fatigue Life Cycle (Equation (11))	Accumulated Damage (Equation (15))
(I)	2162553	1.00
(II)	3671684	0.59
(III)	3481152	0.62
(IV)	3609851	0.60
(V)	3768876	0.57
(VI)	3932370	0.55
(VII)	453810	0.48
(VIII)	5920268	0.37
(IX)	6708234	0.32
(X)	7119810	0.30

Table 2: Fatigue life cycle in phase I = 1.

Coordinate	Fatigue Life Cycle (Equation (11))	Accumulated Damage (Equation (15))
(I)	Crack domain	Crack domain
(II)	2293763	1.00
(III)	3745064	0.87
(IV)	3600837	0.86
(V)	3756262	0.82
(VI)	3950283	0.79
(VII)	4615109	0.68
(VIII)	5911460	0.52
(IX)	6651597	0.46
(X)	7077491	0.44

Table 3: Fatigue life cycle in phase I = 2.

Coordinate	Fatigue Life Cycle (Equation (11))	Accumulated Damage (Equation (15))
(I)	Crack domain	Crack domain
(II)	Crack domain	Crack domain
(III)	2440093	1.00
(IV)	3821137	0.94
(V)	3728167	0.91
(VI)	3913761	0.87
(VII)	4694502	0.70
(VIII)	5902676	0.58
(IX)	6595819	0.51
(X)	7035625	0.48

Fourth Stage

At each node, the damage at each phase is added (Equation (15)). Tables 1-3 show the relationship between the fatigue life cycle and the damage. For example, the accumulated damage of coordinate (II) in Table 2 is calculated as follows:

$$D_{2,(\mathrm{II})} = D_{1,(\mathrm{II})} + \frac{n_{2,(\mathrm{II})}}{N_{2,(\mathrm{II})}} = 0.59 + \frac{9.4\times10^5}{22.9\times10^5} = 1.00 \tag{17}$$

where the subscript indexes represent the phase number and the coordinate number, respectively. Using Equation (16), the parameter $n_{2,(\mathrm{II})}$ is calculated as follows:

$$n_{2,(\mathrm{II})} = N_{2,(\mathrm{II})}\left(1 - D_{1,(\mathrm{II})}\right) = 22.9\times10^5 \times (1-0.59)$$
$$= 9.4\times10^5 \tag{18}$$

The accumulated damage of the coordinate (III) in Table 3 is calculated as follows:

$$D_{3,(\mathrm{III})} = D_{1,(\mathrm{III})} + D_{2,(\mathrm{III})} + \frac{n_{3,(\mathrm{III})}}{N_{3,(\mathrm{III})}} = 0.87 + \frac{3.2\times10^5}{24.4\times10^5} = 1.00 \quad (19)$$

The parameter $n_{3,(\mathrm{III})}$ is calculated as follows:

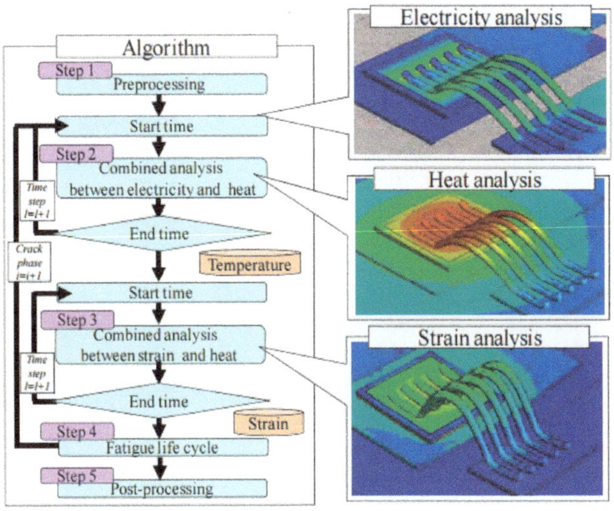

Figure 4: Algorithm.

$$n_{3,(\mathrm{III})} = N_{3,(\mathrm{III})}\left(1 - D_{1,(\mathrm{III})} - D_{2,(\mathrm{III})}\right) = 24.4\times10^5 \times (1-0.87) = 3.2\times10^5 \quad (20)$$

As shown in Figure 4, many iterations from step 1 to step 5 are needed with respect to crack propagation, which increases the total computing time. Therefore, the 1,000 elements ranked as having the highest damage among all the elements of the solder mesh are deleted in every phase.

Fifth Stage

The relationship between the fatigue life cycle and crack length is obtained. In this study, the projection of the crack length on the y axis is shown in Figure 24.

Sixth Stage

The analysis data (stress, thermal, and electrical distribution) are visualized. Figure 4 illustrates the algorithm that uses the power cycle. In the case that the fatigue life cycle is evaluated based upon the thermal cycle, the combined thermal and electrical analysis is not needed.

CALCULATION CONDITIONS

For this paper, ABAQUS[R] was used as a software tool. In the computing mesh, there are 34634 and 169749 elements. The solder thickness is set to 37.5 μm. This solder mesh is separated into 3 blocks layered in the z direction, as shown in Figure 5. In this model, it is assumed that cracks are created only in the intermediate solder layer, and are not created in either the top or bottom solder layers. In the experimental results, cracks tend to occur in the intermediate solder layer. In the electrical analysis, the collector side current on boundary (i) was 60.0 A (as shown in Figure 1). The voltage on the emitter side (ii) was set to 0.0 V. In the thermal analysis, the temperature on the bottom of the heat sink was set to 25°C (the natural cooling condition). The heat transfer coefficient is set to $5.0\times10^{-3}\left(w/\left(mm^2.k\right)\right)$ [17]. In the structural analysis, the displacement with respect to the x, y, and z axes was fixed at the bottom edges of the heat sink. The material properties used in the calculated model are summarized in Table 4. The plastic strain data of the solder with respect to the yield stress and the creep strain data of the solder are shown in Figure 6 and Table 5, respectively. The time step was set to 0.1 s. However, the program automatically subdivides a large time step into several smaller steps if the algorithm has problems.

In order to compare the evaluation of fatigue life based on the power cycle versus evaluation based on the thermal cycle, three different calculation conditions were set in the computing model.

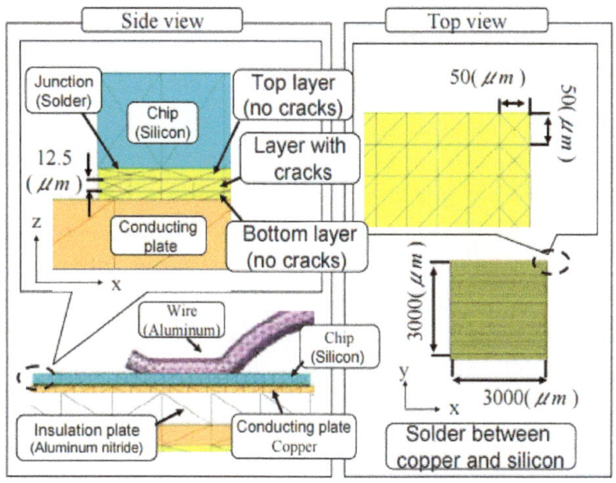

Figure 5: Mesh.

Table 4: Material constant

Material	Thermal conductivity (W/(mmK))	Electrical conductivity (S/mm)	Density (kg/mm^3)	Specific heat (J/(kg·K))	Young modulus (GPa)	Poisson ratio (Nondimension)	Coefficient of thermal expansion (1/K)
Silicon	0.15	0.5	2.33×10^{-6}	700.0	131.0	0.28	4.2×10^{-6}
Aluminum	0.24	37.7×10^3	2.70×10^{-6}	900.0	70.0	0.30	2.15×10^{-5}
Copper	0.40	59.6×10^3	8.96×10^{-6}	380.0	120.0	0.30	1.70×10^{-5}
Solder (Sn-Ag-Cu)	0.05	9.1×10^3	7.40×10^{-6}	234.0	40.0	0.30	2.30×10^{-5}
Aluminum nitride	0.15	1.0×10^{-15} (= 0.0)	3.40×10^{-6}	710.0	320.0	0.24	4.60×10^{-6}

Note: $1.00\ S = 1.00\ \Omega^{-1} = 1.00\ A/V$

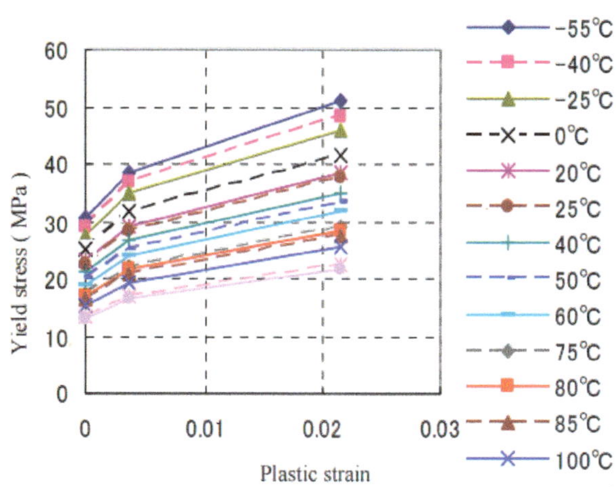

Figure 6: Plastic strain of solder [16].

Table 5: Solder parameter of creep strain in Equation(8) [16]

Temperature	*A [MPa/sec]*	*n*
-40.0	1.1×10^{-25}	13.2
-30.0	7.0×10^{-25}	12.9
-20.0	3.4×10^{-24}	12.5
-10.0	1.7×10^{-23}	12.2
0.0	8.3×10^{-23}	11.9
10.0	4.1×10^{-22}	11.6
20.0	2.0×10^{-21}	11.3
30.0	9.9×10^{-21}	10.9
40.0	4.9×10^{-20}	10.6
50.0	2.4×10^{-19}	10.3
60.0	1.2×10^{-18}	10.0
70.0	5.9×10^{-18}	9.6
80.0	2.9×10^{-17}	9.3
90.0	1.4×10^{-16}	9.0
100.0	7.1×10^{-16}	8.7
110.0	3.5×10^{-15}	8.4
120.0	1.7×10^{-14}	8.0
130.0	8.4×10^{-14}	7.7

The power cycle test is defined as calculation condition 1. By switching the electrical power supply on and off, stress between parts is generated by thermal expansion. The thermal cycle test is defined as calculation condition 2. The power device is gradually heated on its bottom. The temperature rises to a specified point and is then lowered to room temperature. Using experimental equipment, the heat is conveyed from the plate of this to an object through the bottom of the object. This type of experimental equipment is modeled in calculation condition 2.

The thermal shock cycle test is defined as calculation condition 3. The temperature cycle of condition 3 is almost in agreement with the cycle of condition 2. The temperature gradient of calculation condition 3 with respect to time is larger than it of calculation condition 2. The power device is rapidly heated in calculation condition 3. Heat is conveyed from a fluid to the object through all the surfaces of the object by setting a power device in the fluid (air, etc.).

RESULTS

Power Cycle and Thermal Cycle

Power devices are used to adjust currents and voltages. They are installed in hybrid cars, elevators, etc., and do not always work with uniform currents and voltages. An electric current begins to flow at start up, after which it may decrease if the speed of the equipment becomes constant. Such operations are constantly repeated by the starting and the stopping of cars, elevators, etc.

An On-Off control is shown in Figure 7. In the first case, the voltage is adjusted from 10 V to 5 V by placing a series resistance in the circuit. This 5 V drop in voltage causes an energy loss in the circuit. A power device, however, makes it possible to supply an average voltage of 5 V through the use of On-Off switching, which results in a low degree of energy loss. However, when the electrical power supply is switched on and off, the temperature of the power device increases and decreases, respectively. These iterative increases and decreases in temperature naturally cause mechanical stress to occur due to the CTE mismatch.

Temperature cycling is generally applied as a method of evaluating the fatigue life of power devices.

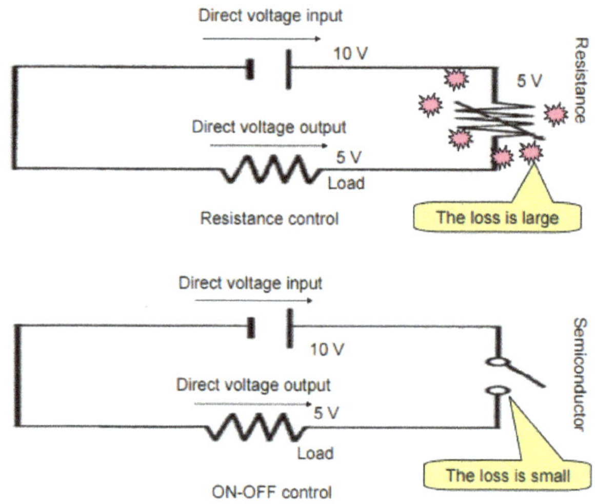

Figure 7: Resistance control and ON-OFF control.

In a combined thermal-mechanical analysis, the temperature is raised to a specified point and then lowered to room temperature. This form of analysis is not considered in an electrical analysis.

As shown in Figure 8, the electricity is turned on until a certain high temperature is reached. After that, the electricity is turned off until the temperature returns to room temperature. These processes together constitute one cycle. The on time and off time are set to 3.5 s and 18 s, respectively. The voltage contour and current density contour are shown in Figures 9 and 10, respectively. An electrical energy loss occurs when the voltage on the collector side of the silicon chip is reduced rapidly. This energy loss is converted to heat energy.

Time History of Temperature (No Cracks (Phase I = 0))

The time history of the maximum temperature with respect to power is shown in Figure 11. In calculation condition 1, the temperature rises during the ON time in the power cycle, as shown in Figure 8. The temperature decreases during the OFF time.

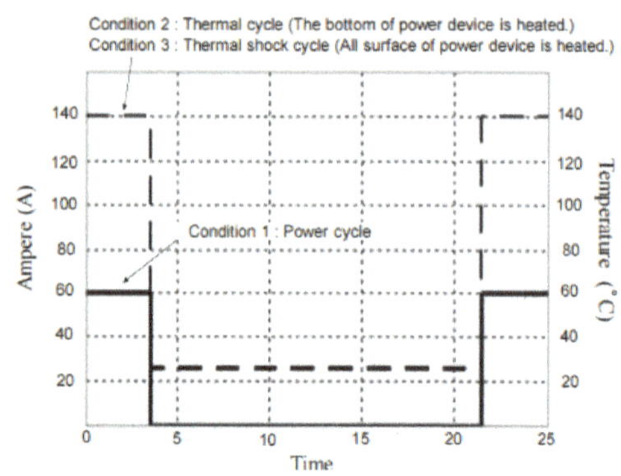

Figure 8: Power cycle and thermal cycle

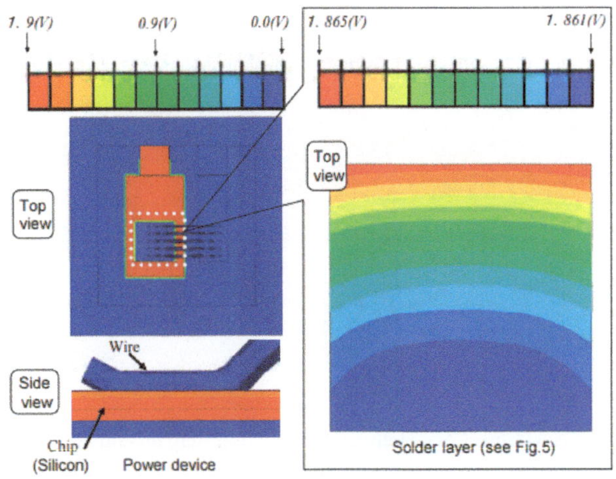

Figure 9: Voltage contour (25 (s)) (phase 0).

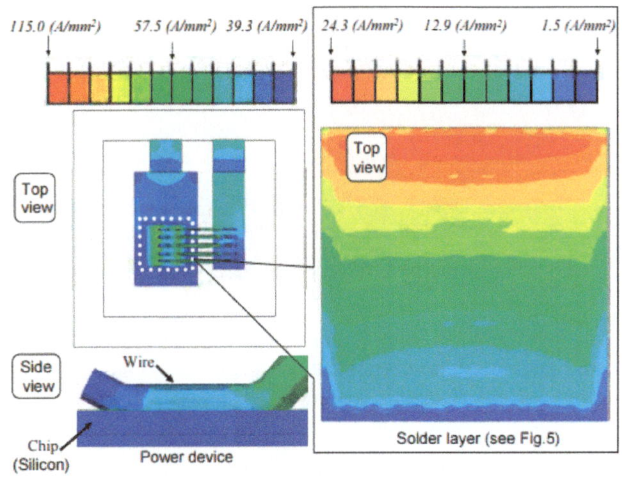

Figure 10: Current density contour (25 (s)) (phase 0).

Heat is essentially not generated on the emitter side connection. Electrical energy losses occur when the voltage on the collector side is rapidly lowered. This energy loss is converted to heat energy by Equation (3). In the wire type power device, the heat concentration occurs in the small connecting space between the wire and the silicon plate, as shown in Figures 12-14. The fatigue life cycle of a power device depends on the maximum temperature it experiences. By adjusting the maximum temperature, the maximum temperature in calculation conditions 2 and 3 is in agreement with that of calculation condition 1.

As shown in Figures 15-17 and in calculation condition 2, heat is gradually transferred from the bottom to the top. As shown in Figures 18-20, heat is transferred from all sides on the surface of the power device. The temperature gradient of calculation condition 2 with respect to time is lower than it is in calculation condition 3.

The temperature distributions on the solder are shown in Figure 21. In this study, the fatigue life of a power device is estimated by analyzing the strain in relation to the solder crack. Fatigue life depends on the temperature distribution in the solder plate.

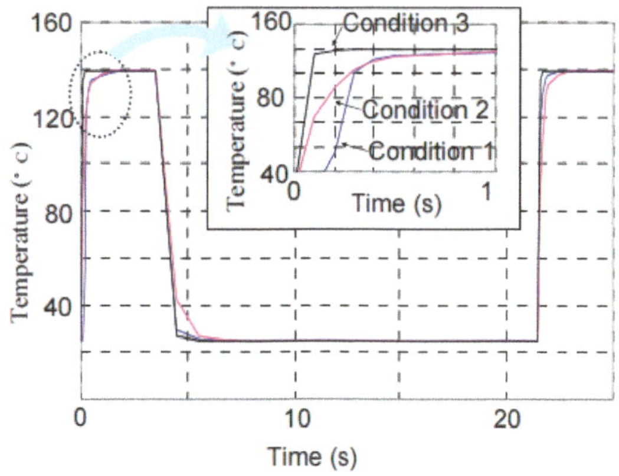

Figure 11: Time history of maximum temperature in power device with respect to power cycle and thermal cycle.

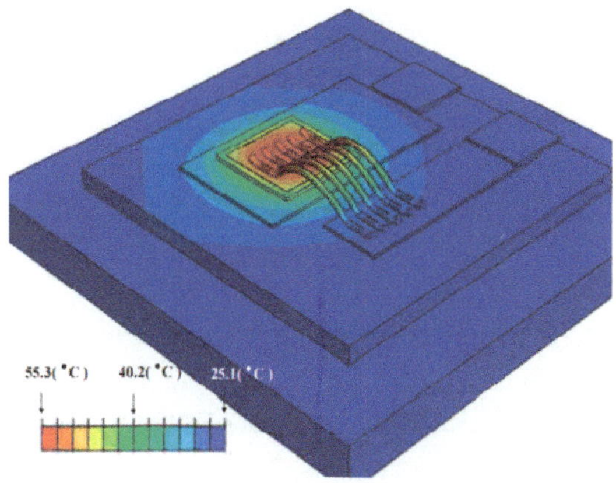

Figure 12: At time 0.1(s), temperature contour map of semiconductor power devices in calculation condition 1 (Power cycle test).

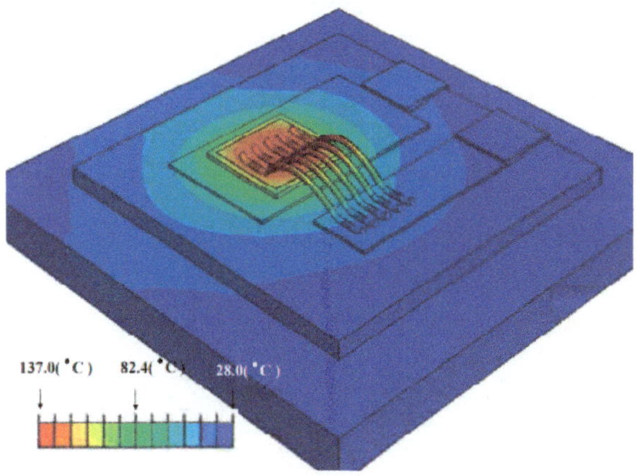

Figure 13: At time 0.5(s), temperature contour map of semiconductor power devices in calculation condition 1 (Power cycle test).

As shown in Figure 21, in calculation condition 1, the temperature distribution on the solder layer becomes non uniform. In the calculation conditions 2 and 3, the temperature distribution on the solder layer becomes non uniform.

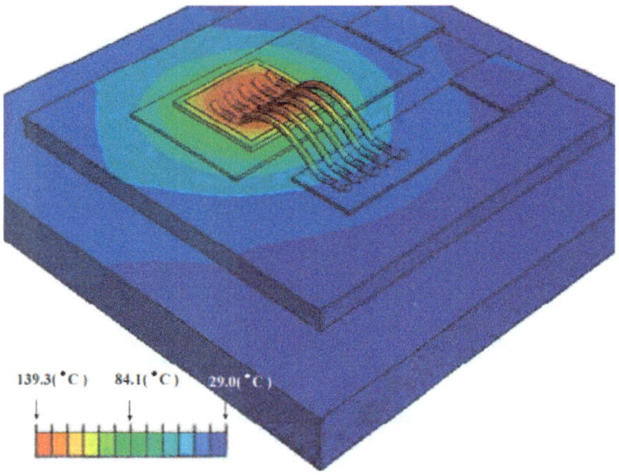

Figure 14: At time 3.5(s), temperature contour map of semiconductor power devices in calculation condition 1 (Power cycle test).

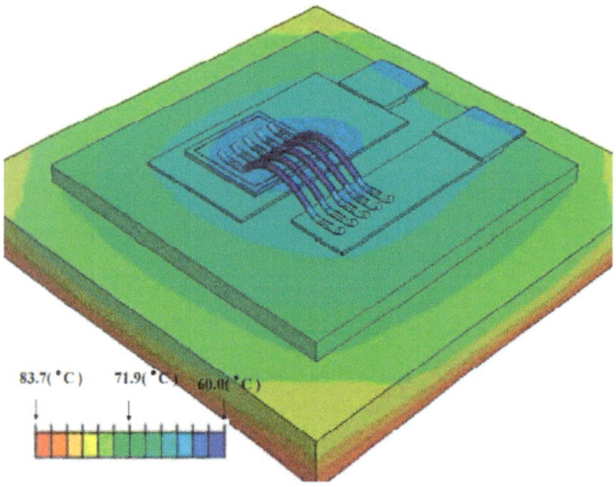

Figure 15: At time 0.1(s), temperature contour map of semiconductor power devices in calculation condition 2 (Thermal cycle test).

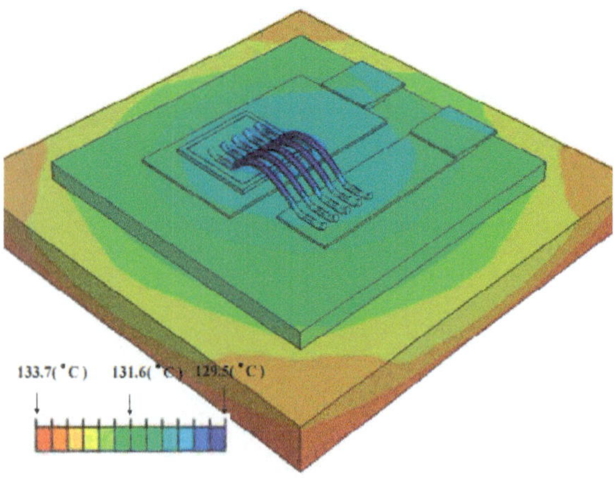

Figure 16: At time 0.5(s), temperature contour map of semiconductor power devices in calculation condition 2 (Thermal cycle test).

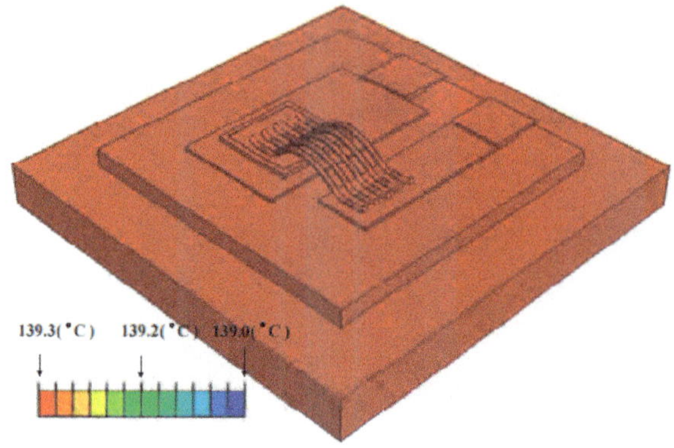

Figure 17: At time 3.5(s), temperature contour map of semiconductor power devices in calculation condition 2 (Thermal cycle test).

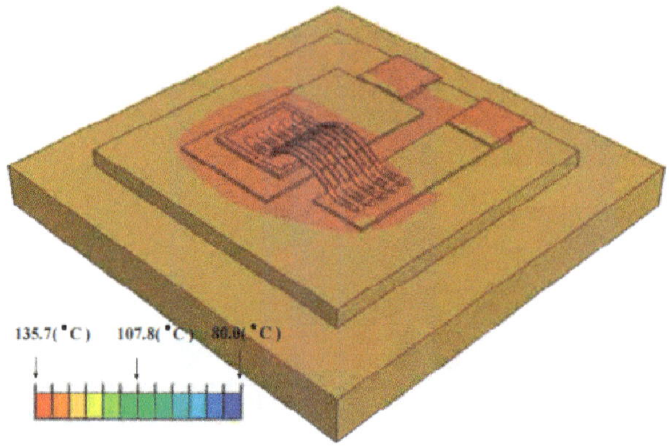

Figure 18: At time 0.5(s), temperature contour map of semiconductor power devices in calculation condition 3 (Thermal shock cycle test).

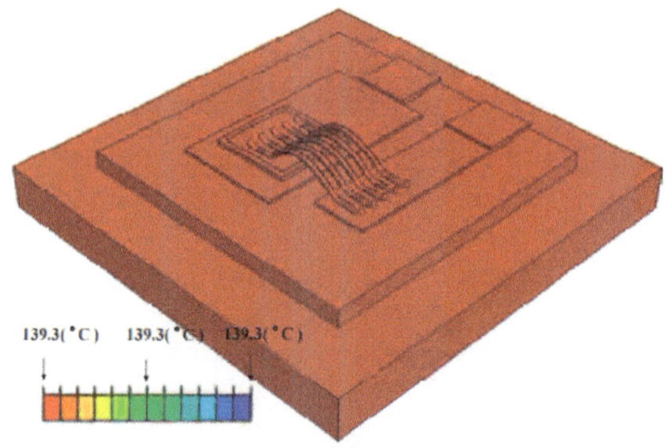

Figure 19: At time 0.5(s), temperature contour map of semiconductor power devices in calculation condition 3 (Thermal shock cycle test).

Strains (No Cracks (Phase I = 0))

The time histories of the maximum plastic strain and maximum creep strain with respect to the power cycle and the thermal cycle (Figure 8) are shown in Figures 22 and 23, respectively. In calculation condition 1 (the power cycle), the maximum strains occur at the center of the solder junction between the silicon plate and the copper plate.

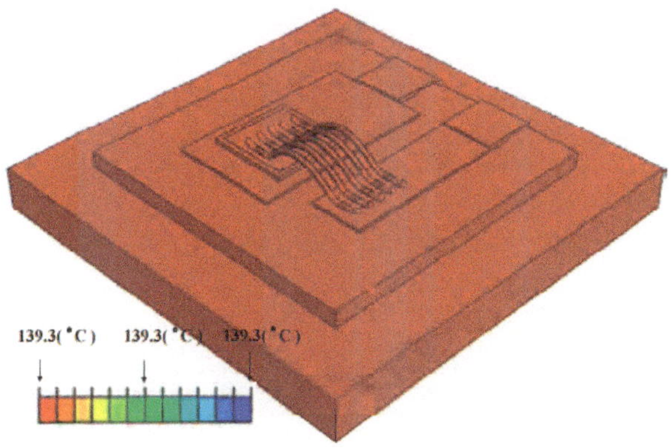

Figure 20: At time 3.5(s), temperature contour map of semiconductor power devices in calculation condition 3 (Thermal shock cycle test).

In calculation condition 2 (the thermal cycle) and calculation condition 3 (the thermal shock cycle), the maximum strains occur over the solder plate. Both plastic strain and creep strain are accumulated during power transitions. In this study, the fatigue life cycle is estimated based upon the ductility of the second ON time between 21.5 s and 25 s. The strain range is defined in Equation (12). It is assumed that there is no difference in the strain range after the second ON time. As shown in Figure 22, the maximum plastic strain of the three calculation conditions is almost in agreement. As shown in Figure 23, the maximum creep strain for calculation condition 1 is larger than it is for calculation conditions 2 and 3.

Crack Propagation

Crack propagation is illustrated in Figure 24. In calculation condition 1 (power cycle), an ellipsoidal crack first begins in the solder below the aluminum wire bonding. No cracks begin in the corners. The size of the crack increases according to the time in the power cycle. In calculation condition 2 (thermal cycle), the crack begins from both the corner and the center of the solder. In calculation condition 3 (thermal shock cycle), the crack begins only from the corner of the solder and then progresses along the edge of the solder.

Fatigue Life Cycle

The relationship between fatigue life cycle and crack length is shown graphically in Figure 25. Crack length is defined as shown in Figure 24. The fatigue life cycles determined by the thermal cycle test and by the power cycle test are 5747 and 4537 for the initial crack length (50 μm), respectively. Therefore, the power device may

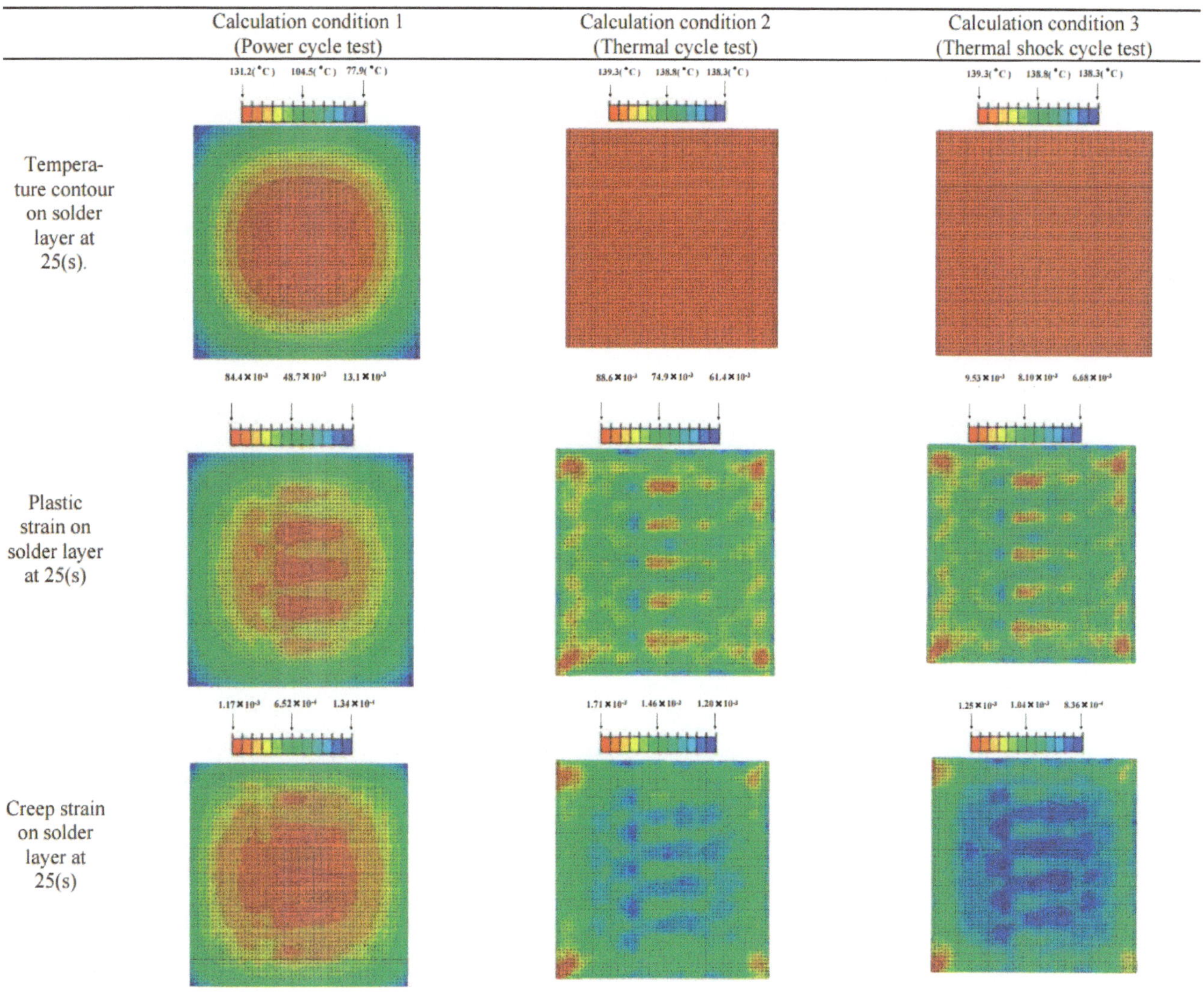

Figure 21: Solder between copper plate and silicon plate.

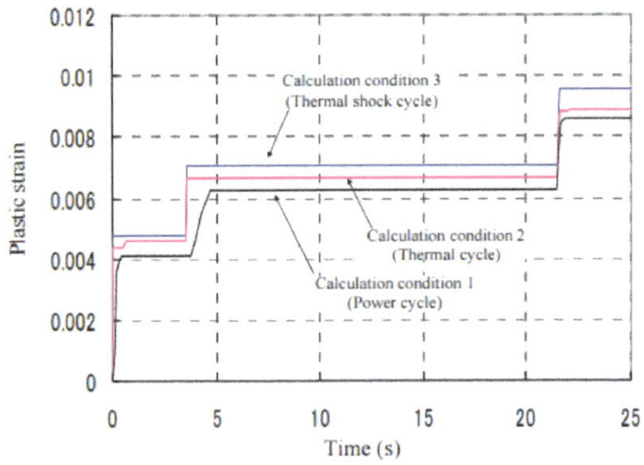

Figure 22: Time history of plastic strain with respect to power cycle and thermal cycle at maximum point on solder.

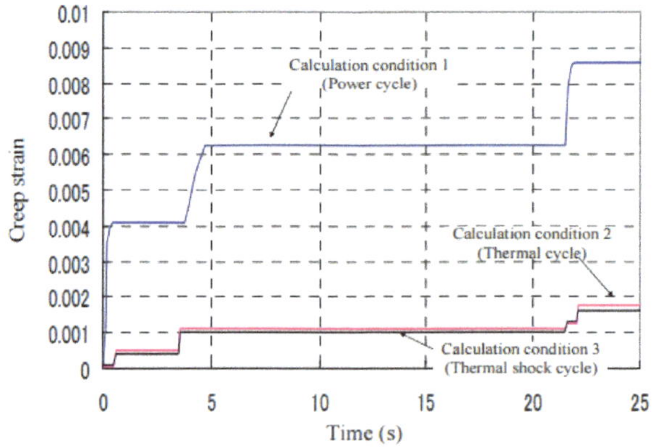

Figure 23: Time history of creep strain with respect to power cycle and thermal cycle at maximum point on solder.

cause failures earlier than was predicted by the simulation results, if the reliability of the power device is evaluated by the thermal cycle test.

CONCLUSIONS

To accurately evaluate the fatigue life of power devices using a power cycle test, a combination of electricity and heat-structure analysis is presented.

By comparing the evaluation of fatigue life obtained by the power cycle test in this study with that obtained by the thermal cycle test in the literature, we arrived at the following conclusions:

In the temperature distribution on the solder junction layer, the heat diffuses radially from the junction of the wire bonding. Therefore, the temperature distribution of the solder junction tends to become non uniform.

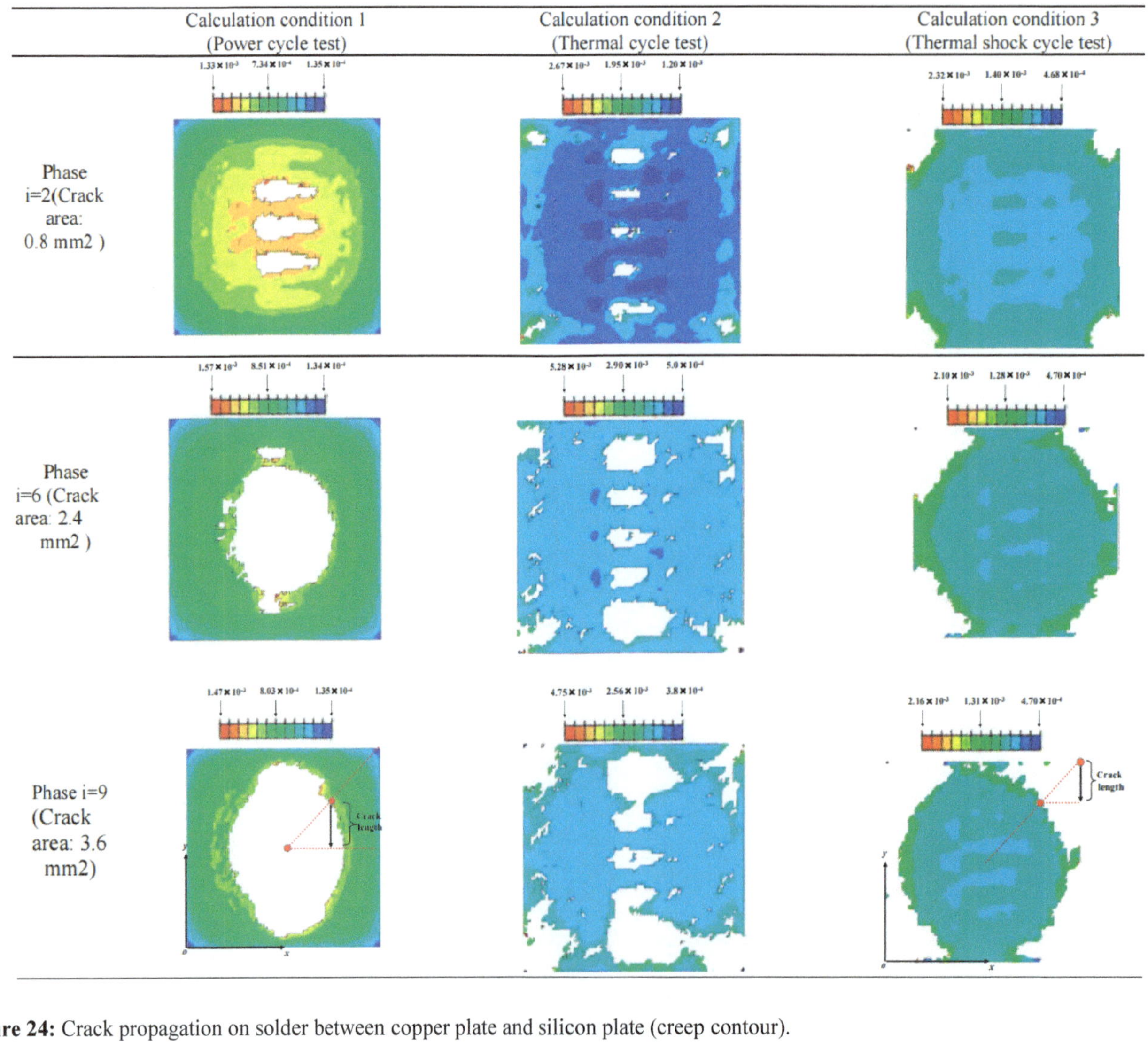

Figure 24: Crack propagation on solder between copper plate and silicon plate (creep contour).

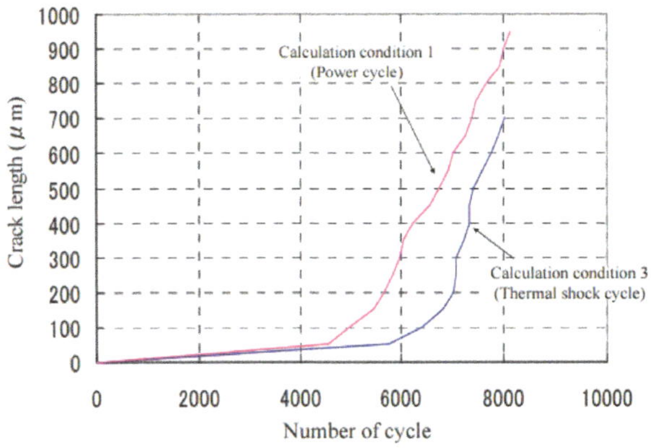

Figure 25: Fatigue life cycle with respect to crack length.

In the power cycle test, an ellipsoidal crack began at the center of the solder below the aluminum wire bonding. No cracks began in the corners. The size of the crack increased with the time of the power cycle. In the thermal shock cycle test, a crack began only from the corner of the solder and then progressed along the edge of the solder.

In the initial crack length (50 μm), the fatigue life cycles obtained by the power cycle test and the thermal cycle test were 5747 and 4537, respectively. The fatigue life cycle determined by the power cycle test was shorter than that determined by the thermal shock cycle test. The actual power device operated by a power cycle may produce failures earlier than is predicted based on simulation results if the reliability of the power device is evaluated by the thermal cycle test. Therefore, the evaluation of fatigue life using the power cycle test is clearly superior, as it produces results that are quite close to those of the actual operating conditions.

REFERENCES

1. B. J. Baliga, "Fundamentals of Power Semiconductor Devices," Springer-Verlag, Berlin, 2008.
2. H. Lu, W-S. Loh, T. Tilford, et al., "Reliability of Power Electronic Modules," InterPACK Conference American Society of Mechanical Engineers, New York, 8-11 July 2007, pp. 883-888.
3. S. Ramminger, N. Seliger and G. Wachutka, "Reliability Model for Al Wire Bonds Subjected to Heel Crack Failures," Microelectronics Reliability, Vol. 40, No. 8-10, 2000, pp. 1521-1525. [Paper reference:1]
4. O. Usui, H. Muto and T. Kikunaga, "Evaluation of Temperature Distribution of a Power Semiconductor Chip Using Electrothermal Simulation," IEEJ Transactions on Industry Applications, Vol. 124, No. 1, 2003, pp. 108- 115.
5. M. Ishiko, M. Usui, T. Ohuchi and M. Shirai, "Design Concept for Wire-bonding Reliability Improvement by Optimizing Position in Power Devices," Microelectronics Reliability, Vol. 37, No. 3, 2006, pp. 262-268.
6. M. O›Keefe and A. Vlahinos, "Impacts of Cooling Technology on Solder Fatigue for Power Modules in Electric traction Drive Vehicles," 2009 IEEE Vehicle Power and Propulsion Conference, Michigan, 7-10 September 2009, pp. 1182-1188.
7. X. Xie, X. Bi and G. Li, "Thermal-mechanical Fatigue Reliability of PbSnAg Solder Layer of Die Attachment for Power Electronic Devices," International Conference on Electronic Packaging Technology & High Density Packaging, Beijing, 10-13 August 2009, pp.1181-1185.
8. L. J. Ladani, "Reliability Estimation for Large-Area Solder Joints Using Explicit Modeling of Damage," Device and Materials Reliability, IEEE Transactions, Vol. 8, No. 2, 2008, pp. 375-386.
9. W. D. Zhuang, P. C. Chang, F. Y. Chou and R. K. Shiue, "Effect of Solder Creep on the Reliability of Large Area Die Attachment," Microelectronics Reliability, Vol. 41, No. 12, 2001, pp. 2011-2021.
10. H. Ye, M. H. Lin and C. Basaran, "Failure Modes and FEM Analysis of Power Electronic Packaging," Finite Elements in Analysis and Design, Vol. 38, No. 7, 2002, pp. 601-612.
11. Y. Nagatomo, T. Nagase and S. Shimamura, "FEM Analysis of Thermal Cycle Properties of the Substrates for Power Modules," Journal of Japan Institute of Electronics Packaging, Vol. 3, No. 2, 2000, pp. 330-334.
12. T. Matsunaga and S. Sudo, "Evaluation of Fatigue Life Reliability and New Lead Bonding Technology for Power Modules," Mitsubishi Electr. Adv., Vol. 113, No. 5, 2006, pp. 13-16. [Paper reference:1]
13. K. Shinohara and Q. Yu, "Fatigue Evaluation of Power Devices,", International Conference on Electronic Packaging Technology & High Density Packaging, 10-13 August 2009, pp. 1277-1283. [Paper reference:1]
14. O. C. Zienkiewicz and R. L. Taylor, "The Finite Element Method," Butterworth-Heinemann, Oxford, 2005. [Paper reference:1]
15. A. Tanaka, Q. Yu, T. Shibutani, Y. Kobayashi and M. Shiratori, "Low Cycle Fatigue Reliability Evaluation of Lead-Free Solders," Nippon Kikai Gakkai Zairyo Rikigaku Bumon Koenkai Koen Ronbunshu, 2005, pp. 475- 476. [Paper reference:2]
16. The Society of Materials Science, "Factual Database on Creep and Creep-Fatigue Properties of Sn-37Pb and Sn-3.5Ag Solders," The Society of Materials Science, Japan, 2004. [Paper reference:2]
17. H. Iwasaki, K. Hisano and T.Takamatsu, "Thermal Prolem of Next Generation Semiconductor Power Devices," Nippon Kikai Gakkai Nenji Taikai Koen Ronbunshu, Vol. 6, 2003, pp. 251-252.

CHAPTER 17

FUTURE SEMICONDUCTOR DEVICES FOR MULTI-VALUED LOGIC CIRCUIT DESIGN

Supriya Karmakar, Faquir C. Jain

Keywords: MVL, Quantum Dots, FET, QDCFET, QDGFET, SWSFET

INTRODUCTION

The basic building block to implement binary logic is the metal-oxide-semiconductor field-effect transistor (MOSFET). The trend predicted by Moore's Law has called for these MOS devices to shrink in half every 18 - 24 months [1,2]. As feature sizes have started to approach sub-22 nm regime, several issues have begun to make further miniaturization difficult [3-5]. As transistors have decreased in size, the thickness of the gate dielectric needs to be decreased to increase the gate capacitance, and thereby the drive current and device performance [6-9]. As the gate dielectric thickness is scaled down below 2 nm, leakage currents due to the tunneling of charge carriers increase drastically, which increases the unwieldy power consumption of the device and reduces device reliability [10-12]. Another approach to increase integration is to increase the bit handling capacity of a fabricated device. Traditional MOSFETs can only process one bit at a time.

Higher level of bit density can be achieved using multi-valued logic (MVL) [13]. Multi-value logic is defined as a non-binary logic and involves the switching between more than 2 states.

Research is ongoing towards the development of devices that are suitable for multi-valued operation. Resonant tunneling transistors (RTTs) [14-16] and resonant tunneling diodes (RTDs) [17,18] are the major promising semiconductor devices for MVL applications. All these devices produce negative differential resistance (NDR) in the specific voltage range of operations.

All the devices that have been proposed so far however, until now are not matured yet for MVL circuit implementation. These devices have more serious problems than complementary metal-oxide-semiconductor (CMOS) devices, such as excess leakage current, being unable to operate at room temperature, or requiring complex fabrication processes. Research is ongoing to find a better combination of stack materials for resonant tunneling devices. A common problem in the NDR devices is to reduce the valley current, which is due to the unintentional excess leakage current. In this paper we propose some new devices which can be promising circuit element in future to implement multi-valued logic with better performance. Before discussing new devices, we will highlight other two existing devices like resonant tunneling diode (RTD) and resonant tunneling transistor (RTT) in Sections 2 and 3. Three new devices, quantum dot gate field effect transistor (QDGFET), quantum dot channel field effect transistor (QDCFET) and spatial wave-function switched field effect transistor (SWSFET) are discussed in Section 4 which is followed by conclusions in Section 5.

RESONANT TUNNELING DIODE (RTD)

A resonant tunneling device is similar in construction to a p-n junction diode with two terminals [19]. The device has a quantum well layer separated from two conducting regions by quantum barriers (on the order of 50 Å to 100 Å thick). The barriers can be higher bandgap AlGaAs, for example, while the well and conduction regions can be lower bandgap GaAs.

Electrons tunnel through two barriers separated by a well when flowing from source to drain in a resonant tunneling diode, which is also known as quantum mechanical tunneling [20]. The flow of electrons is controlled by diode bias. This

Citation: S. Karmakar and F. Jain, "Future Semiconductor Devices for Multi-Valued Logic Circuit Design," Materials Sciences and Applications, Vol. 3 No. 11, 2012, pp. 807-814. doi: 10.4236/msa.2012.311117.

matches the energy levels of the electrons in the source to the quantized level in the well so that electrons can tunnel through the barriers. The energy level in the well is quantized because the well is small. When the energy levels are equal, a resonance occurs, which allows electron flow through the barriers as shown in Figure 1(b). No bias or too much bias, in Figures 1(a) and (c) respectively, yields an energy mismatch between the source and the well, and electrons cannot conduct.

When the bias is increased from zero across the RTD, the current increases, then decreases, correspond to off, on and off states. Figure 2 shows the I-V characteristic of a resonant tunneling diode. This makes simplification of conventional transistor circuits possible by substituteing a pair of RTD for two transistors. The potential application of RTD is to reduce the component count, area, and power dissipation of conventional transistor circuits by replacing some, though not all, transistors.

RTDs have allowed us to realize certain applications that will be beyond the capability of CMOS technology [21,22]. These low-power, high speed, and small devices are especially important as we continue to scale down to the size of atoms, where heat and parasitic effects are a major problem. However, in order for RTD to reach its full potential, more mature fabrication techniques are necessary. Precise barrier thickness control is required to ensure uniformity across the whole wafer. Also, the output power of RTDs is limited. More research is required to realize RTD circuits without an amplifier or other drivers. This will minimize the power and area of the integrated circuit (IC).

RESONANT TUNNELING TRANSISTOR (RTT)

A Resonant Tunneling Transistor [23] is similar to the resonant tunneling device which has source, drain and a gate electrode attached to its well region [24].

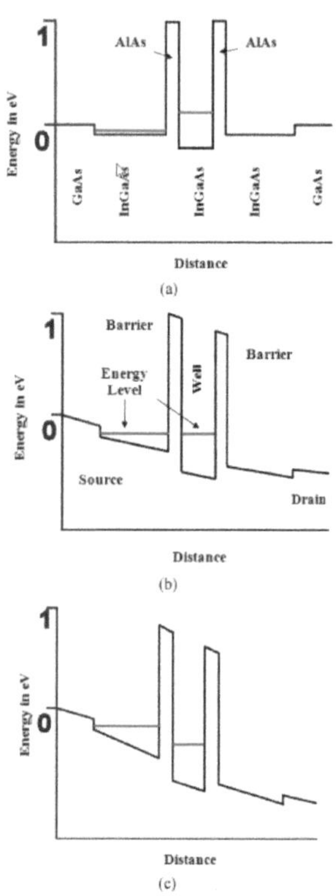

Figure 1: Resonant tunneling diode (RTD): (a) No bias, source and well energy levels not matched, no conduction. (b) Small bias causes matched energy levels (resonance); conduction results. (c) Further bias mismatches energy levels, decreasing conduction.

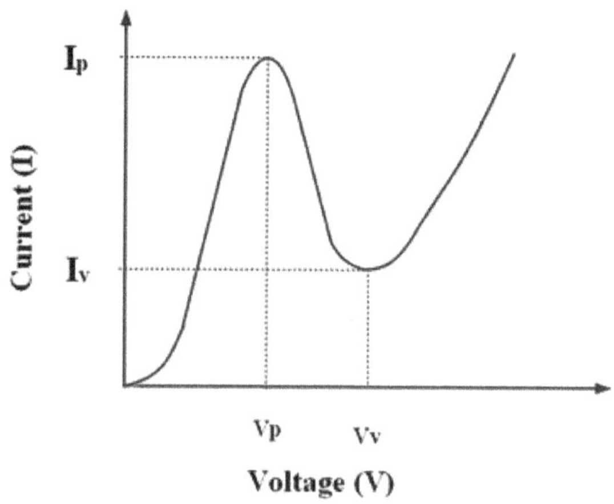

Figure 2: RTD I-V characteristic.

Electrons are injected into the device through the source electrode. An applied gate voltage adjusts the energy of the well levels with respect to the energy range of the occupied conduction states in the source region. Carriers tunnel through the barrier into the well from the source region when the energy of a well state falls within the energy range of the occupied conduction states in the source [25, 26]. With a voltage applied from the source region to drain region, the charge then flows into the drain region.

NEW PROPOSED DEVICES

Quantum Dot Gate Field Effect Transistor (QDGFET)

The device structure of a quantum dot gate field effect transistor (QDGFET) is shown in Figure 3(b). Unlike conventional FET (**Figure 3**(a)), a quantum dot layer is deposited on top of the gate insulator in QDGFET.

FEI Strata 400S Dual-Beam FIB instrument was used to prepare the sample for cross-sectional TEM imaging. A 1 μm thick membrane having dimension of 10 μm across and 5 μm in height was cut through the sample using Ga+ ion beam. Platinum layer was deposited onto the surface of the FET to minimize the ion beam damage to the device during FIB machining.

A JEOL 2010 FasTEM was used to examine the TEM specimens. This instrument is equipped with a high-resolution objective lens pole-piece and operate at 200 kV and produce a point-to-point resolution of <0.19 nm in phase contrast images. Chemical microanalysis was also performed in situ using an EDAX Phoenix atmospheric thin window EDXS. TEM image (Figure 4) obtained from the experiment shows the deposition of two layers of SiO_x cladded-Si quantum dots in the gate region of the QDGFET.

QDGFETs exhibit an intermediate state between "ON" and "OFF" states in their transfer characteristic because of stored charge in the quantum dots in the gate region.

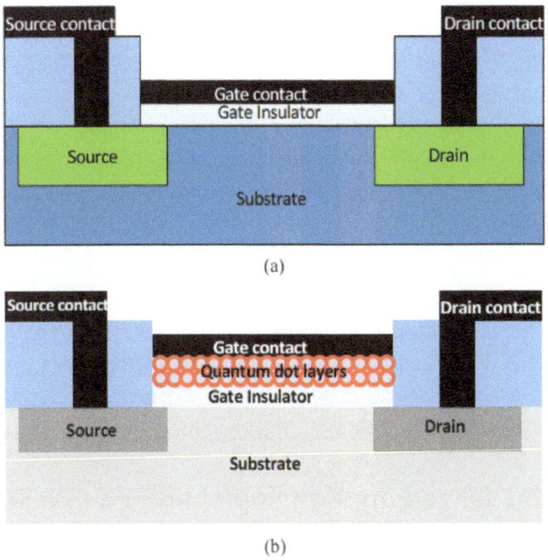

Figure 3: Cross-sectional schematic of (a) conventional FET, (b) quantum dot gate FET (QDGFET).

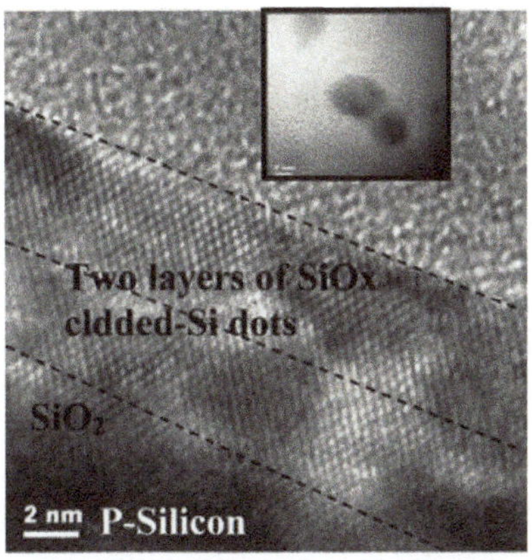

Figure 4: HRTEM picture of fabricated QDGFET, (inset) TEM image of fabricated SiO_x-Si quantum dots.

In the intermediate state the drain current (I_D) is independent over a range of gate voltages, which is determined by the thicknesses of tunnel gate insulator, quantum dot cladding as well as their core diameter [27-29].

This novel three-state characteristic of QDGFET provides new versatility in designing MVL implementation with significantly reduced FET counts. In previous work, we have shown the usage of these novel transistors in the construction of three state complement functions and memory cells [30]. Different ternary logics can be implemented using this three state property of the QDGEFT.

Theory of Operation

Figures 5(a)-(d) show the energy band diagrams and the carrier concentration plots in the quantum well channel and two quantum dot gate layers at various gate voltages. Figure 5(a) shows the state of the device at a gate voltage before the dots are charged. In Figure 5(b), the first layer of dots closest to the channel is charged via tunneling of charge carriers from inversion channel. The buildup of the charge in the first quantum dot layer as the energy levels in the second quantum dot layer are not lined up is shown in Figure 5(c). In Figure 5(d), the resonant tunneling occurs to the second quantum dot layer due to alignment of the bound states in the first quantum dot layer and the second quantum dot layer. Table 1 shows different simulation parameters for device simulation.

Quantum Dot Channel Field Effect Transistor (QDCFET)

Quantum dot channel FET (QDCFET) is another promising circuit element to implement MVL. Jain et al. [31, 32] already discussed the fabrication and theory of operation of QDCFET. The structural difference between a QDCFET and a conventional FET is the presence of quantum dot layers between the source and drain region in the QDCFET.

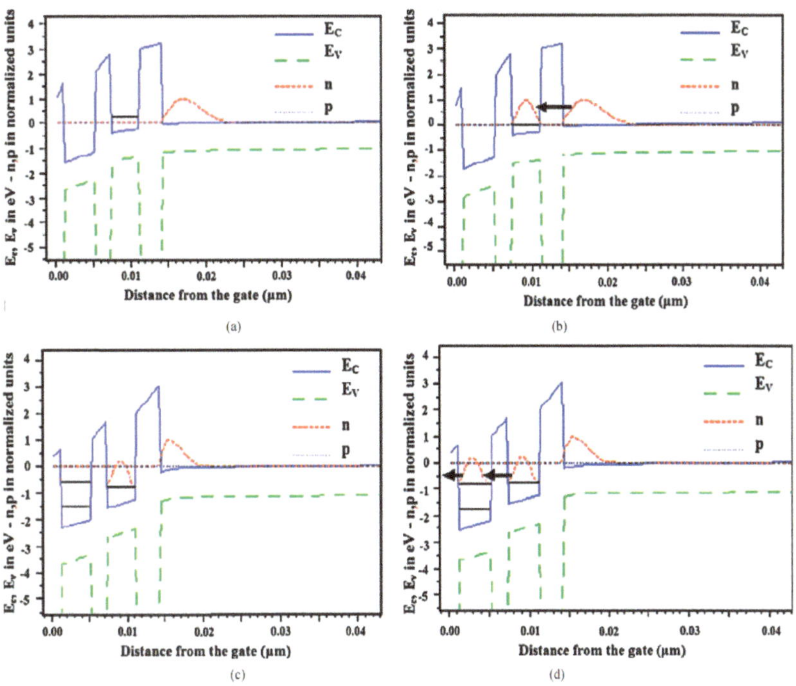

Figure 5: Device simulation shows the tunneling of electron wave function from inversion channel to different quantum dot layers when gate voltage increases from (a) to (d) gradually [30].

Table 1: QDGFET parameters for charge control simulation.

Parameters	Value
Channel Width (W)	15 μm
Channel Length (L)	5 μm
QD gate capacitance (C_{ox})	4.4 × 10^{-7} F/cm^2
Drain-source voltage (V_{DS})	0.5 V
Mobility (μ_n)	650 cm^2/V-s
Si dot core diameter	12 nm
Si-SiO$_x$ Barrier height	3.27 eV
Threshold voltage without QD gate charge (V_{th0})	−2 V
Total gate layer	26 nm
Effective dielectric constant ε_r	8
Center of QD1 from gate	~18 nm
Center of QD2 from gate	~6 nm

The presence of quantum dots in the channel region produces four states in the transfer characteristic of the QDCFET which can be used to implement four state inverter and MVL.

The device structure of a QDCFET is shown in Figure 6(a). The presence of quantum dots in the channel region is shown in Figure 6(b).

The current transport in a QDCFET depends on the carrier concentration in the inversion layer and their transport in the quantum dot channel as a function of the gate voltage [32]. Quantum dots inside the channel formed like a super lattice. The transfer of the charge carriers from the source to the drain region of QDCFET depends on the mini band structure formed by overlapping of energy bands of different quantum dots in the channel region. These mini-bands are occupied by charge carriers based on the drain-to-source voltage (V_{DS}). As V_{DS} is increased, the next level of mini-band is filled and charge flows through both mini-bands.

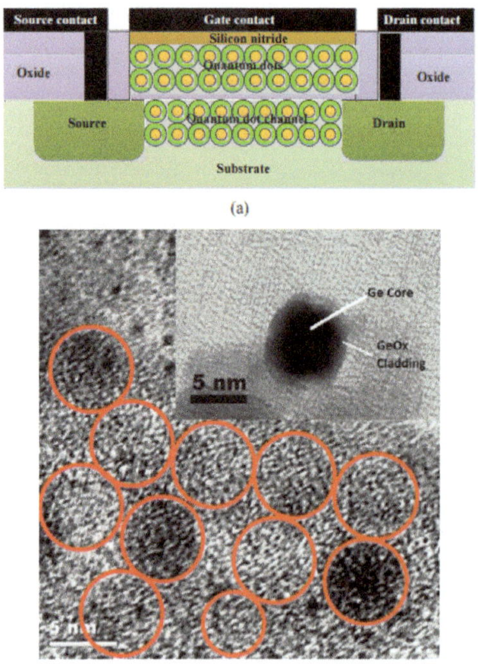

(a)

(b)

Figure 6: (a) The device structure of a QDCFFET, (b) HRTEM image of quantum dots in the channel region of QDCFET, (Inset) TEM image of GeOx cladded-Ge quantum dots.

The quantum dots in the gate region change the threshold voltage of the device because of tunneling of the charge carriers from the quantum dot channel to the quantum dot layer in the gate region. The generation of the steps in the transfer characteristic of QDCFET is the cumulative effects of the quantum dots in the channel region as well as in the gate region of the device. Figure 7 shows the transfer characteristic of a fabricated QDCFET having four states in its transfer characteristic. The two intermediates states are at around 1.6 V and 2.15 V of gate voltage which is shown in Figure 7.

The generation of four states in the transfer characteristic of a QDCFET can be used to implement quaternary logic in future. More research is necessary to explore this device as a circuit element to implement multi-valued logic in future.

Spatial Wave-Function Switched FET (SWSFET)

Spatial wave-function switched (SWS) FET is a promising circuit element for digital integrated circuits.

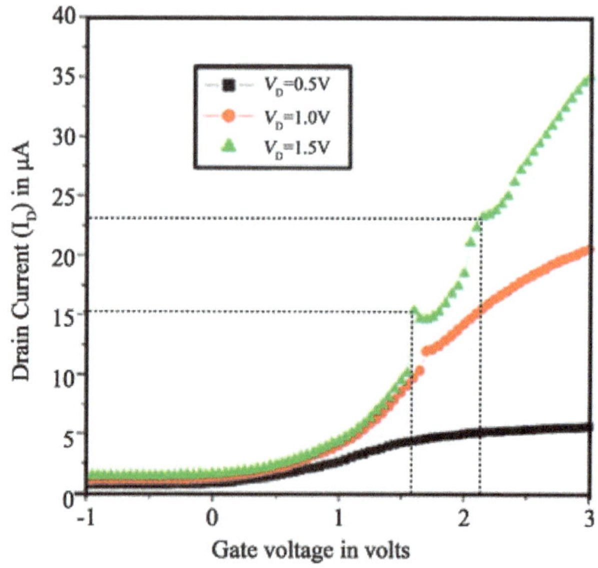

Figure 7: I_D vs. V_{GS} characteristic of a fabricated QDCFET.

Based on the applied gate voltage, the charge flows through different channels in a SWSFET. When the gate voltage is low but above the threshold voltage (V_{th1}) of the device, charge carriers are confined in the lower quantum well channel and flows from source to drain region. As the gate voltage is increased (V_{th2}), charge carriers transfer from the lower channel to the upper quantum well channel and current flows through the upper channel of the device. The device structure of a SWSFET is shown in Figure 8. Based on the applied gate voltage, the electron concentration as well as the electron wave function switches from one channel to the other channel and current flows through different drain terminals.

The energy band-diagram of the SWSFET is schematically shown in Figure 9 [33]. Figure 10 shows the capacitance-voltage (C-V) characteristics of a fabricated two quantum well channel InGaAs-AlInAs SWSFET [33]. The C-V curves show distinct peak before the accumulation regime (gate voltage less than −2V) where the capacitance becomes constant. The capacitance of the SWS device reaches to first maximum value at a gate voltage ~−0.1 V. This occurs when the holes are in the lower quantum well channel #1. When the gate voltage is reduced further, holes are transferred from the first quantum well layer (channel #1) to the second quantum well layer (channel #2). The capacitance increases as the carriers are in the proximity of gate until total accumulation is reached. The detailed device operation is already published elsewhere [33].

The circuit model of a SWSFET is developed in Berkeley short channel IGFET model (BSIM 3). In the circuit model, SWSFET effect is introduced by modifying the threshold voltage of conventional FET model in BSIM. The simulated transient characteristic of the SWSFET model is shown in Figure 11 [34,35]. Because of two drain terminals, a two-channel SWSFET is also known as a twin-drain SWSFET.

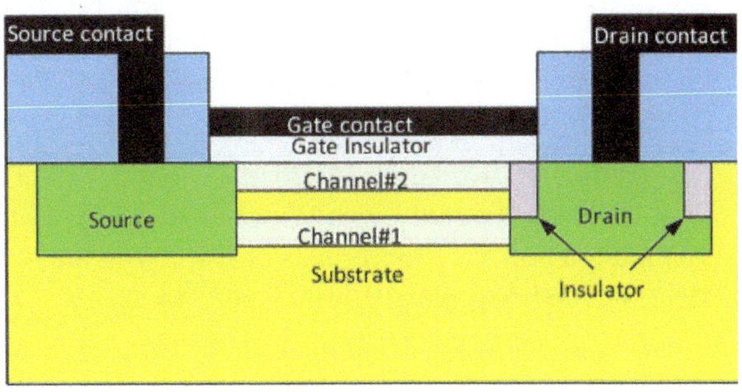

Figure 8: Device structure of SWSFET.

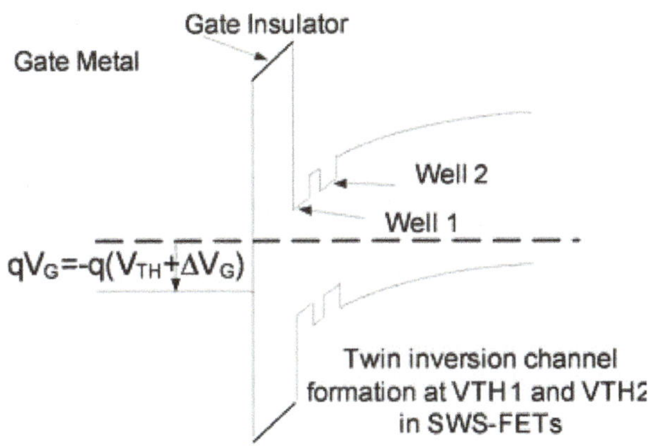

Figure 9: Energy band diagram of a SWSFET [33].

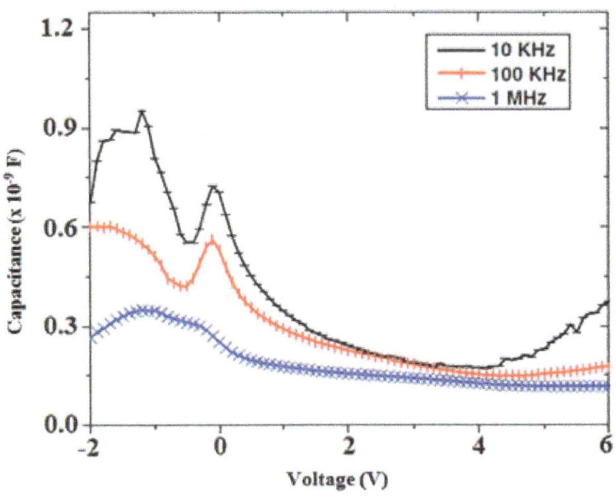

Figure 10: C-V characteristics of SWSFET [35].

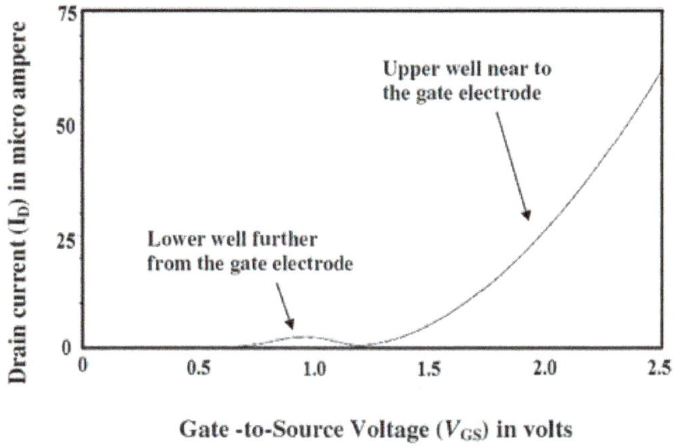

Figure 11: Transfer characteristic of the SWSFET model [34,35].

In this simulation we connect two drains of the twin SWSFET and simulate the transfer characteristic (I_D vs. V_{GS}) of the device. The simulation result shows that the drain current initially flows through the lower channel which is farther from the gate terminal. When the gate voltage increases further, charge carriers transfer from the lower channel to the upper channel (closer to the gate terminal) and current flows decreases in the lower channel and increases in the upper channel and ultimately all current flows through the upper channel.

CONCLUSION

In this paper, we introduce three new devices to implement multi-valued logic in future. These devices produce more stable output because of the cladded quantum dots in their structure. All these devices work based on different mechanism of charge storage in their device structure. In QDGFET, three states generate because of stored charge in the quantum dots in the gate region. Whereas in QDCFET, the multi-states generate because of formation of mini band structure in the channel region and the stored charge in the gate region. In the SWSFET, the multi valued logic can be implemented base on the charge in different quantum well channel in their device structure. The improved noise margin and the existing CMOS fabrication mechanism can make these quantum dot based devices as promising circuit element for multi valued logic implementation in future.

REFERENCES

1. G. E. Moore, "Cramming More Components onto Integrated Circuits," Electronics, Vol. 38, No. 8, 1965, pp. 114-117.
2. G. E. Moore, "No Exponential Is Forever: But 'Forever' Can Be Delayed!" Solid-State Circuits Conference, IEEE International Digest of Technical Papers, Vol. 1, 2003, pp. 20-23. doi:10.1109/ISSCC.2003.1234194
3. E. J. Nowak, "Maintaining the Benefits of CMOS Scaling When Scaling Bogs Down," Journal of Research and Development, Vol. 46, No. 2-3, 2002, pp. 169-180.
4. T. N. Theis, "Beyond the Silicon Transistor: Personal Observations," Computing in Science & Engineering, Vol. 5, No. 1, 2003, pp. 25-29.
5. S. Borkar, "Design Perspectives on 22 nm CMOS and beyond," Design Automation Conference, San Francisco,

26-31 July 2009, pp. 93-94.

6. M. Goto, S. Kawanaka, S. Inumiya, N. Kusunoki, M. Saitoh, K. Tatsumura, A. Kinoshita, S. Inaba and Y. Toyoshima, "The Study of Mobility-Tin, Trade-Off in Deeply Scaled High-k/Metal Gate Devices and Scaling Design Guideline for 22 nm-Node Generation," 2009 Symposium on VLSI Technology, Kyoto, 16-18 June 2009, pp. 214-215.

7. R. Huang, H. M. Wu, J. F. Kang, D. Y. Xiao, X. L. Shi, X. An, Y. Tian, R. S. Wang, L. L. Zhang, X. Zhang, et al., "Challenges of 22 nm and beyond CMOS Technology," Science in China Series F: Information Sciences, Vol. 52, No. 9, 2009, pp. 1491-1533.

8. S. Thompson, et al., "A 90 nm Logic Technology Featuring 50 nm Strained Silicon Channel Transistors, 7 Layers of Cu Interconnects, Low k ILD, and 1 μm^2 SRAM Cell," International Electron Devices Meeting Technical Digest, Vol. 6, 2002, pp. 61-64.

9. K. Mistry, et al., "A 45 nm Logic Technology with High-k + Metal Gate Transistors, Strained Silicon, 9 Cu Interconnect Layers, 193 nm Dry Patterning, and 100% Pb-Free Packaging," International Electron Devices Meeting Technical Digest, Washington DC, 10-12 December 2007, pp. 247-250.

10. P. Bai, et al., "A 65 nm Logic Technology Featuring 35 nm Gate Lengths, Enhanced Channel Strain, 8 Cu Interconnect Layers, Low-k ILD and 0.57 μm^2 SRAM Cell," International Electron Devices Meeting Technical Digest, San Francisco, 13-15 December 2004, pp. 657-660.

11. H. K. Lim and J. G. Fossum, "Threshold Voltage of ThinFilm Silicon-on-Insulator (SOI) Mosfet's," IEEE Transactions on Electron Devices, Vol. 30, No. 10, 1983, pp. 1244-1251.doi:10.1109/T-ED.1983.21282

12. J. P. Colinge, "Transconductance of Silicon-on-Insulator Mosfets," IEEE Electron Device Letters, Vol. EDL-6, 1985, pp. 573-574. doi:10.1109/EDL.1985.26234

13. L. J. Micheel, A. H. Taddiken and A. C. Seabaugh, "Multiple-Valued Logic Computation Circuits Using Microand Nanoelectronics Devices," Proceedings of 23rd IEEE International Symposium on Multiple-Valued Logic, Sacramento, 24-27 May 1993, pp. 164-169. .

14. A. C. Seabaugh, W. R. Frensley, J. N. Randall, M. A. Reed, D. L. Farrington and R. J. Matyi, "Pseudomorphic Bipolar Quantum Resonant-Tunneling Transistor," IEEE Transaction on Electron Devices, Vol. 36, No. 10, 1989, pp. 2228-2234.

15. J. Stock, J. Malindretos, K. M. Indlekofer, M. Pottgens, A. Forster and H. Luth, "A Vertical Resonant Tunneling Transistor for Application in Digital Logic Circuits," IEEE Transaction on Electron Devices, Vol. 48, No. 6, 2001, pp. 1028-1032.

16. F. Capasso and R. A. Kiehl, "Resonant Tunneling Transistor with Quantum Well Base and High-Energy Injection: A New Negative Differential Resistance Device", Journal of Applied Physics, Vol. 58, No. 3, 1985, pp. 1366-1368.

17. H. C. Lin, "Resonant Tunneling Diodes for Multi-Valued Digital Applications," Proceedings of 24th IEEE International Symposium on Multiple-Valued Logic, Boston, 25- 27 May 1994, pp. 188-195.

18. A. Forster, "Resonant Tunneling Diodes: The Effect of Structural Properties on Their Performance," Advances in Solid State Physics, Vol. 33, 1993, pp. 37-62.

19. T. Waho, K. J. Chen and M. Yamamoto, "ResonantTunneling Diode and HEMT Logic Circuits with Multiple Thresholds and Multi-Level Output," IEEE Journal of Solid-State Circuits, Vol. 33, No. 2, 1998, pp. 268-274. .

20. H. C. Lin, "Resonant Tunneling Diodes for Multi-Valued Digital Applications," Proceedings of 24th IEEE International Symposium on Multiple-Valued Logic, Boston, 25- 27 May 1994, pp. 188-195. .

21. P. Mazumder, S. Kulkarni, M. Bhattacharya, J. P. Sun and G. I. Haddad, "Digital Circuit Applications of Resonant Tunneling Diodes," Proceedings of IEEE, Vol. 86, No. 4, 1998, pp. 664-686. doi:10.1109/5.663544

22. J. P. A. van der Wagt, H. Tang, T. P. E. Broekaert, A. C. Seabaugh and Y.-C. Kao, "Multibit Resonant Tunneling Diode SRAM Cell Based on Slew-Rate Addressing," IEEE Transactions on Electron Devices, Vol. 46, No. 1, 1999, p. 5562. doi:10.1109/16.737441

23. T. Waho, "Resonant Tunneling Transistor and Its Application to Multiple-Valued Logic Circuits," Proceedings of 25th IEEE International Symposium on Multiple-Valued Logic, 1995, pp. 130-138. .

24. W. L. Chen, G. O. Mums, L. Davis, P. K. Bhattacharya and G. I. Haddad, "The Growth of Resonant Tunneling Hot Electron Transistors Using Chemical Beam Epitaxy," The 4th International Conference in Chemical Beam Epitaxy, Section S-7, Nara, 21-23 July 1993. .

25. T. Futatsugi, Y. Yamaguchi, K. Ishii, K. Imamura, S. Muto, N. Yokoyama and A. Shibatomi, "A Resonant Tunneling Bipolar Transistor (RBT): A Proposal and Demonstration for New Functional Device with High Current Gains," International Electron Devices Meeting Technical Digest, Washington DC, 7-10 December 1986, p. 286.

26. A. C. Seabaugh, W. R. Frensley, Y. C. Kao, J. N. Randall and M. A. Reed, "Quantum-Well Resonant Tunneling Transistors," The Proceedings of the 1989 IEEE Conference, Ithaca, 1989, p. 255.

27. S. Karmakar, M. Gogna, E. Suarez, F. Alamoody, E. Heller, J. chandy and F. Jain, "3-State Behavior of Quantum Dot Gate FETs with Lattice Matched Insulator," Proceedings of 2009 Nanoelectronic Devices for Defense and Security, Fort Lauderdale, 28 September-2 October 2009.

28. S. Karmakar, J. A. Chandy and F. C. Jain, "Application of 25 nm Quantum Dot Gate FETs to the Design of ADC and DAC Circuits," International Journal of High Speed Electronics and Systems, Vol. 20, No. 3, 2011, p. 653.

29. S. Karmakar, A. P. Suresh, J. A. Chandy and F. C. Jain, "Design of ADCs and DACs Using 3-State Quantum DOT Gate FETs," Proceedings of International Semiconductor Device Research Symposium, College Park, 9-11 December 2009.

30. S. Karmakar, E. Suarez and F. Jain, "Quantum Dot Gate Three State FETs Using ZnS-ZnMgS Lattice-Matched Gate Insulator on Silicon," Journal of Electronic Materials, Vol. 40, No. 8, 2011, pp. 1749-1756. [Paper reference:2]

31. F. Jain, S. Karmakar, R. A. Croce Jr., M. Gogna, E. Suarez, J. Chandy and E. Heller, "Multi-State Quantum Dot Channel (QDC) Field-Effect Transistors (FETs): A New Paradigm in Circuit Design," 2011 Nanoelectronic Devices for Security Conference, Brooklyn, 29 August-1 September 2011.

32. F. Jain, S. Karmakar, P.-Y. Chan, E. Suarez, M. Gogna, J. Chandy and E. Heller, "Quantum Dot Channel (QDC) Field-Effect Transistors (FETs) Using II-VI Barrier Layers," Journal of Electronic Materials, Vol. 41, No. 10, 2012, pp. 2775-2784. doi:10.1007/s11664-012-2161-z .

33. F. C. Jain, B. Miller, E. Suarez, P.-Y. Chan, S. Karmakar, F. Al-Amoody, J. Chandy and E. Heller, "Spatial Wavefunction Switched (SWS) InGaAs FETs with II-VI Gate Insulators," Journal of Electronic Materials, Vol. 40, No. 8, 2011, pp. 1717-1726. doi:10.1007/s11664-011-1667-0 .

34. S. Karmakar, F. C. Jain, J. A. Chandy and E. Heller, "Circuit Model of SWSFET to Implement Multi-Valued Logic," 21st CMOC Symposium, University of Connecticut, Storrs, 9 April 2012.

35. S. Karmakar, J. A. Chandy and F. C. Jain, "Implementation of Unipolar Inverter Based on Spatial Wave-Function Switched (SWS) FET," IEEE Lester Eastman Conference on High Performance Devices, Brown University, Providence, 7-9 August 2012.

CHAPTER 18

SEMICONDUCTOR OPTICAL AMPLIFIER (SOA)-FIBER RING LASER AND ITS APPLICATION TO STRESS SENSING

Yoshitaka Takahashi, Shinji Sekiya, Tatsuro Suemune

Keywords: Fiber Laser, Fiber Sensor, Ring Laser, Semiconductor Optical Amplifier, Optical Sensor

INTRODUCTION

Optical fiber sensors are widely used in various uses since it has many advantages. Most of them detect an optical intensity change as a sensing signal caused by change in polarization, phase, loss, and fluorescence. In such kinds of sensors, however, fluctuation of the source intensity and/or a propagation loss will cause a measurement error. Sensors which are not influenced by the fluctuation have been required and studied, e.g. optical heterodyne, a frequency-domain sensor, and so on. For the application to a frequency-domain sensor, the authors have studied a novel optical fiber ring laser [1-3]. In general cases birefringence applied to a ring laser is reciprocal effect and no phase difference generates between two counter-propagating lights. Fiber ring lasers for frequency-domain sensors were proposed, e.g., an optical fiber gyro [4,5] using Sagnac effect and an optical current sensor [6,7] using Faraday effect. These effects are non-reciprocal effect. But in the present study reciprocal effect, i.e., birefringence was used, and making good use of the gain anisotropy of SOA the authors developed the SOA-fiber ring laser in which the phase difference occurred by introducing a birefringent medium in the ring cavity between two counter-propagating oscillations. To investigate the performance of the present laser, we introduced a photoelastic material in the ring laser cavity and confirmed the frequency difference of the counter-propagating oscillations changed by applying stress to the material proportionally.

Citation: Y. Takahashi, S. Sekiya and T. Suemune, "Semiconductor Optical Amplifier (SOA)-Fiber Ring Laser and Its Application to Stress Sensing," Optics and Photonics Journal, Vol. 1 No. 4, 2011, pp. 167-171. doi: 10.4236/opj.2011.14027.

PRINCIPLE OF OPERATION

The operating principle of the SOA ring laser [1] is shown in Figure 1. SOA is the gain medium of the laser. It has polarization anisotropy of its gain and is regarded as an amplifier only for TE-polarization, which is assumed to be in accordance with the horizontal plane. The ring cavity contains Faraday rotators R_1 and R_2 which generate +45° and –45°-rotation of the state of polarization (SOP) of light propagating in the clockwise (cw) direction. The cavity also contains a medium S which has birefringence (retardation R) and works as a sensing element when the laser is applied to a fiber sensor. First, consider light which propagates in the cw direction in the ring cavity. Horizontally polarized light emitted from SOA is coupled into the cavity and passes through R_1, and, as a result, SOP of the cw light rotates 45° (an arrow in a solid circle shown in Figure 1). The dielectric axes (F-axis and S-axis) of the birefringent medium S are at an angle of 45° to the horizontal plane.

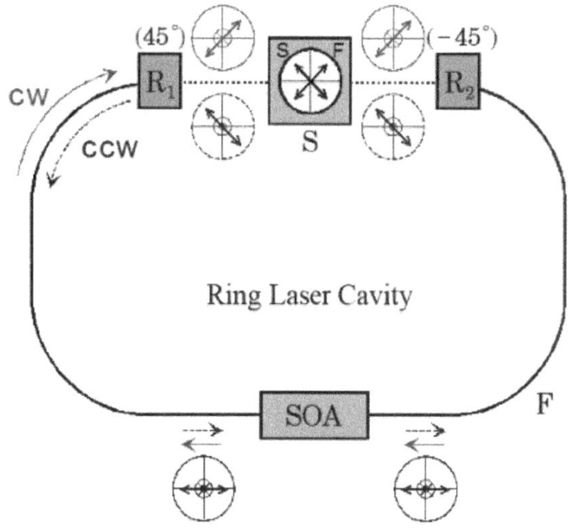

Figure 1: Schematic diagram of SOA-fiber ring laser.

The cw light, therefore, passes through S with the polarization plane in accordance with the F-axis. The cw light turns horizontally polarized after passing through another Faraday rotator R_2, and returns to SOA with the polarization corresponding to its TE-polarization. On the other hand, the counterclockwise (ccw) light, whose SOP's are denoted as arrows in a doted circle, is emitted from SOA and passes through S with the polarization plane in accordance with the S-axis because of the non-reciprocity of Faraday effect. It returns to SOA with the polarization corresponding to its TE-polarization in a similar way. That is, the cw and ccw lights have orthogonal polarization to each other in passing through S, while they have the same polarization in other regions and are both amplified by SOA. Thus phase difference is brought about between the cw and ccw lights by retardation in S, and causes them to oscillate in different frequencies. The frequency difference f_B is proportional to the retardation R (nm) and given by [1]

$$f_B = \frac{c}{\lambda L} R = \frac{R}{\lambda} f_{FSR} \qquad (1)$$

where l nm is the lasing wavelength, c is the velocity of light, L and f_{FSR} = c/L is the optical path length and the free spectral range of the ring cavity, respectively. If S is the medium which causes the retardation change by the physiccal phenomenon to be measured, the change can be detected by measuring f_B and the laser in this study can be applied to a sensor which detects the signal in a frequency domain.

CONFIGURATION

Figure 2 shows the configuration of the fiber ring laser. The emitted lights at 1.55 mm from both sides of SOA (Anritsu Corp.) were coupled into singlemode fibers (SMF) with singlemode ball-lens fibers (BLF), collimated with fiber-collimators (C_1, C_2) and then passed through a sensing element (S) in the space-propagating region. To reduce the lasing bandwidth 150-mm-thick glass plate E was also placed as etalon. At the ends of the two fiber-collimators Faraday rotators (R_1, R_2) 0.7 mm thick were attached. They were magnetized and generated 45°-rotation of the polarization angle of the 1.55 mm light without an external magnet. Both the cw and ccw lights of the ring laser were output via a 10 dB-singlemode fiber directional coupler (FC_1), and in order to examine the frequency difference they were combined with each other via a 3 dB-singlemode fiber directional coupler (FC_2) and detected by an avalanche photodiode (APD). Using two fiber polarization controllers (FPC) the SOP of both cw and ccw lights were adjusted in accordance with the TE-polarization of SOA and to be linearly polarized at the end of the fiber-collimators. Two Faraday rotators (R_1, R_2) were placed complementarily as described before and thus the cw and ccw lights propagated in the orthogonal polarization to each other in passing through S. The frequency difference f_B is measured by APD with a spectrum analyser as a beat signal.

Experiment and Results

The lasing characteristics of SOA-fiber ring laser was investigated without briefringent medium in the resonator first and the power spectrum is shown in Figure 3. The laser operated in multimode in spite of introducing an etalon, then in Figure 3 the spectrum corresponding to the signals of the free spectral range f_{FSR} and its harmonic are shown. The center of the lasing wavelength was 1.562 mm.

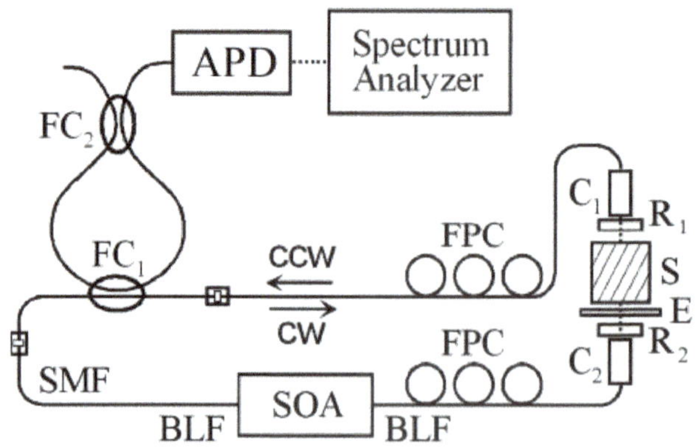

Figure 2: Configuration of SOA-fiber ring laser.

The optical path length of the ring cavity L was 16.1 m and the calculated value of f_{FSR} = c/L is 18.6 MHz, which corresponds to the one shown in Figure 3. Various wave plates were inserted in the laser cavity instead of a sensing element as known birefringent media to change the frequency difference f_B and the result is shown in Figure 4. The dashed line is calculated value with Equation (1) and the observed power spectrum at R of 387.5 nm is also shown in Figure 4 with a quarter-wave plate of 1550 nm. The result shows that retarda tion can be determined by measuring the shift of f_B and the present fiber ring laser would be a sensor of birefringence in a frequency domain.

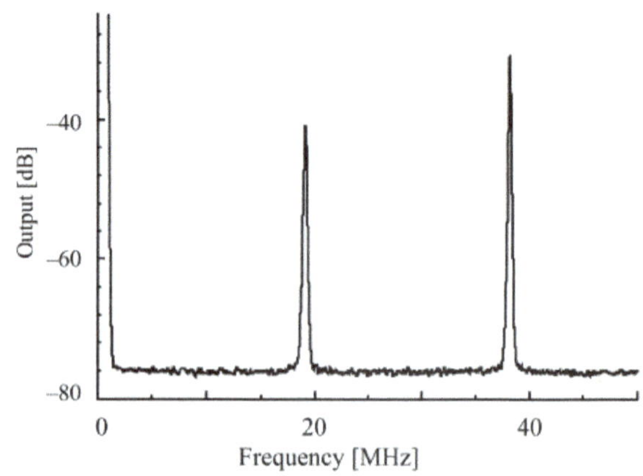

Figure 3: Power spectrum of SOA-fiber ring laser without birefringent media.

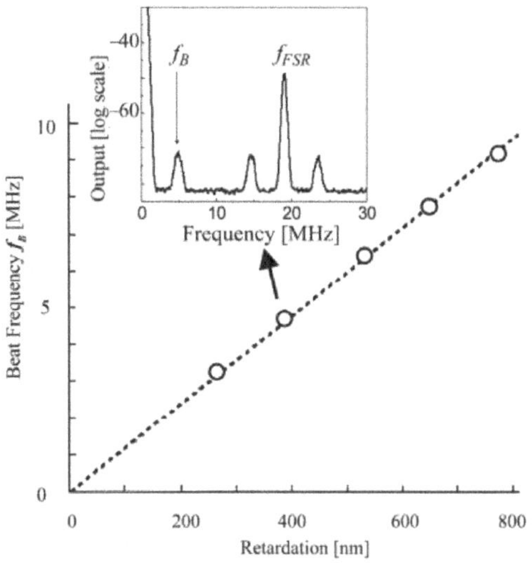

Figure 4: Change of frequency difference f_B with applied retardation. Inset power spectrum is the case of R = 387.5 nm.

In a ring laser when the phase difference between counter-propagating lights is small, oscillating frequencies of them do not differ because of lock-in phenomenon. Then if the inserted birefringence is small, f_B remains zero. Even if it is non-zero, a sign of the birefringence is uncertain. To avoid these problems a bias element is inserted in the space-propagating region in addition to a sensing element S. As the case of small birefringence, a sheet of PET film for viewgraph was inserted. Figure 5 shows the signal of $f_{FSR} + f_B$ with and without PET film. The traces are shifted up and down to make the difference clear. The retardation was calculated as 17.2 nm from the frequency shift 0.20 MHz, which means the laser can measure such a small birefringence with a bias element. As the experiment for the application to stress sensing a fused silica substrate was used as photoelastic crystal. It is 5 × 10 × 10 mm^3 and the applied stress was monitored an attached strain gauge. It was inserted in the space-propagating region as well as a quarter-wave plate. The collimated beam passes through the 10 × 10 mm^2 square plane and the stress was applied by a vise in the direction perpendicular to the beam. The dependence of the beat frequency shift on the pressure is shown in Figure 6. The dotted line denotes calculated value from Brewster coefficient of fused silica at 633 nm [8] as reference. The beat frequency shift changes with the pressure proportionally and closely to the dotted line. Since dispersion of Brewster coefficient supposes to be small the present ring laser can be applied to a fiber stress sensor in a frequency domain.

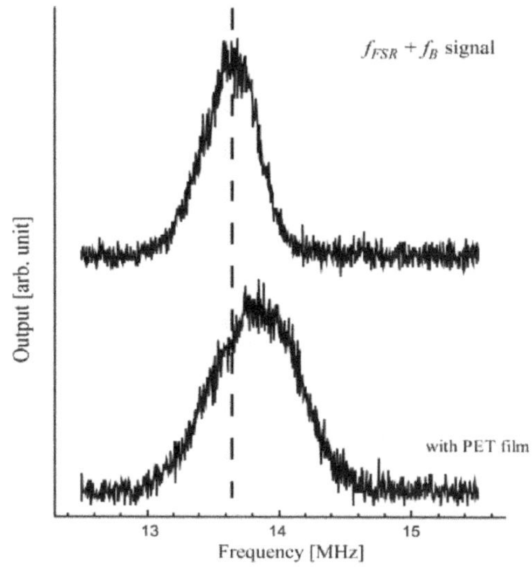

Figure 5: Power spectrum of $f_{FSR} + f_B$ signal. The upper and the lower traces show without and with PET film.

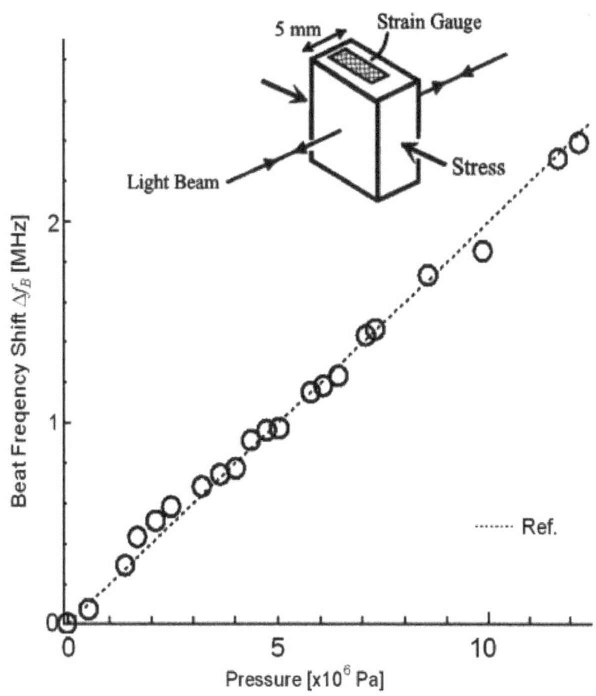

Figure 6: Beat frequency difference on applied pressure on fused silica.

DISCUSSION

In Figure 6 there are fluctuations and the accuracy is degraded, which must be the problem in developing a practical sensor. These might be caused by errors of a sensing element and instability of the laser oscillation.

a) Errors of a sensing element

The applied stress might cause nonuniform strain in the crystal. And it was difficult to attach a strain gauge on the fused silica and the adhesion might be imperfect, which causes errors of the measured pressure.

b) Instability of the laser oscillation

The polarization of the propagating light is not constant because singlemode fibers were incorporated into the ring laser cavity, and it is probable to cause the instability. In addition the difference between the designed wavelength (1.55 mm) and the lasing wavelength, which centered at 1.562 mm, will affect the oscillation. It causes errors of both the polarization and the phase difference. The former reduces the SOA gain and the latter induces errors of the retardation. Assuming circular birefringence of a Faraday rotator and linear birefringence of a sensing element has no dispersion, the rotation angle and the retardation changes in proportion to the ratio of the wavelengths. With this assumption the retardation error was calculated and the result is plotted in Figure 7. From Figure 7 the difference of the wavelength does not seem to affect the oscillation significantly

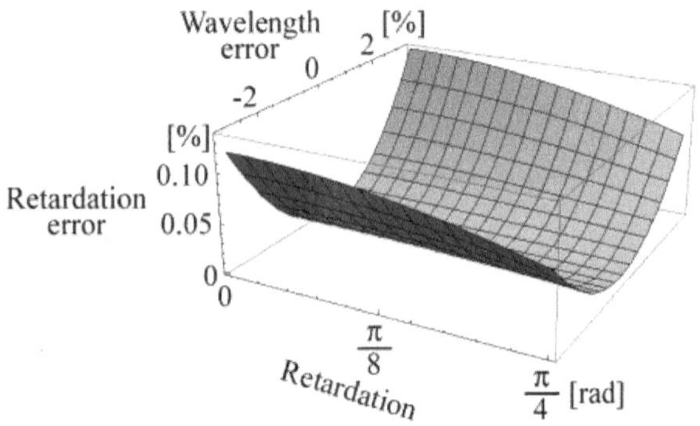

Figure 7: Retardation errors dependence on wavelength deviation.

CONCLUSIONS

The SOA-fiber ring laser has been developed and as the application to stress sensing by introducing a photoelastic material in the cavity the beat frequency changed as applying pressure proportionally. In the present laser phase difference is generated by birefringence which is reciprocal effect in the ring cavity. Since birefringence is induced by many phenomena such as electromagnetic field, pressure, and temperature, etc., sensors utilizing the present fiber ring lasers are expecting to detect such physical quantity in a frequency domain. As described in the previous section, because of using singlemode fiber the polarization of the propagating light might vary and should be stabilized. Using polarization-maintaining fiber instead of singlemode fiber the authors have studied the stabilization of the fiber ring laser. The improvement would make the practical fiber sensor.

REFERENCES

1. Y. Takahashi, S. Sekiya and N. Iwai, "Semiconductor Optical Amplifier(SOA)-Fiber Ring Laser for Sensor Application," Optical Review, Vol. 10, No. 4, 2003, pp. 315-317. doi:10.1007/s10043-003-0315-1
2. Y. Takahashi, H. Kuroda and T. Suemune, "Semiconductor Optical Amplifier-Fiber Laser and Its Application for Temperature Sensing," The Review of Laser Engineering, Vol. 33, No. 5, 2005, pp. 329-332.
3. T. Suemune and Y. Takahashi, "SOA-fiber ring laser and its application to electric field sensing in frequency domain," Optics and Lasers in Engineering, Vol. 45, No. 7, 2007, pp. 789-794. doi:10.1016/j.optlaseng.2006.12.002
4. R. K. Kadiwar and I. P. Giles, "Optical Fibre Brillouin Ring Laser Gyroscope," Electronics Letters, Vol. 25, No. 25, 1989, pp. 1729-1731. doi:10.1049/el:19891157
5. S. K. Kim, H. K. Kim and B. Y. Kim, "Er3+-Doped Fiber Ring Laser for Gyroscope Applications," Optical Letters, Vol. 19, No. 22, 1994, pp. 1810-1812. doi:10.1364/OL.19.001810
6. A. Kung, P.-A. Nicati and P. A. Robert, "Reciprocal and Quasi-Reciprocal Brillouin Fiber-Optic Current Sensors," IEEE Photonics Technology Letters, Vol. 8, No. 12, 1996, pp. 1680-1682. doi:10.1109/68.544717
7. Y. Takahashi and T. Yoshino, "Fiber Ring Laser with Flint Glass Fiber and Its Sensor Applications," Journal of Lightwave Technology, Vol. 17, No. 4, 1999, pp. 591-597. doi:10.1109/50.754788
8. D. E. Gray, "American Institute of Physics Handbook," 3rd Edition, McGraw-Hill, New York, 1972, pp. 233-236.

CHAPTER 19

ZNSE/ZNS QUANTUM-DOT SEMICONDUCTOR OPTICAL AMPLIFIERS

K. H. Al-Mossawi

Keywords: Linear Absorption Coefficient, Linear Emission Coefficient, Rate Equations (Res), Noise Figure

INTRODUCTION

Quantum dot photonic materials have attracted much attention in recent years as they have the potential to deliver the stability and coherence of atomic sources within a compact and efficient semiconductor device. characteristics such as reduced sensitivity to optical feedback makes such materials attractive as laser sources. In addition, suppression of pattern effects in QD semiconductor optical amplifiers (SOAs) shown to be promises for high speed application. The understanding of the high speed carrier dynamics of these materials is crucial for their optimization and exploitation. To address this issue, time resolved spectroscopy has been used to investigate the fundamental carrier decay time scales of SOA structures and determine their suitability for high-speed applications. Such pump-probe studies are usually performed using pulse width of a few hundred femtoseconds to picoseconds in order to sufficiently resolve the relaxation dynamics of high-speed devices. [1] Quantum dot (QD) semiconductor optical amplifiers (SOAs) demonstrate best features when compared with other SOAs based on bulk or quantum well materials. As a result, QD SOAs are very promising for applications in high-speed optical communications. One of the most important features of QD materials results in these best performances is the discrete structure of their energy levels [2]. In this work, ZnSe/ZnS QD-SOAs are studied. First, energy levels are calculated using quantum box model where the dot is assumed in the form a cube with quantum dot size is $10 \times 10 \times 10$ nm^3. net gain is then calculated by using linear Absorption coefficient and linear emission coefficient components in the QD.

Citation: K. Al-Mossawi, "ZnSe / ZnS Quantum-Dot Semiconductor Optical Amplifiers," Optics and Photonics Journal, Vol. 1 No. 2, 2011, pp. 65-69. doi: 10.4236/opj.2011.12010.

GAIN OF QDS

Gain and threshold current models are calculated with quasi-Fermi levels energies E_{fc} and E_{fv}, which are directly controlled by the doping concentration or injected current. Given that w_x, w_y and w_z are the lengths in each dimension of a quantum box, see Figure 1, and E_{cnml} and E_{vnml} are allowed quantized electron and hole energies, respectively, the corresponding carrier densities, n and p(~n) may be expressed as [3]

$$n = \sum_{nml} \frac{2 f_c(E_{cnml})}{w_x w_y w_z} = \sum_{nml} \frac{2}{\left[1 + \exp\left(\frac{E_{cnml} - E_{f_c}}{kT}\right)\right] w_x w_y w_z} \quad (1)$$

$$p = \sum_{nml} \frac{2\left(1 - f_v(E_{vnml})\right)}{w_x w_y w_z} = \sum_{nml} \frac{2}{\left[1 + \exp\left(\frac{E_{f_v} - E_{cnml}}{kT}\right)\right] w_x w_y w_z} \quad (2)$$

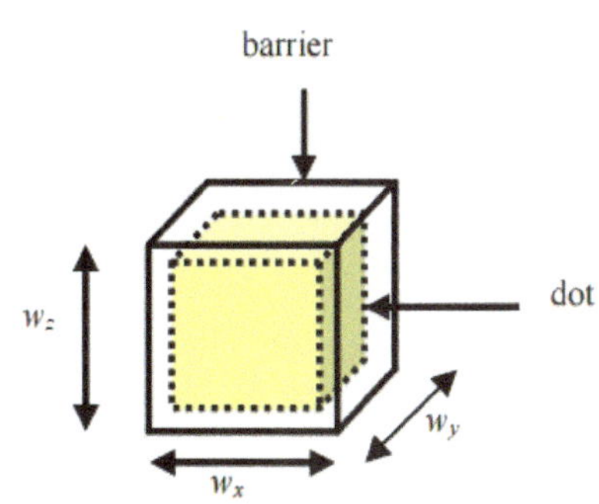

Figure 1: Diagram of dot/barrier quantum box.

Where f_c and f_v are the corresponding Fermi functions for electrons in the conduction and valence bands. The factor of two takes into account the electron spin. The summation is over all the confined energy states, which depends on the QD material and the dot/barrier valence and conduction band offsets (ΔE_c and ΔE_v) that enable electron and hole confinement. Solving known E_{fc} and E_{fv} are ed for directly and applied to find the linear absorption coefficient α(w) and linear emission coefficient e(w) components in the QD [3]

$$a(w) = \frac{w}{n_r}\sqrt{\frac{\mu_o}{\varepsilon_o}}\sum_{nml}\int_{E_g}^{\infty}\langle R_{cv}^2\rangle \times \frac{g_{cv} f_v(E_1)\left[1-f_c(E_2)\right]\left(\frac{\hbar}{\tau_{in}}\right)}{(E_{cv}-\hbar w)^2+\left(\frac{\hbar}{\tau_{in}}\right)^2} dE_{cv} \tag{3}$$

$$e(w) = \frac{w}{n_r}\sqrt{\frac{\mu_o}{\varepsilon_o}}\sum_{nml}\int_{E_g}^{\infty}\langle R_{cv}^2\rangle \times \frac{g_{cv} f_c(E_2)\left[1-f_v(E_1)\right]\left(\frac{\hbar}{\tau_{in}}\right)}{(E_{cv}-\hbar w)^2+\left(\frac{\hbar}{\tau_{in}}\right)^2} dE_{cv} \tag{4}$$

Where R_{cv} is the dipole moment, n_r is the refractive index and g_{cv} is the density of states for the QD, given by [3]

$$g_{cv} = \frac{2\delta\left(E_{cv}-E_{cnml}-E_g\right)}{w_x w_y w_z} \tag{5}$$

Where τ_{in} the intraband relaxation time. Equations (3) and (4) signify the rates of absorption and emission per unit length with E_1 and E_2 representing the hole energy level in the valence band and the electron energy level in the conduction band, respectively. The net gain of the system is then [3]

$$G(w) = e(w) - a(w) \tag{6}$$

To determine the optimal operating region for an amplifier system, the gain coefficient, expressed as [3]

$$\text{gain coefficient} = \frac{G(w)}{P_{in}} \tag{7}$$

QD-SOA

Based on the rate equations used in [4] for the wetting layer (WL) carrier density N_w, occupation probabilities for ground and exited states (f and h), respectively one can write.

$$\frac{\partial N_w}{\partial t} = \frac{J}{qL_w} - \frac{N_w(1-h)}{\tau_{w2}} + \frac{N_w h}{\tau_{2w}} - \frac{N_w}{\tau_{wR}} \tag{8}$$

$$\frac{\partial h}{\partial t} = \frac{N_w L_w(1-h)}{N_Q\tau_{w2}} - \frac{N_w L_w h}{N_Q\tau_{2w}} - \frac{(1-f)h}{\tau_{21}} + \frac{f(1-h)}{\tau_{12}} \tag{9}$$

$$\frac{\partial f}{\partial t} = \frac{(1-f)h}{\tau_{21}} - \frac{f(1-h)}{\tau_{12}} - \frac{f^2}{\tau_{1R}} - \frac{g_p L}{N_Q}(2f-1)S_{av}\frac{c}{\sqrt{\varepsilon_s}}\Gamma \tag{10}$$

Where J is the current density, q is the electron charge, L_w is the effective thickness of the active layer, τ_{w2} is the carrier relaxation time from the two-dimensional WL to the ES, τ_{2w} is the carrier escape time from the ES to the WL, τ_{wR} is the spontaneous radiative lifetime in WL, τ_{1R} is the spontaneous radiative lifetime in the QDs, τ_{21} is the carrier relaxation time from ES to GS, τ_{12} is the carrier relaxation time from the GS to the ES, N_Q is the surface density of QDs, ε_s is the dielectric constant of the QD material and Γ is the optical confinement factor. g_p is the peak material gain. According to Pauli's exclusion principle, the occupation probability in the quantum-dot ground state relates to the carrier density N in the QD by f = $N/2(N_Q/w_x)$ [5] where wx is the size of the QD. The average signal photon density Sav is given by [6]

$$S_{av} = \left[\frac{(G-1)(1-R_f)(1+R_b G)}{\left(1-\sqrt{R_f R_b}G\right)^2 + 4\sqrt{R_f R_b}G\sin^2\Phi}\right]\frac{P_{in} n_g}{\hbar w_p L_w D L g_p c} \tag{11}$$

Where P_{in} is the input signal power, n_g is the group refractive index, ħ is the normalized Plank's constant, w_p is the peak frequency, D is the width of the active layer, L is the cavity length, c is the free space light speed. Note that the Fibry-Perot amplifier gain G is given by [7]

$$G = \frac{(1-R_f)(1-R_b)G_S}{\left(1-\sqrt{R_f R_b}G_s\right)^2 + 4\sqrt{R_f R_b}G_s\sin^2\Phi} \tag{12}$$

R_f is the front mirror reflectivity, R_b the back mirror reflectivity. Φ is the phase angle while G_S is the single-pass gain of the structure, given by [7]

$$G_s = \exp\left[\left(g_p\Gamma - a_{\text{int}}\right)L\right] \tag{13}$$

Note that αint is the loss coefficient. The carrier dynamics described by the rate Equations are related only to electrons in the conduction band while hole dynamics can be neglected due to their larger effective mass [4]. In addition, typical values of the material parameters in Table 1 [5] for ZnSe/ZnS system where (E_g = 2.8 eV, m_e = 0.17 m_o and m_v = 0.60 m_o) for ZnSe and (E_g = 3.68 eV, m_e = 0.39 m_o and m_v = 0.49 m_o) for ZnS.

Rate equations model are solved at steady state case using the expression of the average signal photon density S_{av}, Equation (11). The noise added to the signal during the amplification process is a fundamental property for every kind of amplifiers. The noise characteristics of an amplifier are quantified by a parameter called noise figure, when the shot noise part is taken into account, the amplifier noise figure can be written as [7].

Table 1: Parameters used in the Calculations [5]

Parameter	Value	Unit
L	200	μm
L_w	0.2	μm
D	10	μm
N_Q	5×10^{10}	cm^{-2}
τ_{w_2}	3	ps
τ_{2w}	1	ns
τ_{12}	1.2	ps
τ_{21}	0.16	ps
τ_{1R}	0.4	ns
τ_{wR}	1	ns
Φ	0	
$R_f = R_b$	10^{-4}	
α_{int}	3	1/cm
Γ	0.006	
L_b	20	nm
L_c	75	nm
P_{in}	1	μW

$$F_n = 2n_{sp}(G-1)/G+(1/G) \tag{14}$$

The spontaneous emission factor n_{sp} is given by [8]

$$n_{sp} = \frac{\Gamma e(w)}{[\Gamma G(w)] - a_{\text{int}}} \tag{15}$$

RESULTS AND DISCUSSION

Using quantum box model we calculate QD energy levels for conduction and valence bands and then it is used to calculate linear absorption coefficient, linear emission coefficient and net gain. The barrier layer thickness is 20 nm while the thickness of the clad layer is 75 nm, see Figure 2. All calculations are done at $\tau_{in} = 0.1$ ps [3] and carrier density 2×10^{24} m^{-3}. Also we calculate the small signal gain for structures contain one and six layers. Gain coefficients are calculated using Equation (7) for several values of input power. Figure 3 shows a comparison between the linear absorption and emission coefficients where a multi-peaks appear for ground and excited state transitions. One can refers to the higher emission value compared to that of absorption which results from lower transparency value for this structure. The net gain is shown in Figure 4, where a double peak can also be seen. This is very important in some applications. For example, this can be used in long haul optical transmission. The reason for this double emission is reasoned to finite ground state relaxation time, which brings ground state emission to a constant value after the excited state threshold [10]. Figure 5 shows the effect of intraband relaxation time τin where a higher gain is obtained at longer intraband time. Gain of the highest peak increases by ~ 4.5 time when τ_{in} increases from 0.1 ps to 0.5 ps. Really, intraband relaxation time controlled the transition between intersubband thus one can use this property to work with the same structure at higher gain by choosing the subbands that occur between them. Also, one can refers to the somewhat increase between the ground and excited peaks with inraband relaxation time. Rate equations are then solved at steady state, using the parameters listed in Table 1, the QD-SOA characteristics are examined. Small-signal gain spectrum of this QD-SOA is shown in Figure 6. These gain values are comparable to that obtained by Ben-Azra [9]. Note that it is obtained here at a lower current value. So, this is a good modification. The effect of QD layer is also examined where the gain increases with QD layers as shown in Figure 7. Figure 8 shows the output power calculated as a function of carrier density with input power is taken as a parameter. Including the shot noise part, the noise figure spectrum is shown in Figure 9, while the noise figure curve at different carrier densities is shown in Figure 10 from the last two figures one can refers to the lower noise accompanied with this amplifier while the last three figures shows the properties of ZnSe/ZnS QD-SOA which is a good results to use it in applications.

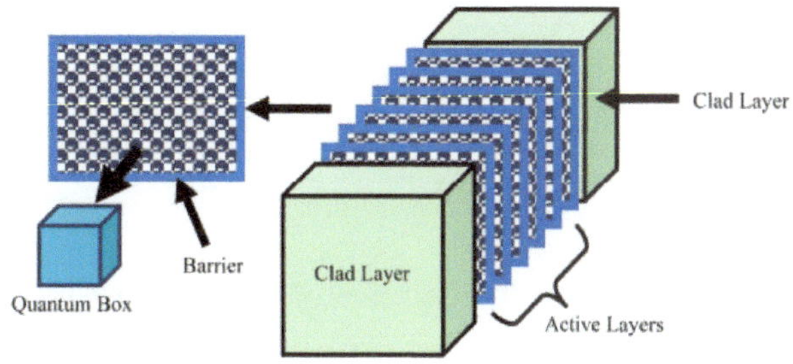

Figure 2: QD system contain six active layers.

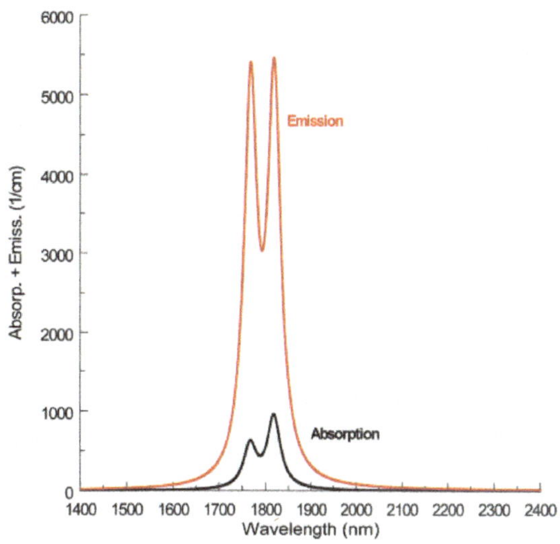

Figure 3: Comparison between linear absorption and emission coefficients.

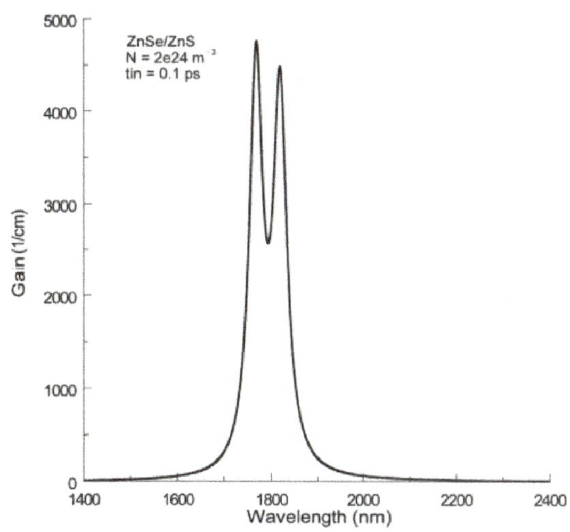

Figure 4: Gain of ZnSe/ZnS QDs.

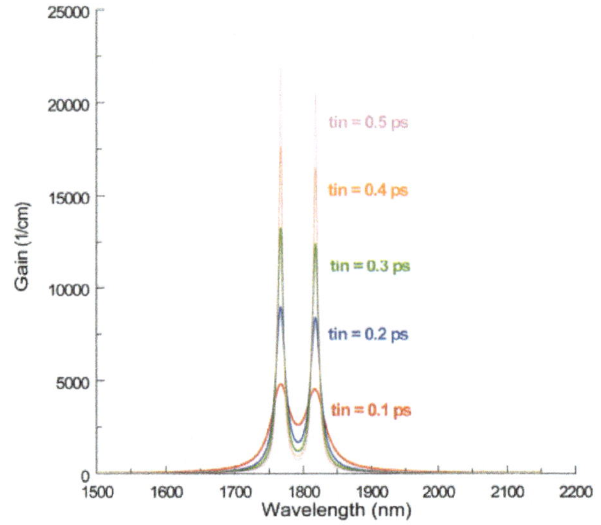

Figure 5: Gain versus wavelength at several values of intraband relaxation time.

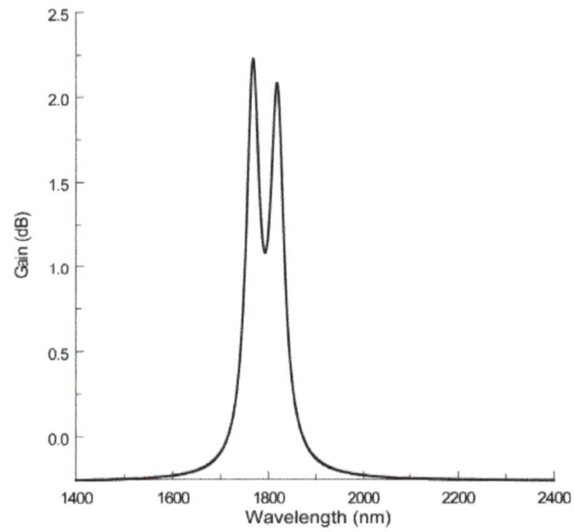

Figure 6: Small signal gain spectrum.

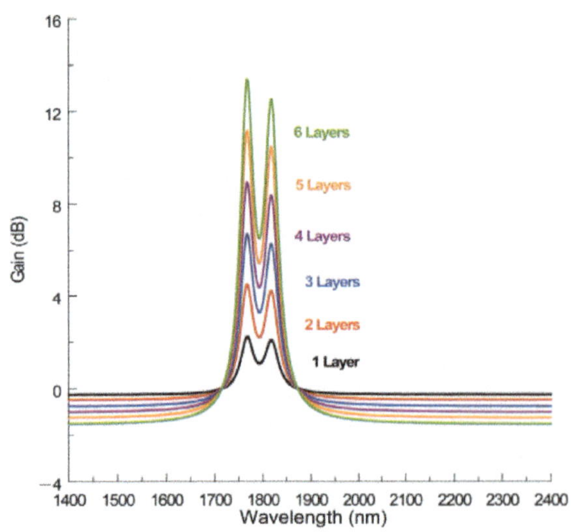

Figure 7: Small-signal gain spectra with QD layer as a parameter.

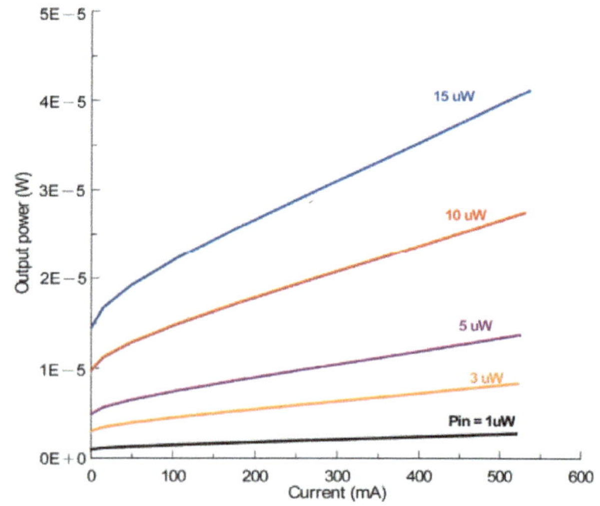

Figure 8: Output power versus current at values several from input power.

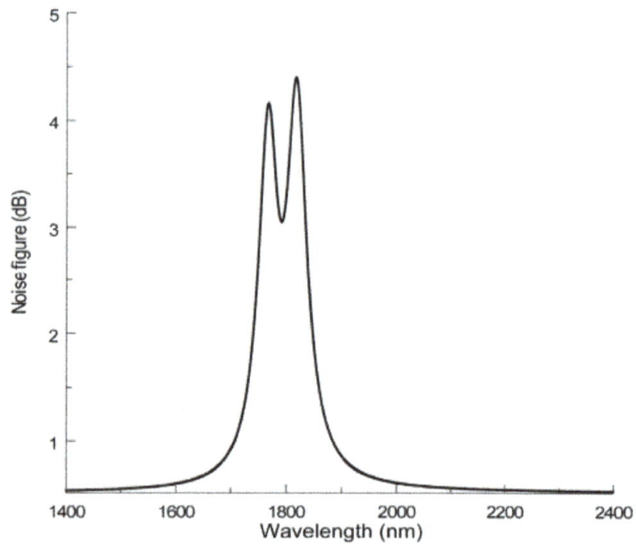

Figure 9: Noise figure with shot noise included in the calculations.

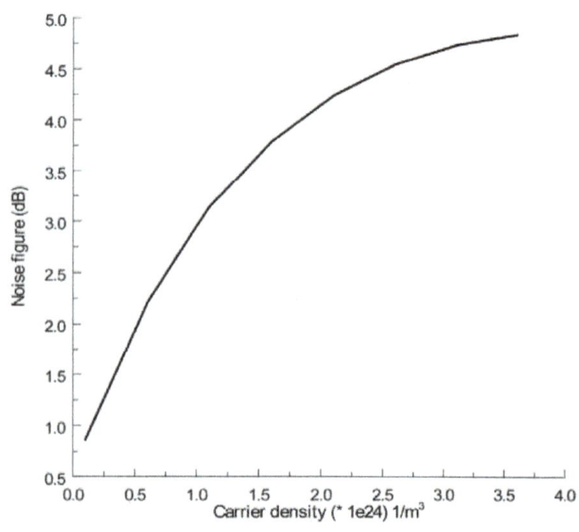

Figure 10: noise figure with shot noise versus carrier density.

REFERENCES

1. T. Piwonski, I. O'Driscoll, J. Houlihan, G. Huyet and J. Manning, "Carrier Capture Dynamics of InAs/GaAs Quantum Dot," Applied Physics letters, Vol. 90, No. 12, 2007, pp..
2. T. Erneux, E. A. Viktorov, P. Mandel, T. Piwonski, G. Huyet and J. Houlihan, "The Fast Recovery Dynamics of a Quantum Dot Semiconductor Optical Amplifier," Applied Physics Letters, Vol. 94, No. 11, 2009, pp. 113501. doi:10.1063/1.3098361
3. C.-J. Wang, L. Y. Lin and B. A. Parviz, "Modeling and Simulation for a Nano-Photonic Quantum Dot Waveguide Fabricated by DNA-Directed Self-Assembly," IEEE Journal of Selected Topics in Quantum Electronics, Vol. 11, No. 2, 2005, pp.500-509. doi:10.1109/JSTQE.2005.845616
4. Y. Ben-Ezra, M. Haridim and B. I. Lembrikov, "Theoretical Analysis Of Gain-Recovery Time and Chirp in QD-SOA," IEEE Journal of Selected Topics in Quantum Electronics, Vol. 17, No. 9, 2005, pp. 1803-1805.
5. M. A. Al-Mossawi, A. H. Al-Khursan, R. A. Al-Ansari, "ZnO-MgZnO Quantum-Dot Semiconductor Optical Amplifiers," Recent Patents on Electrical Engineering, Vol. 2, No. 3, 2009, pp. 226-238. doi:10.2174/1874476110902030226
6. M. Vasileiadis, D. Alexandropoulos, M. J. Adams, H. Simos and D. Syvridis, "Potential of InGaAs/GaAs Quantum Dots for Applications in Vertical Cavity Semiconductor Optical Amplifiers" IEEE Journal of Selected Topics in Quantum Electronics, Vol. 14, No. 4, 2008, pp. 1180-1187. doi:10.1109/JSTQE.2007.915517
7. E. Forestieri, "Optical Communication Theory and Techniques" Ch. 8, Springer, Pisa, 2005.
8. Y. Ben-Ezra, B. I. Lembrikov and M. Haridim, "Acceleration of Gain Recovery and Dynamics of Electrons in QD-SOA," IEEE Journal of Selected Topics in Quantum Electronics, Vol. 41, No. 10, 2005, pp. 1268-1273. doi:10.1109/JQE.2005.854131
9. K. Veselinv, F. Grillot, C. Cornet, J. Even, A. Bekiarski, M. Giannini and S. Loualiche, "Analysis of the Double Laser Emission Occurring in 1.55 mm InAs-InP (133) B Quantum-Dot Lasers," IEEE Journal of Selected Topics in Quantum Electronics, Vol. 43, No. , 2007, pp. 810-.

CHAPTER 20

HARMONIC CURRENTS OF SEMICONDUCTOR PULSE SWITCHING CONVERTERS

Petr Bilik, Jan Zidek, Vaclav Kus, Tereza Josefova

Keywords: Rectifiers; Harmonic Analysis, Harmonic Distortion, Power System Harmonics, Current Measurement

INTRODUCTION

Conventional controlled rectifiers designed for speed regulation of DC motors were characterized by both the consumption of harmonic currents and, especially, the variable power factor (proportional to the steering angle of the converter). There were mostly high power converters, so the negative effects on the power network were dealt with by experts. In the spectrum of the current, there mostly occurred so called characteristic harmonic ones and their level was computed using the amplitude principle. When mentioning small power sources for the use in electronics (charging of capacitors), there were, on the contrary, very low powers and very small amount of appliances, so that negative effects were ignored. A significant increase of negative effects of the converters on the power network occurs with the increase of frequency converters with a DC voltage intermediate circuit. These converters usually contain uncontrolled rectifiers at the input. The converter thus works with a negligible phase shift between the voltage and the current, and the power factor (fundamental harmonic cos φ_1) is close to 1. However, the current harmonic ones become a problem. The very first converters even worked without sufficient circuits for limitation of the harmonic ones. In principle, it is a rectifier working to capacitive load; the values of harmonic currents are very high. In a simplified calculation, there can be used a so-called "generalized amplitude principle" (1) [1,2].

$$I_h = \frac{1}{h} I_1 . \frac{\sin\frac{hd}{2}}{\sin\frac{d}{2}} \tag{1}$$

where: h – harmonic order, I1 – magnitude of fundamental harmonic, d – diode conduction time. Detailed information about calculations of characteristic harmonics, non-characteristic harmonics and interharmonics can be found in [2-5]. Topic of harmonics elimination and power factor improvement of three-phase rectifier is described in [6]. Equation (1) advantage is that it also applies to conventional controlled rectifiers, to calculation of the harmonic ones with intermittent currents, to calculation of harmonic voltage converters as well as to frequency converters or other rectifiers working to capacitive loads. See the calculated shape of current in Figure 1. The unambiguous result of this relation is the fact that the values of harmonic ones depend on the time during which the diode is open (the capacity is recharged). As for large capacities, the time is very short and thus the harmonic ones are even higher. This, paradoxically, results in a situation where the load requires large voltage smoothing (and thus a large capacitor) and the character of the current consumption requires, on the contrary, a capacitor as small as possible. A large capacitor means low d factor in (1). The value of the harmonic one then reaches up to 80-90% of the value of the fundamental harmonic one of the current, even for the nominal load. With the increasing improvement of quality of the components and control electronics, the number of converters is increasing rapidly and thus the need to deal with the limitations of harmonic currents. Besides the methods consisting of deployment of filters, protected compensation etc., solutions appear directly in the construction and the control methods of converters. One of the ways is the use of pulse rectifiers.

Citation: P. Bilik, J. Zidek, V. Kus and T. Josefova, "Harmonic Currents of Semiconductor Pulse Switching Converters," Energy and Power Engineering, Vol. 5 No. 4B, 2013, pp. 1120-1125. doi: 10.4236/epe.2013.54B214.

THE PRINCIPLE OF OPERATION OF PULSE RECTIFIERS

The basic diagram of a pulse rectifier is shown in Figure 2. The connection does not practically differ from a single-phase voltage inverter. The principle of operation consists of suitable switching of transistors. Approximately sinusoidal current consumption (with very little distortion) is achieved this way. At the same time, it is necessary to guarantee current consumption of the first harmonic one with a very good power factor. The phasor diagram shows that it is necessary to ensure power supply voltage of the bridge Uv that is delayed behind the consumed current Ia by the angle. More information about pulse rectifiers and its control principles can be found in [7-9].

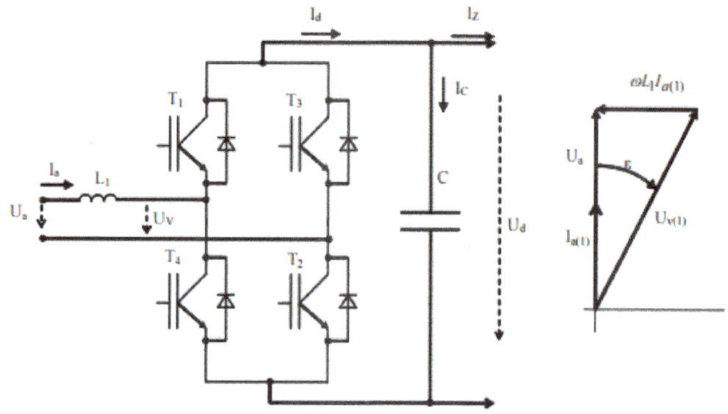

Figure 2: The block diagram of the pulse rectifier and the corresponding phasor diagram.

HARMONIC CURRENTS OF PULSE RECTIFIERS

Figure 2 shows that this rectifier is supposed to meet the basic requirement: consumption of a sinusoidal current without a phase shift. To determine the harmonic currents, an analytical calculation of the shape of the current that this converter takes from the power network was performed (in Figure 3, it is the waveform in the lower part, illustrated with the solid line). For simplicity, modulation frequency f_m=1kHz was selected. From this simple figure and the low modulation frequency, it is obvious that the shape of the current taken by the converter from the power network is little distorted and the power factor is very good as well. The determination of the higher harmonic ones was then performed numerically. Let observe the dependence of percentage values of the harmonic currents as a function of the ε angle In fact, ε triangle is therefore a relative unit of the load. For ε = 0, the converter is close to the idle run and vice versa. The unambiguous result from Figure 4 is that for the frequencies near the modulation frequency, there are high percentage values of the harmonic currents.

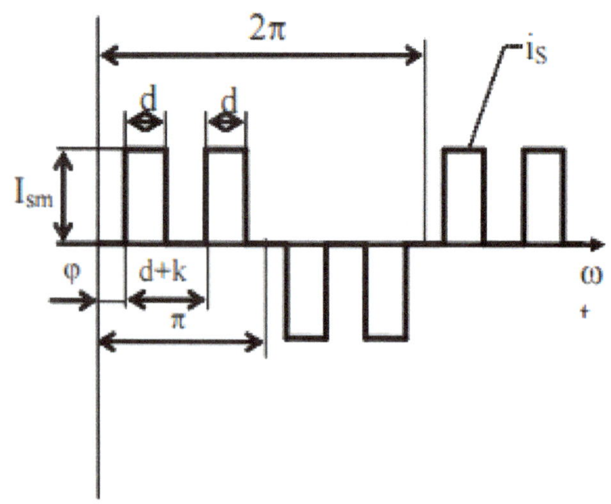

Figure 1: Idealized shape of the current taken by the converter from the power network.

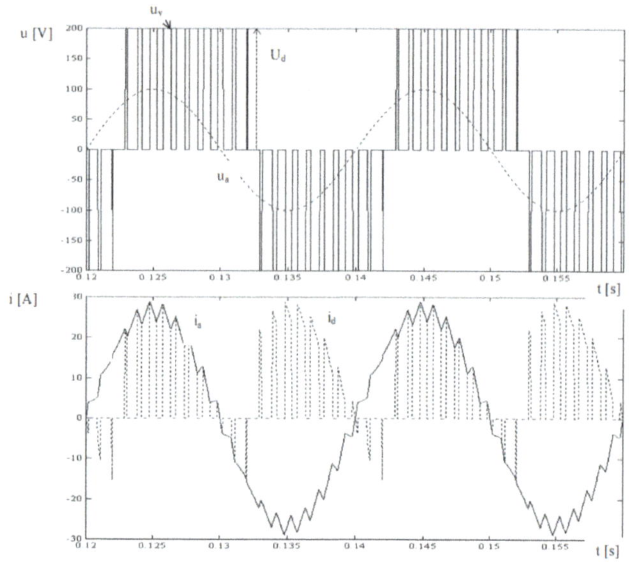

Figure 3: Voltage and current ratios of the pulse voltage rectifier.

If this graph is not complemented with comments on the size of the absolute values of the current or dependence of the fundamental harmonic current as a function of angle ε, this result is highly misleading. The first impression may be that the operation of this converter is construed as inappropriate. Presentation of absolute values, which will be dealt with in Chapter 4, is much more suitable.

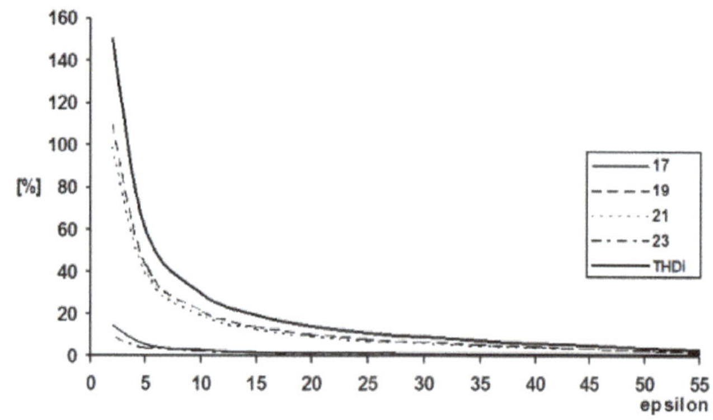

Figure 4: Percentage values of harmonic currents and THDi (from the computed waveform).

MEASUREMENT OF PULSE RECTIFIERS IN THE LABORATORY

The measurement was performed in the laboratories of electric drive of Department of Electromechanics and Power Electronics in The University of West Bohemia in Pilsen. It was a single-phase pulse rectifier of the voltage type with a rated current of 6A. The modulation frequency was f_m = 1kHz. As it was mentioned in the previous chapters, the characteristic harmonic ones that lie closest to the modulation frequency appear. In our case, they will be the 19th and the 21st harmonic ones. The measurement was performed for various loads and for the operation of both drive and brake. In order to conduct the first assessment of the measured value, Figure 5 shows two types of current waveforms - one for the nominal values and the other for values nearing the idle run. As for the rated current, the waveform is of high quality and low distortion can be expected. As for the small load, the current is rippled, but the amplitude of the current measured is very low. In the Figure 6, can be seen an incorrectly represented graph. Here is shown the percentage value of the current of the 3rd and 5th harmonic ones and the current of the modulation frequency depending on the load. From this graph can be seen again that when the load is small, the percentage values are high. Therefore, there is a risk that the converter will be considered inappropriate from the point of view of the harmonic content. The graph in Figure 7 is much more apposite. There were produced graphs for the waveforms from Figure 5 for the values of the harmonic ones, now in the absolute values. In the graph, we can see that due to the rated input current of the converter I_N = 18A (the fundamental harmonic), the values of the harmonic ones are negligible, even in the mode of a very small load. For example, at the load (on the DC side) of 1.4A, the value of the 19th harmonic one is 0.25A, which is approximately 1.4% of the rated current of the fundamental harmonic. There is then a sharp contrast with Figure 6, which is technically flawless, but otherwise unsuitable. Notice: beware of the scale - in the headlines of the figures there are values of the output current (the load of the converter), but the graph shows the input currents.

MEASUREMENT OF PULSED SWITCHING CONVERTERS IN PRACTICE

The practical measurements were performed with a measuring apparatus using the analyzer BK-ELCOM ENA440, see Figure 8. As for the analyzer, there were used input modules with A/D converters with the 24-bit resolution. The currents higher than 5A (the photovoltaic power plant and the converter) were measured by means of current clamps while taking advantage of the compensation of amplitude and phase errors. As for currents of under 5A, the direct current input was used. The sampling was the same in all channels and the sampling frequency was 50kS/s for each channel.

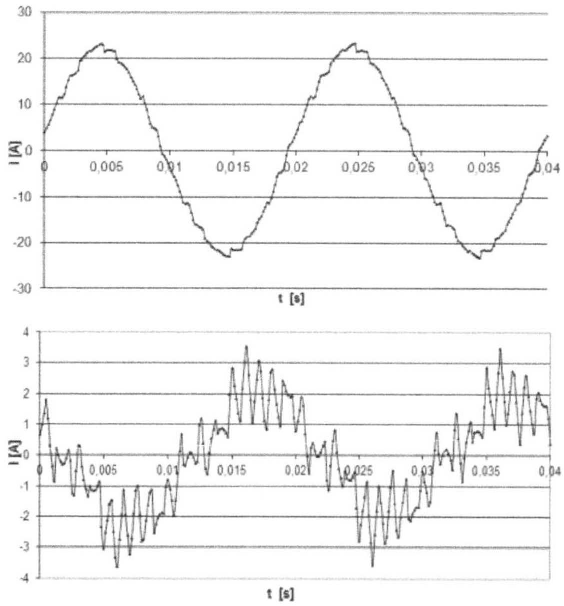

Figure 5: Waveforms of the current measured for various loads.

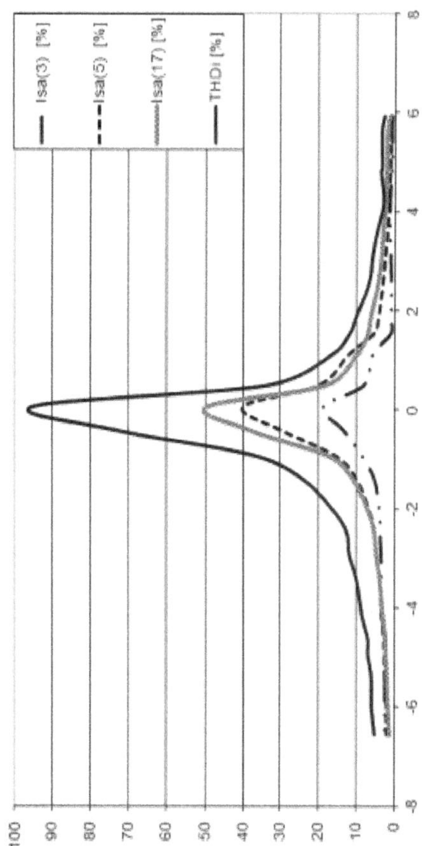

Figure 6: Percentage values of harmonics depending on the load of the converter.

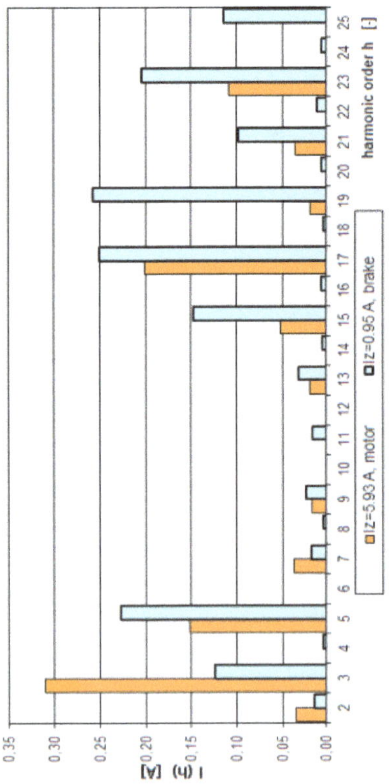

Figure 7: A correctly used figure for documentation of harmonic pulse converters.

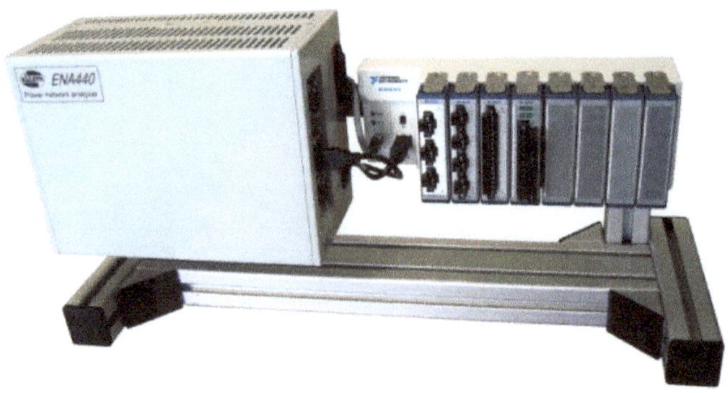

Figure 8: Analyzer BK-ELCOM in version ENA440x.

As technical interest and for the notion of the dynamic range of the input modules used with regard to the very low system noise, there are also stated the minimum detectable amplitudes of the harmonic ones:

- The voltage input of the range of 300Vrms is able to evaluate correctly, in the band of 0-25kHz, the signal of the minimum level of 300 μV (0.0001% of the range).
- The direct current input of the range of 5Arms is able to evaluate correctly, in the band of 0-25 kHz, the signal of the minimum level of 70 μA (0.0001% of the range)

The First Measurement

The first measurement was performed at the photovoltaic power plant (PVPP) with a nominal output of 20kW, see Figures 9 and Figure 10. The measurement at the PVPP was performed on a sunny day in the middle of September, at about 09:30. The PVPP uses several single-phase converters for each phase. The non-equivalent load of the individual phases may be caused by the malfunction of one of the converters. The most distorted waveform of the current appeared in phase L3 where there were found out the following parameters: U=226V, I=7,12A, P= 1596W, PF=0,992, cos φ = 0.998. From the waveforms in Figure 9 as well as from the basic data of the measurement, it is obvious that the converter works with a very good power factor. The distortion of the current is minimal. Although there is used the pulse modulation of the inverter, the current is also (perhaps by means of the influence of the high modulation frequency) at the frequency of the modulation frequency very low. It is confirmed by the results of the harmonic analysis shown in Figure 10. More about disturbing impact of grid connected converters can be found in [10].

The Second Measurement

The second measurement was carried out on a frequency converter with so called "Active Front End" in a laboratory at VSB-TU Ostrava (see Figure 11 to 14). Therefore, it is also a pulse rectifier connected to the power network. The maximum load of approximately 3kW, which could have been reached, was limited by the equipment of the laboratory. When the converter was loaded, the following parameters were obtained in phase L3: U=228 V, I=4.9 A, P=963W, PF=0.85, cosφ=0.865. When the converter was iddle, the following parameters were obtained in phase L3: U=224 V, I=7.15 A, P=384W, PF=0.24, cosφ=0.26. From the current waveforms and the results of the harmonic analysis, it is obvious that the distortion of the input current is low due to the activity of the pulse rectifier. When the load is small, a wide spectrum of harmonic of the input current appears due to the activity of the pulse rectifier.

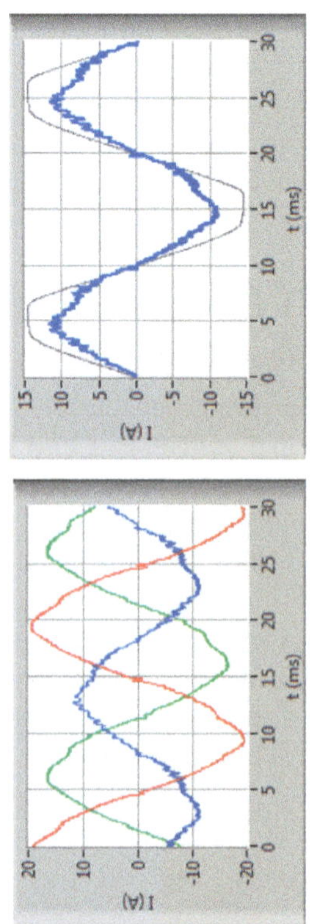

Figure 9: Left: Currents in PVPP. Right: relation of U and I of phase L3.

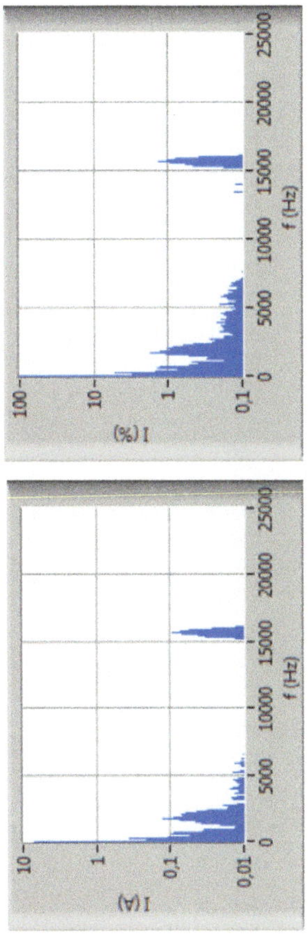

Figure 10: The spectrum of the current of phase L3. Left: graph in Amps. right: graph in percent.

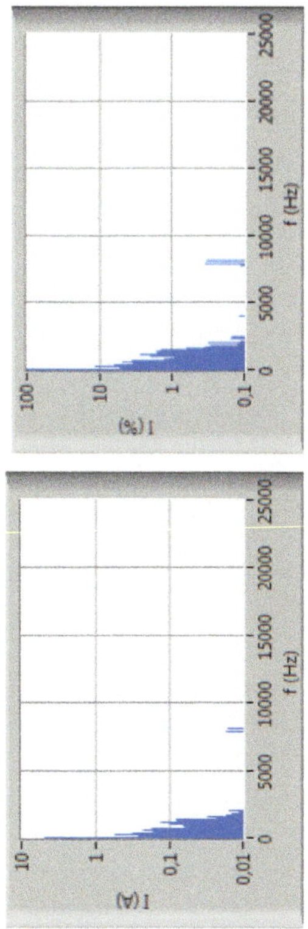

Figure 12: The spectrum of the current of phase L3. Left: graph in Amps. Right: graph in percentage.

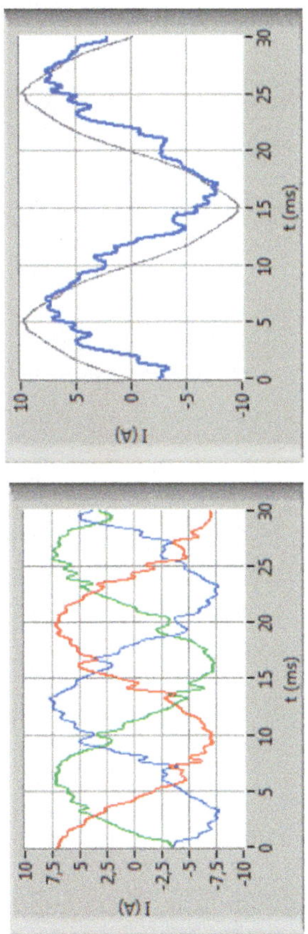

Figure 11: Left: Currents of the converter under a load. Right: relation of voltage and current of phase L3.

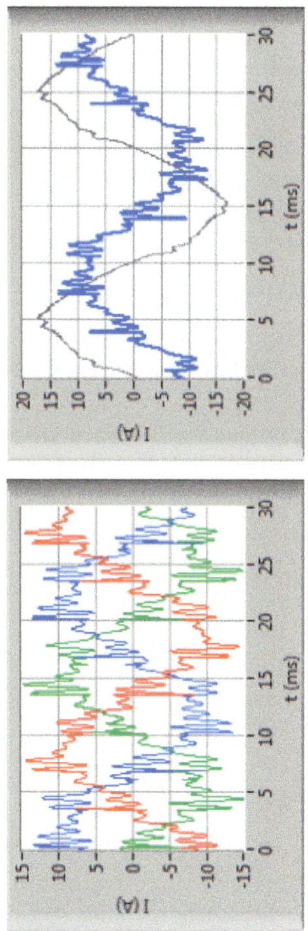

Figure 13: Left: Currents of the converter at the idle run. Right: relation of voltage and current of phase L3.

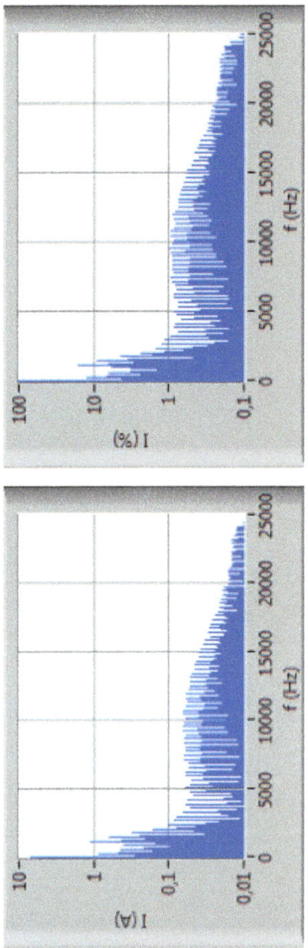

Figure 14: The spectrum of the current of phase L3. Left: graph in Amps. Right: graph in percentage.

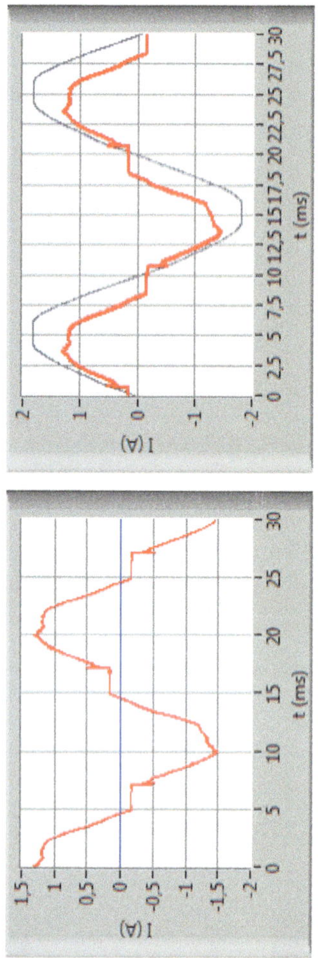

Figure 15: The waveform of the current of a plasma TV. Right: relation of voltage and current.

However, the absolute value of the harmonic ones is low, especially if a comparison is done with the fundamental harmonic of the current at the nominal load.

The Third Measurement

The third measurement was performed on a typical household appliance - a television, as a demonstration of the fact that producers of consumer electronics try to improve the parameters of their products with respect to the reverse impact on the feed array. The measurement was carried out on a plasma TV with a diagonal of 42", model 2010, see Figures 15 and 16. The following parameters were obtained: U= 230 V, I=0.85 A, P=182 W, PF= 0.932, cosφ= 0.959. Despite the fact that the diagram of the connection of the feed source of the television was not available, the result of the measured courses of the current and its harmonic analysis is that there is no occurrence of higher harmonic ones that would have a dangerous impact on the power network.

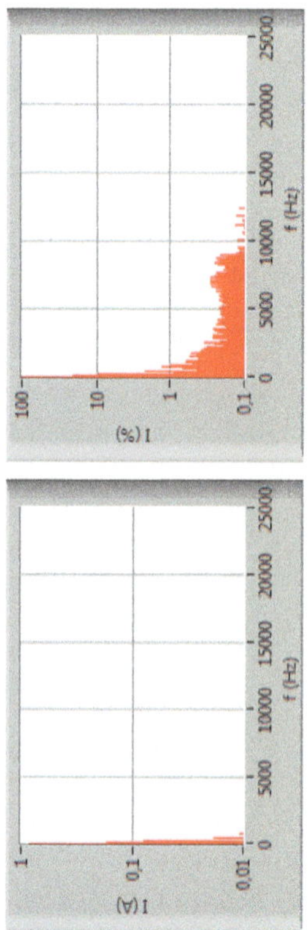

Figure 16: The spectrum of the current. Left: graph in Amps. Right: graph in percentage.

CONCLUSIONS

The article shows the development of semiconductor converters with respect to higher harmonic currents taken by the converter from the feed array. If modern methods of pulse switching are used elimination of these currents, the characteristic harmonic ones of these currents are low. Currents of higher harmonic ones, that are identical to the frequencies of modulation frequencies, newly appear in the spectrum.

However, these currents are not significant in their absolute values. The graphs show that when the loads are small, the percentage values of the currents of modulation frequencies can be high, but with respect to the low current consumption (in relation to the nominal values), the absolute values are low. The article also presents results of several practical measurements. The article does not feature the aspects of largeness, i.e. concurrence of a high number of small household appliances, including modern sources of light, on the feed array. Generally, there is no need to be afraid of converters with pulse modulation. From the point of view of their effect on the power network, the use of these converters can even be considered beneficial.

ACKNOWLEDGEMENTS

This paper was supported by the project No. GACR 102/09/1164, by the European Regional Development Fund and the Ministry of Education, Youth and Sports of the Czech Republic under the Regional Innovation Centre for Electrical Engineering (RICE), project No. CZ.1.05/2.1.00 /03.0094. This work was supported in part by The Ministry of Education, Youth and Sports of Czech Republic under the project KONTAKT II registration number LH12183.

REFERENCES

1. V. Kus and P. Drabek, “The Education of EMC at the UWB,” Electrical Power Quality and Utilisation (EPQU), 2011 11th International Conference on, 2011, pp. 1-6. doi:10.1109/EPQU.2011.6128811
2. V. Kus, “Impact of Power Electronic Converters on Power Distribution Network. Book edited in BEN-technická literature, Praha, Czech Republic, 2002.
3. V. Kus, P. Drabek and J. Fort, “Negative Influence of Controlled Rectifiers on Power Distribution Network,” Proceedings international exhibition and conference: Nuernberg, PCIM Europe 2007.
4. V. Kus, Z. Peroutka and P. Drabek, “Non-characteristic Harmonics and Interharmonics of Power Electronic Converters,” 18th International Conference and Exhibition on Electricity Distribution, 2005 CIRED, Turin, Italy.
5. M. Bauta and M. Grotzbach, “Noncharacteristic Line Harmonics of AC/DC Converters with High DC Current Ripple,” Power Delivery, IEEE Transactions on, Vol. 15, No. 3, 2000, pp. 1060-1066. doi:10.1109/61.871375
6. C. Filote, C. Ciufudean, S. Alaei and A. M. Cozgarea, “Harmonic Elimination and Power Factor Improvement of Three-phase Rectifier Using RNSIC Variant,” Clean Electrical Power (ICCEP), 2011 International Conference on, pp. 40-45, 14-16 June 2011.
7. V. Blahnik, Z. Peroutka, J. Molnar and J. Michalik, “Control of Primary Voltage Source Active Rectifiers for Traction Converter with Medium-frequency Transformer,” Power Electronics and Motion Control Conference, 2008. EPE-PEMC 2008, 13th, pp. 1535-1541, 1-3 Sept. 2008. doi:10.1109/EPEPEMC.2008.4635485
8. M. Malinowski, M. P. Kazmierkowski, S. Hansen, F. Blaabjerg and G. D. Marques, “Virtual-flux-based Direct Power Control of Three-phase PWM Rectifiers,” Industry Applications, IEEE Transactions on, Vol. 37, No. 4, 2001, pp. 1019-1027. doi:10.1109/28.936392
9. P. Verdelho and G. D. Marques, “DC Voltage Control and Stability Analysis of PWM-voltage-type Reversible Rectifiers,” Industrial Electronics, IEEE Transactions on, Vol. 45, No. 2, pp. 263-273.
10. O. Prochazka, J. Gurecky and S. Misak, “Feedback Effects Definition of Alternative Electric Energy Sources Connected to the Distribution Power Network,” 11th International Scientific Conference on Electric Power Engineering EPE 2010, Brno, 2010, pp. 243 - 248.

INDEX

M

N

O

P

Q

R

S

T

U

V

W